LEGUME-BASED FERMENTED FOODS

Editors

N. R. Reddy, Ph.D.
Research Associate
Department of Food Science and Technology
Virginia Polytechnic Institue and State University
Blacksburg, Virginia

Merle D. Pierson, Ph.D.
Professor and Head
Department of Food Science and Technology
Virginia Polytechnic Institute and State University
Blacksburg, Virginia

D. K. Salunkhe, Ph.D.
Vice-Chancellor
Mahatma Phule Agricultural University
Rahuri, India

CRC Press, Inc.
Boca Raton, Florida

Library of Congress Cataloging-in-Publication Data
Main entry under title:

Legume-based fermented foods.

Bibliography: p.
Includes index.
1. Legumes as food. 2. Food, Fermented.
I. Reddy, N. R. II. Pierson, Merle D.
III. Salunkhe, D. K.
TX558.L4L444 1986 641.3'65 85-28003
ISBN 0-8493-6286-5

Direct all inquiries to CRC Press, Inc., 2000 Corporate Blvd., N.W., Boca Raton, Florida, 33431.

International Standard Book Number 0-8493-6286-5

Library of Congress Card Number 85-28003
Printed in the United States

PREFACE

Legume-based fermented foods are a significant part of the human diet in Southeast Asia, the Near East, and parts of Africa. In these areas of the world, legume-based fermented foods serve as a significant source of protein, calories, and certain vitamins. Furthermore, these foods are becoming popular in the westerm world, due to their high nutritive value and organoleptic characteristics. Preparation of legume-based fermented foods has remained to some extent an art, and their nutritional quality is of interest to both professionals and consumers. This book provides a comprehensive review of published data on the production of various legume-based fermented foods and critically examines their nutritional quality. The book consists of 14 chapters where the preparation, composition, nutritional quality, and food safety of individual legume-based fermented foods are discussed in detail. This treatise on legume-based fermented foods will be helpful to food scientists and nutritionists in improving the nutritional quality, organoleptic quality, and safety of these foods. It will also serve as an important reference book for scientists and technologists involved in fermented foods research.

THE EDITORS

N. R. Reddy, Ph.D., Research Associate, Department of Food Science and Technology, Virginia Polytechnic Institute and State University, Blacksburg, Virginia.

Dr. Reddy obtained his B.Sc. (Agric.), and M. Sc. (Agric.) degrees from Andhra Pradesh Agricultural University, Hyderabad, India in 1970 and 1972, respectively. He graduated in 1976 from Alabama A & M University, Normal, Alabama with an M.S. in Food Science and received a Ph.D. degree in Nutrition and Food Science in 1981 from Utah State University, Logan, Utah.

Dr. Reddy is a member of American Association of Cereal Chemists, American Oil Chemists Society, Institute of Food Technologists, and Sigma Xi.

Dr. Reddy has published more than 25 scientific papers, reviews, and book chapters.

M. D. Pierson, Ph. D., is Professor and Head, Department of Food Science and Technology, Virginia Polytechnic Institute and State University, Blacksburg, Virginia.

Professor Pierson graduated in 1964 from Iowa State University, Ames, Iowa with a B.S. degree in Biochemistry and obtained his M.S. and Ph.D. degrees in Food Science from University of Illinois, Urbana, Illinois in 1969 and 1970, respectively. He also served as a Research Chemist with George A. Hormel and Co., Austin, Minnesota.

Professor Pierson is a member of American Association for the Advancement of Science; American Society for Microbiology; The Society for Applied Bacteriology; American Meat Science Association; Institute of Food Technologists; International Association of Milk, Food, and Environmental Sanitarians; Gamma Sigma Delta; Phi Kappa Phi, Phi Tau Sigma, and Sigma Xi. He received the Gamma Sigma Delta VPI & SU Faculty Research Award in 1984 and serves as chairman-elect of the Institute of Food Technologists Food Microbiology division.

He is presently a member of the Editorial Board of *Journal of Food Protection* and served on the Editorial Board of *Journal of Food Science*. Professor Pierson has served on numerous advisory panels and as a consultant to several food processing companies.

Under Professor Pierson's guidance, 25 graduate students received their M.S. or Ph.D. degrees. He has published more than 50 scientific papers, reviews, and book chapters.

D. K. Salunkhe, Ph.D., is Vice-chancellor, Mahatma Phule Agricultural University, Rahuri, Maharashtra State, India. He was formerly professor of Nutrition and Food Science at Utah State University, Logan, Utah.

He graduated with a B.Sc. (Agric.) degree with honors from Poona University, Poona, India in 1949; in 1951 and 1954 he received his M.S. and Ph.D. degrees from Michigan State University.

He was an Alexander Humbolt Senior Fellow and Guest Professor at the Technical University in Karlsruhe, West Germany. He was a Guest Lecturer at the Technological Institute, Moscow, U.S.S.R. and exchange scientist to Czechoslovakia, Rumania, and Bulgaria on behalf of the National Academy of Sciences, National Research Council, as well as advisor to the U.S. Army Food Research Laboratories and many food storage, processing, and consumer organizations.

Professor Salunkhe was Sigma Xi president, Utah State University Chapter; Fellow of Utah Academy of Sciences, Arts, and Letters; Fellow of the Institute of Food Technologists and Danforth Foundation Faculty Associate. He delivered the Utah State University's 50th Faculty Honor Lecture, ''Food, Nutrition and Health: Problems and Prospect'', on the basis of his creative activities in research and graduate teaching.

Professor Salunkhe is presently a member of the editorial boards of *Journal of Food Quality,* and *International Journal Qualitas Plantarum/Plant Food for Human Nutrition* and served on the editorial boards of *Journal of Food Science* and *Journal of Food Biochemistry.*

Under Professor Salunkhe's guidance, 80 graduate students received their M.S. or Ph.D. degrees. He has authored more than 300 scientific papers, reviews, book chapters, and reports. Some of his papers have received recognition and awards as outstanding articles in biological journals. He has authored and/or co-authored fifteen

CONTRIBUTORS

Larry R. Beuchat, Ph.D.
Professor
Department of Food Science
University of Georgia
Agricultural Experiment Station
Experiment, Georgia

Chandramohan B. Chawan, Ph.D.
Assistant Professor
Department of Food Science and Animal Industries
Alabama A&M University
Normal, Alabama

B. B. Desai, Ph.D.
Associate Professor
Mahatma Phule Agricultural University
Rahuri, India

Hideo Ebine, Ph.D.
Director
Central Miso Research Institute
Tokyo, Japan

Nobutake Nunomura, Ph.D.
Research Scientist
Soy Sauce Science Research Laboratory
Kikkoman Corporation
Noda, Chiba, Japan

S. Ayo Odunfa, Ph.D.
Senior Lecturer
Department of Botany
University of Ibadan
Ibadan, Nigeria

Teruo Ohta, Ph.D.
Director
Applied Microbiology Division
National Food Research Institue
Ministry of Agriculture, Forestry, and Fisheries
Tsukuba, Ibaraki, Japan

Sudhanand R. Pulusani, M.S.
Instructor
Department of Food Science and Animal Industries
Alabama A&M University
Normal, Alabama

D. Ramkishan Rao, Ph.D.
Professor
Department of Food Science and Animal Industries
Alabama A&M University
Normal, Alabama

Masaoki Sasaki, Ph.D.
Senior Research Scientist
Soy Sauce Science Research Laboratory
Kikkoman Corporation
Noda, Chiba, Japan

Shridhar K. Sathe, Ph.D.
Research Associate
Department of Foods and Nutrition
Purdue University
West Lafayette, Indiana

S. R. Shurpalekar, Ph.D.
Project Coordinator
Flour Milling and Baking Technology Discipline
Central Food Technological Research Institute
Mysore, India

Yuan-Chi Su, Ph.D.
Dean and Professor
College of Agriculture
National Taiwan University
Taipei, Taiwan

F. G. Winarno, Ph.D.
Director, Food Technology Development Center
Professor and Head, Food Science Department
Bogor Agricultural University
Bogor, Indonesia

TABLE OF CONTENTS

Chapter 1

INTRODUCTION

N. R. Reddy, M. D. Pierson, and D. K. Salunkhe

Fermentation is one of the oldest and most economical methods of producing and preserving food. Various types of fermentation have been used by nearly every civilization since prehistoric times. Foods are fermented for many reasons including enhancement of nutritive value and improvement in sensory characteristics like flavor and taste.[1] The increased nutritive value of fermented foods is due to the breakdown of complex components, such as carbohydrates, proteins, and lipids, to more easily digested sugars, free fatty acids, amino acids, as well as synthesis of certain vitamins.[2] In some parts of the world, large amounts of fermented foods are produced and serve as an essential part of the diet.[2] Many of the food fermentations are natural and/or mixed culture fermentations consisting of different species and genera of yeasts, fungi, and/or bacteria.

Food legumes represent a diverse group of plants. These are found worldwide and consumed in one form or other in practically every country of the world. About 20% of the protein currently available to man is derived from food legumes in developing countries. Fermented foods derived from legumes and cereals are an important part of the human diet in Southeast Asia, the Near East, and parts of Africa. Legume-based fermented foods originated centuries ago in these countries and are often referred to as indigenous fermented foods.[3] Some of legume-based fermented foods are also becoming popular in the Western world.[4] The popularity of legume-based fermented foods is due to desirable changes in the legume that include texture and organoleptic characteristics (flavor, aroma, and appearance or consistency), especially elimination of beany flavors, improvement in digestibility, enhanced keeping-quality of the product, improved safety (absence of toxins and partial and/or complete elimination of antinutritional factors), increased nutrition, and reduced cooking time. The organoleptic characteristics make fermented legumes more attractive to the consumer than the raw beans or legumes.

In many instances, the final legume-based fermented product is an important source of protein, calories, and certain vitamins. Soybeans, black gram, bengal gram, and mung beans are the four principal food legumes used in the preparation of legume-based fermented foods. Recently, several other food legumes have been substituted for legumes in some legume-based fermented foods. Food legumes can be fermented either alone or in combination with cereals. Legume-based fermented foods of different countries are presented in Table 1. The preparation, composition, nutritional quality, and food safety of the commonly fermented legumes are presented in this volume. Legume-based fermented foods are used in the human diet for various purposes[1-5] such as:

1. A main course (tempe and oncom in Indonesia and idli in India)
2. A flavoring agent for food (soy sauce in Japan, kecap in Indonesia, and dawadawa in Central and West Africa)
3. A soup base (miso in Japan and China)
4. To add color to the food and change physical state of the substrate

Fermentation of legume-based foods may even have a more important role in the future in providing essential food nutrients and new foods as well as being a means to

Table 1
LEGUME-BASED FERMENTED FOODS

Name	Substrate(s)	Microorganisms involved	Use of food	Areas
Dawadawa (Iru)	Locust beans	*Bacillus pumilus* *Bacillus licheniformis* *Bacillus subtilis* *Bacillus* spp.	Supplement to soups and stews Condiment	West and central Africa
Dhokla	Bengal gram and rice or wheat	*Leuconostoc mesenteroides* *Lactobacillus fermenti* *Streptococcus faecalis*	Breakfast or snack food	India (North India)
Dosai	Black gram and rice	*Leuconostoc mesenteroides* *Lactobacillus delbrueckii* *Lactobacillus fermenti* *Streptococcus faecalis* *Bacillus* spp. Yeasts	Breakfast or snack food	India
Idli	Black gram and rice	*Leuconostoc mesenteroides* *Lactobacillus fermenti* *Lactobacillus delbrueckii* *Lactobacillus lactis* *Streptococcus lactis* *Pediococcus cerevisiae* *Streptococcus faecalis* Yeasts	Breakfast food	India, Sri Lanka
Kecap	Soybeans and wheat	*Aspergillus oryzae* *Lactobacillus* spp. *Hansenula* spp. *Saccharomyces* spp.	Flavoring agent	Indonesia and vicinity
Kenima	Soybeans	Acid-producing bacteria	Snack food or condiment	Nepal, Sikkim, Darjeeling districts of India
Ketjap	Black soybeans	*Aspergillus oryzae*	Seasoning	Indonesia
Khaman	Bengal gram	*Leuconostoc mesenteroides* *Lactobacillus fermenti* *Lactobacillus lactis* *Pediococcus acidilactici* *Bacillus* spp.	Breakfast or snack food	India (North India)
Meitauza	Soybean press cake	*Mucor meitauza* *Actinomucor elegans*	Snack food	China, Taiwan
Meju	Black soybeans	*Aspergillus oryzae* *Rhizopus* spp.	Seasoning agent	Korea
Milk and related products	Soybeans	Lactic acid bacteria *Lactobacillus* spp.	Drink	China, Taiwan, Japan
Miso	Soybeans and wheat or rice or barley	*Aspergillus oryzae* *Streptococcus faecalis* *Pediococcus acidilactici* *Pediococcus halophilus* *Micrococcus* spp. *Bacillus* spp. *Saccharomyces rouxii* Other yeasts	Flavoring agent Soup base	East Asia Japan, China, U.S.
Natto	Soybeans	*Bacillus natto*	Cake as a meat substitute	Japan (northern part)
Oncom	Peanut press cake	*Neurospora intermedia* *Rhizopus oligosporus*	Snack	Indonesia

Table 1 (continued)
LEGUME-BASED FERMENTED FOODS

Name	Substrate(s)	Microorganisms involved	Use of food	Areas
Papadam, Papad	Black gram	*Saccharomyces* spp.	Condiment	India, Pakistan
Soy sauce (Shoyu)	Soybeans and wheat	*Aspergillus oryzae* *Saccharomyces rouxii* *Pediococcus halophilus* *Lactobacillus delbrueckii*	Seasoning	Japan, China, Taiwan, U.S.
Sufu (Chinese soybean cheese)	Soybeans	*Actinomucor elegans* *Mucor hiemalis* *Mucor silvaticus* *Mucor* spp.	Cheese	China, Taiwan
Tao-si	Soybeans and wheat flour	*Aspergillus oryzae*	Seasoning	Phillippines
Tao-tjo	Soybeans and roasted wheat meal or glutinous rice	*Aspergillus oryzae*	Condiment	East Indies
Tauco	Soybeans and cereals	*Rhizopus oligosporus* or *Aspergillus oryzae*	Seasoning	Indonesia, West Java
Tempe	Soybeans	*Rhizopus oligosporus* *Aspergillus oryzae*	Cake used as snack	Indonesia, New Guinea, Surinam, Netherlands, Canada, U.S.
Ugba	African oil bean	*Leuconostoc mesenteroides* *Bacillus* spp. *Staphylococcus* spp. *Micrococcus* spp.	Condiment	West and central Africa
Waries	Black gram or bengal gram	*Saccharomyces cerevisiae* *Candida krusei* Acid-producing bacteria	Condiment	India, Pakistan

preserve food.[1] Legume-based foods form an excellent vehicle for utilization of dry beans and providing adequate amounts of food energy, protein, other nutrients, and vitamins to the people of underdeveloped and developed countries. Further, these foods can also help eliminate protein-calorie malnutrition and other nutritional deficiencies in the developing countries.

REFERENCES

1. Hesseltine, C. W., The future of fermented foods, *Nutr. Rev.*, 41, 293, 1983.
2. Wang, H. L. and Hesseltine, C. W., Use of microbial cultures: legume and ceral products, *Food Technol.*, 35, 79, 1981.
3. Hesseltine, C. W. and Wang, H. L., Fermented foods, *Chem. Ind.*, 12, 393, 1979.
4. Reddy, N. R., Pierson, M. D., Sathe, S. K., and Salunkhe, D. K., Legume-based fermented foods: Their preparation and nutritional quality, *CRC Crit. Rev. Food Sci. Nutr.*, 17, 335, 1982.
5. Beuchat, L. R., Fermented soybean foods, *Food Technol.*, 38, 64, 1984.

Chapter 2

SOY SAUCE

N. Nunomura and M. Sasaki

TABLE OF CONTENTS

I. INTRODUCTION

Soy sauce is probably man's oldest prepared seasoning. Its exact origin is prehistoric and it probably was used to prevent food spoilage and preserve food. In the sixth century Buddhism came to Japan from China. Along with this new religion came a vegetarian seasoning, a salty paste of fermented beans. This is the earliest direct ancestor of soy sauce. In the 14th century, Japan broke away from China's influence and entered a period during which the distinctive Japanese culture developed. During this period, soy sauce in Japan underwent considerable development. The result was a seasoning with a fine balance of the proper concentrations of the major flavor-enhancing proteins, amino acids, sugars, and alcohols.

Soy sauce is known as "Shoyu" in Japan, "Chiang-yiu" in China, "Kecap" in Indonesia, "Kanjang" in Korea, "Toyo" in the Philippines, and "see-ieu" in Thailand. It is mainly used as an all-purpose seasoning in Japan, China, Taiwan, Malayasia, Indonesia, and other countries. Furthermore, it is also becoming popular in the U.S. as a sauce for barbecue, steaks, hamburgers, etc. Average consumption of shoyu in Japan is about 10 ℓ per capita per year. On an average, the Japanese consume about 34.1 g of shoyu daily, which supplies 2.4 g protein, 0.2 g fat, 5.8 g salt, and 14 Kcal.[1] Since less-technical information is available in the literature on Chinese soy sauces, this chapter will mainly describe the fermentation, types, composition, flavor, and other characteristics of Japanese soy sauces.

II. TYPES OF SOY SAUCE

The soy sauces from China differ in organoleptic qualities and composition from that of Japan. Chinese Chiang-yiu is dark in color and has a higher specific gravity, viscosity, and nitrogen content. Japanese shoyu is lower in viscosity, nitrogen, lighter red in color than Chiang-yiu, and has high glutamic acid content. The difference in aroma distinguishes the types of soy sauce from China and Japan. Differences in aroma are due to the use of different proportions of wheat and soybeans during the sauce preparation. Japanese soy sauce is usually pasteurized, while Chinese soy sauce is unpasteurized. The characteristic aroma compounds in the Japanese soy sauce are produced mainly from wheat constituents during koji fermentation, alcoholic fermentation, and aging.

The Japanese Agricultural Standards (JAS) define the qualities of various shoyu which are classified into five types: (1) koikuchi, (2) usukuchi, (3) tamari, (4) shiro, and (5) saishikomi. The production of these shoyu produced in Japan for 1982 is presented in Table 1. About 84% of the shoyu produced in Japan is of koikuchi type. It is prepared from equal parts of soybeans and wheat. Usukuchi amounts to less than 13% of the production. Tamari shoyu represents the Chinese style of soy sauce and originated in China. It represents about 2% of the total production and is made primarily of soybeans (90%) and small proportions of wheat (10%). Tamari shoyu is consumed in parts of Nagoya, Japan. The other two types, shiro and saishikomi, account for less than 1% of production. Shiro is primarily prepared from wheat with small amounts of soybeans. Saishikomi is made from enzymatically degraded soybeans and wheat, instead of the usual salt water.

The typical composition of the five types of shoyu are presented in Table 2. All types of shoyu have relatively high levels of salt, in the range of 14 to 19% (Table 2) and all are used as seasoning agents to enhance the flavor of meats, seafoods, and vegetables. Koikuchi is an all-purpose seasoning, characterized by strong aroma, myriad flavor, and deep-brown color. Usukuchi shoyu is used mainly for cooking, where one desired to preserve the original flavor and color of the foodstuff, and is characterized by a

Table 1
ANNUAL PRODUCTION OF DIFFERENT TYPES OF SHOYU IN JAPAN, 1982

Type	Production (kℓ)	Percent of total production
Koikuchi	902,862	84.3
Usukuchi	138,261	12.9
Tamari	20,885	2.0
Shiro	5,042	0.5
Saishikomi	3,130	0.3
Total	1,070,180	100.0

Table 2
TYPICAL COMPOSITION OF DIFFERENT TYPES OF SHOYU

Types	Salt %(w/v)	Total nitrogen %(w/v)	Formol nitrogen %(w/v)	Reducing sugars %(w/v)	Alcohol %(w/v)	pH
Koikuchi	16.95	1.58	0.92	2.99	2.45	4.79
Usukuchi	18.95	1.21	0.82	3.86	2.87	4.79
Tamari	16.85	2.35	1.14	6.98	0.68	4.85
Shiro	17.86	0.51	0.28	17.21	0.11	4.74
Saishikomi	14.30	2.35	1.17	9.43	1.33	4.84

light brown color and milder flavor. Tamari shoyu is distinguished by a deep dark color, thick consistency, and distinctive aroma. Shiro shoyu is characterized by a high level of reducing sugars and yellowish-tan color. Saishikomi is dark brown in color.

III. PREPARATION OF SHOYU

A. Raw Materials

Whole soybeans or defatted soybean flakes or meal and whole wheat (soft wheat) are mainly used as substrates for shoyu preparation. These substrates are used in various proportions to make different types of shoyu. For example, equal amounts of soybeans and wheat are used in preparation of kiokuchi shoyu. Generally defatted soybean flakes are used in the shoyu preparation because of their lower cost. Recently Ogawa and Fujita[2] successfully employed puffed granules of soybeans instead of cooked in the preparation of shoyu. Soybeans and wheat serve as a source of protein and starch, respectively, in shoyu fermentation.

B. Preparation

There are two distinct processes that are in use for shoyu preparation. One process involves fermentation with microorganisms, and the other is a chemical method that involves the use of acids to promote hydrolysis of substrate constituents. The chemical method is not considered traditional or indigenous, and the end product is considered to be inferior.[3] The preparation of koikuchi shoyu is described in detail here because it is the major type of shoyu produced. Soy sauces native to other countries are reviewed in detail by Steinkraus et al.[1]

Japanese fermented shoyu (koikuchi type) preparation is outlined in Figure 1. The

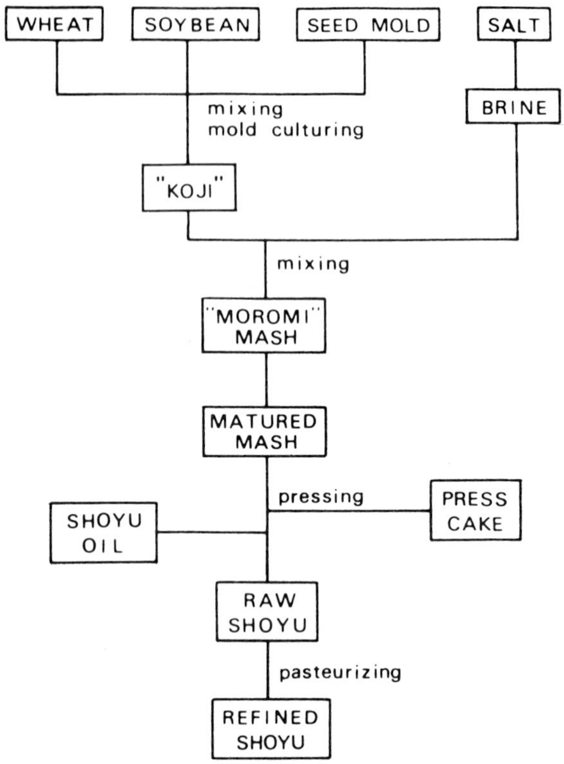

FIGURE 1. Preparation of Japanese fermented koikuchi shoyu.

process includes (1) the treatment of raw materials, (2) koji making, (3) mash production and aging, (4) pressing, and (5) refining and pasteurization.

1. Treatment of Raw Materials

Cleaned soybeans or soybean meal are soaked for 10 to 15 hr in water at room temperature during which a 2.0- to 2.2-fold increase in weight occurs. The water is changed every few hours to prevent acidification by bacteria and growth of undesirable spore-forming bacilli which may eventually be deleterious to the final end product. The hydrated beans are cooked under pressure at 10 to 14 psi for 1 hr in a cooker. The cooked beans are cooled rapidly. In addition, whole wheat kernels are roasted at 170 to 180°C for several minutes and coarsely crushed into four or five pieces. Wheat flour or bran may sometimes be used instead of whole wheat kernels. Roasting causes the breakdown and formation of various conversion products and a series of guaiacyl compounds such as vanillin, vanillic acid, ferulic acid and 4-ethylguiacol. Degradation of lignin and glycosides result in an increase in free phenolic compounds. Further roasting also causes the formation of an array of brown-colored reaction products which contribute to the desired visual properties of shoyu.

2. Koji Making

Equal parts of the resultant cooked soybeans or defatted soybean meal and roasted cracked wheat are mixed together and inoculated with 1 to 2% w/w of the seed koji or a pure culture of *Aspergillus oryzae* or *Aspergillus sojae*. Seed koji is produced by culturing selected strains of *A. oryzae* or *A. sojae* on steamed polished rice. The best shoyu is prepared from a soybean to wheat mixture of 50:50 w/w or 52:48 v/v.[1] The selection of a pure koji mold strain is important in the preparation of shoyu because of its influence on the quality of the final product, handling of koji making, activity of the enzymes, color of the product, etc.

The inoculated mixture is spread on large perforated stainless steel or wooden trays to a depth of 30 to 40 cm and incubated in a room at 25 to 30°C with humidity control for 2 to 3 days. During this period, the temperature, moisture, and aeration are controlled to allow the seed koji to grow on the mixture, to prevent development of *Mucor* spp. or bacteria, and to enhance production of proteolytic and amylolytic enzymes. A rather high moisture level (27 to 37%) is necessary for good mycelial growth and enzyme activity. The resulting end product (clear yellow to yellowish-green in color), is called "koji" which is a mixture of hydrolytic enzymes and the substrate.

3. Mash (Moromi) Production and Aging

The koji is mixed in deep fermentation tanks with an equal amount or more (up to 130% by volume) of salt water (22 to 23% w/v sodium chloride) to form the mash or moromi. The sodium chloride content of the mash should range from 17 to 19%. Concentrations of less than 16% salt may enable growth of undesirable putrefactive bacteria during subsequent fermentation and aging. A concentration of salt in excess of 23% may retard the growth of desirable osmophilic yeasts and halophilic bacteria. The mash or moromi is held in fermentation tanks for up to 8 months (depending upon its temperature) with occasional agitation with compressed air to mix the contents and stimulate microbial growth. The mycelium of koji mold is killed during very early stages of mash preparation.

During the early stages of fermentation, the enzymes from the koji hydrolyze proteins to yield low molecular weight peptides and free amino acids. Starch is converted to simple sugars which in turn are fermented to yield lactic, glutamic, and other acids, as well as alcohols and carbon dioxide, by specific microorganisms. The pH drops from an initial value of 6.7—7.0 to value of 4.7—4.9 during the brine fermentation period and aging. Elevated levels of carbon dioxide will enhance the growth of certain anaerobic microorganisms which in turn may impart undesirable flavor and aroma to the finished product. On the other hand, excessive aeration will hinder proper fermentation. To offset this, stirring must be correlated with the rate of carbon dioxide production. Often, pure cultures of *Pediococcus halophilus* and *Saccharomyces rouxii* are added to the mash at the start, and after 1 month, to promote the desirable fermentation and obtain good quality of shoyu. It has been reported that various groups of bacteria and yeasts predominate in sequence during the fermentation and aging of mash. In the first stage, *Pediococcus halophilus* grows and converts simple sugars to lactic acid, resulting in a pH drop. In the second stage, *Saccharomyces rouxii* and other yeasts grow and, as a result, vigorous alcoholic fermentation occurs. The salt concentration is changed to around 17 to 18% (w/v) after one or two months of fermentation, which restricts microbial growth to a few desirable osmophilic microorganisms. The pleasant aroma and flavor developed in the shoyu are largely due to the activities of microorganisms, especially lactic acid bacteria.

4. Pressing

Once aging of the mash is complete, the liquid (raw shoyu) is separated from the mash by a hydraulic filter press in commercial operations or by a simple mechanical press on a domestic level. Sometimes, fresh salt water is added to the press cake and a second fermentation is allowed to proceed for 1 to 2 months before a second pressed shoyu is produced which is of lower quality than the first. The final moisture content of the press cake is less than 25%. The press cake is called "shoyu cake" which is used as animal feed.

5. Refining and Pasteurization

Raw shoyu separates into three layers (sediment at the bottom, clear shoyu in the middle, and an oily layer at the top) upon standing in tanks. The oil layer is called "shoyu oil" and is removed by decantation. The middle layer, clear shoyu, is pasteurized at 70 to 80°C in a kettle or in a heat exchanger, thus killing vegetative microbial cells, denaturing enzymes, and coagulating proteins. Sometimes a filter aid is added to the pasteurized shoyu to enhance clarification. The clear shoyu is filtered, bottled, and marketed. Preservatives such as benzoic acid or propyl- or butyl *p*-hydroxy benzoate are sometimes added to the filtered shoyu before bottling. Steps for the industrial preparation of shoyu in Japan are mechanized which prevents contamination by undesirable microorganisms.

C. Production of Chemical and Semichemical Shoyu

Nonfermented chemical shoyu is prepared by acid hydrolysis of a mixture of wheat and soybean meal. The mixture is refluxed with constant boiling hydrochloric acid for 12 to 16 hr. The hydrolyzate is then neutralized with sodium hydroxide to pH 4.0 to 4.5, aged in hardwood casks, pasteurized, and bottled. This process offers a cost advantage over traditional preparation. During acid hydrolysis of the mixture, carbohydrates may be converted to undesirable compounds (dark humins, levulinic acid and formic acid) which are absent in traditionally fermented shoyu. Some of the amino acids and sugars are also destroyed during this chemical hydrolysis. The resulting compounds give undesirable flavor to the shoyu.

Compounds such as dimethylsulfide, hydrogen sulfide, and furfural, which have strong off-odors, are derived from methionine, sulfur-containing amino acids, and pentose sugar, respectively. The nutritionally important amino acid tryptophan is completely destroyed or denatured during chemical hydrolysis. Chemical shoyu lacks some of the flavors of traditionally fermented shoyu and has a very limited market.

In order to improve the flavor of chemical shoyu, a semifermented shoyu preparation method has been developed. In this method, raw soybeans are hydrolyzed with 7.0 to 8.0% hydrochloric acid and the hydrolyzate is neutralized with sodium hydroxide. Wheat bran or koji is added to the neutralized mixture. The resultant mixture is inoculated with yeasts and bacteria and allowed to ferment. The semifermented shoyu still contains undesirable compounds responsible for off-flavors. The nonfermented and semifermented soy sauces are mainly used as extenders of fermented shoyu in Japan.

To overcome the differences in soy sauces, the Japanese government has assigned three grades (special, upper, and standard) to soy sauce. These grades are distinguished by organoleptic evaluation, total nitrogen content, soluble solids other than sodium chloride, and color.

IV. COMPOSITION

A. Chemical Composition:

Compositional data of koikuchi and usukuchi shoyu are presented in Tables 2, 3, and 4. Both koikuchi and usukuchi shoyu contain appreciable amounts of total nitrogen in the range of 1.2 to 1.6% (Table 2), of which 80 to 90% are lower peptides, peptones, and amino acids. Ammonia makes up the remaining nitrogen. Some of the soluble protein is coagulated and removed during pasteurization.[1] Color development in shoyu is due in part to browning reactions between amino acids and sugars. Lactic, pyroglutamic, acetic, succinic, and formic acids are present in the koikuchi and usukuchi shoyu. These organic acids contribute to the aroma, flavor, color, and storage quality. Lactic acid is the major organic acid in these shoyu. There are 17 and 18

Table 3
ORGANIC ACID CONTENT (mg/mℓ) OF KOIKUCHI AND USUKUCHI SHOYU

Organic acid	Koikuchi	Usukuchi
Succinic acid	0.30	0.18
Lactic acid	6.84	7.66
Formic acid	0.11	0.02
Acetic acid	1.30	0.90
Pyroglutamic acid	3.09	1.11

Table 4
AMINO ACID CONTENT (mg/mℓ) IN KOIKUCHI AND USUKUCHI SHOYU

Amino acid	Koikuchi	Usukuchi
Aspartic acid	6.05	3.00
Threonine	2.95	2.21
Serine	4.22	3.20
Glutamic acid	11.76	12.13
Proline	4.20	3.34
Glycine	2.40	1.78
Alanine	5.64	4.32
Valine	4.45	3.40
Methionine	0.77	0.80
Isoleucine	4.22	3.07
Leucine	6.29	4.78
Tyrosine	1.09	0.43
Phenylalanine	3.40	2.73
Lysine	3.89	2.79
Arginine	3.00	0.79
Ornithine	1.29	2.10
Cystine	0.00	0.67
Histidine	1.07	0.41

individual amino acids, respectively, in koikuchi and usukuchi (Table 4), glutamic acid being the major amino acid in both. Glutamic acid and its salt are very important in flavoring. Shoyu contains neutral sugars such as glucose, galactose, arabinose, and xylose. Recently neutral sugars in the koikuchi shoyu were determined by liquid chromatography and found to be in the following range: glucose 11 to 13 mg/mℓ, galactose 4 to 6 mg/mℓ, arabinose 1 to 1.5 mg/mℓ, and xylose 0.5 to 1.0 mg/mℓ.[4]

B. Flavor Compounds

1. Japanese Shoyu Flavor Components

Research on Japanese shoyu flavor has been in progress since Tawara's study[5] in 1887. Obata and Yamanishi[6] and Yamanishi et al.[7] investigated the compounds which possessed shoyu flavor. Yokotuska[8] and Yokotuska et al.[9] attempted to find new groups of shoyu flavor components and studied mainly acetals, mercaptals, phenols, dicarbonyls, and other compounds by various chemical methods. Their study[8,9] placed emphasis on guaiacyl compounds especially 4-ethyl-2-methoxyphenol (4-ethylguaiacol), which had an important effect on the Japanese shoyu flavor. Despite these studies, some components that decisively characterize shoyu flavor still were not known.

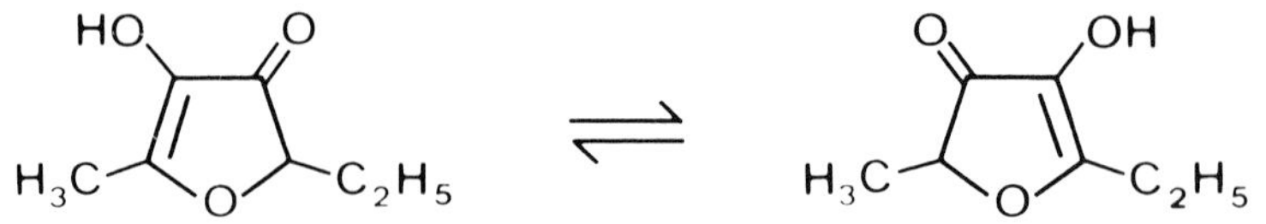

FIGURE 2. Structure of shoyu sweet flavor component, 4-Hydroxy-2 (or 5)-ethyl-5(or 2)-methyl-3(2H)-furanone (HEMF).

Recent improvements in isolation techniques and development of instrumental analysis made it possible to detect trace compounds in shoyu flavor. Nunomura et al.[10-14] separated shoyu flavor concentrates into basic, acidic, and neutral fractions and identified 168 new flavor components in shoyu by combined gas chromatography-mass spectrometry. Nunomura et al.[15] further isolated and identified for the first time an important unstable tautomeric sweet flavor component from shoyu, namely, 4-hydroxy-2(or 5)-ethyl-5(or 2)-methyl-3(2H)-furanone (HEMF) (Figure 2). HEMF possesses a shoyu-like and strong caramel-like flavor. A total of 271 flavor components have been found in Japanese shoyu until now (Table 5). The structures of caramel-like aroma compounds identified from Japanese shoyu are presented in Figure 3. These compounds are generally considered to be very important as flavor components of various foods, especially maltol which is well known as a flavor enhancer. 4-Hydroxy-2,5-dimethyl-3(2H) furanone (HDMF) was first isolated from pineapple.[16] The quantity of HDMF in shoyu is very small. HDMF and 4-hydroxy-5-methyl-3(2H)-furanone (HMF) are probably formed from sugars during heating processes such as the heat treatment of raw materials and the pasteurization of the liquid part of mash. 2-Hydroxy-3-methyl-2-cyclopenten-1-one (cyclotene), 2-acetyl-3-hydroxydifuran (isomaltol), and maltol have been found in various foods and are reported to be formed during the Maillard reaction. 3,5-Dihydroxy-2-methyl-4H-pyran-4-one has been previously isolated from roasted barley[17] and has a weak maltol-like odor. This compound has also been isolated from the degradation of D-glucose by acetic acid and methylamine.[18] The caramel-like aroma compounds in Figure 3 contain hydroxy group of enol in their structures and therefore are considered to exhibit acidic properties and to be water-soluble.

The weight ratios of the basic, acidic, and neutral fractions in the total flavor concentrate are as follows: basic, 27%; acidic, 57%, and neutral, 16%. The acidic fraction amounts to more than 50% of the total flavor concentrate. The weak acidic fraction imparts the characteristic shoyu flavor and this fraction also has an intense caramel-like aroma.

Among the caramel-like aroma compounds, HEMF, HMF, and maltol are present in high levels in shoyu. HEMF is by far the most abundant compound present at levels of 200 to 300 ppm. The odor of HEMF is reported to be more intense than that of HDMF[19] in shoyu. The odor threshold value of HDMF is 0.04 ppb in water.[20] The odor value of HEMF is calculated to be at more than five million. Among all the caramel-like aroma compounds present in the acidic fraction, HEMF has the greatest contribution to the good aroma of this fraction.[13]

The aroma of the fraction containing the higher-boiling compounds of the volatile flavor concentrate constitutes the base of shoyu flavor, and HEMF is one of the major and principal components of the fraction.[15] HEMF itself has an intense sweet and fruity aroma. HEMF is the major and most important constituent of the characteristic good flavor of shoyu.

Another important property of HEMF is its influence on the taste of shoyu. Addition of 0.1 ppm HEMF to shoyu makes the taste milder. The mitigative effect of HEMF on salty taste is very remarkable. Therefore, the taste of shoyu may become

Table 5
FLAVOR COMPOUNDS FOUND IN SHOYU[6,8-13,25,31-67]

I. Hydrocarbons (37)

Benzene
Toluene
Styrene
o-Xylene
m-Xylene
p-Xylene
Ethylbenzene
Mesitylene
1,2,3-Trimethylbenzene
1,2,4-Trimethylbenzene
1-Ethyl-2-methylbenzene
Cumene
Naphthalene
4-Methylindan
5-Methylindan
1,2,3,4-Tetrahydronaphthalene
1-Ethyl-2,3-dimethylbenzene
1-Ethyl-2,4-dimethylbenzene
1-Ethyl-3,5-dimethylbenzene
2-Ethyl-1,3-dimethylbenzene
2-Ethyl-1,4-dimethylbenzene
4-Ethyl-1,2-dimethylbenzene
1-Methyl-2(or 4)-propylbenzene
1,2,3,5-Tetramethylbenzene
1,2,4,5-Tetramethylbenzene
1,2-Diethybenzene
1,3-Diethylbenzene
1,4-Diethylbenzene
Butylbenzene
Cyclohexylcyclohexane
1-Methylnaphthalene
2-Methylnaphthalene
2,3,5(or 6)-Trimethylnaphthalene
Tetradecane
Pentadecane
Hexadecane
5-Phenyldodecane

II. Alcohols (30)

Methanol
Ethanol
2-Propen-1-ol
1-Propen-3-ol
1-Propanol
2-Propanol
2-Methyl-1-propanol
1-Butanol
2-Methyl-2-buten-1-ol
1-Penten-3-ol
3-Penten-2-ol
2-Methyl-1-butanol
3-Methyl-1-butanol
1-Pentanol
3-Pentanol
3-Buten-1,2-diol
2-Ethoxyethanol
L-2,3-Butanediol
meso-2,3-Butanediol
(E)-2-hexen-1-ol
1-Hexanol
Benzyl alcohol
2,3-Dimethyl-2-pentanol
2,4-Dimethyl-3-pentanol
3-Methyl-3-hexanol
2-Phenylethanol
1-Octen-3-ol
5-Nonanol
2-Phenyl-1-butanol
2-Undecanol

III. Esters (41)

Methyl acetate
Ethyl formate
Ethyl acetate
2-Oxopropyl acetate (Acetol® acetate)
Ethyl propionate
Butyl formate
1-Methylpropyl acetate
2-Methylpropyl acetate
Butyl acetate
Ethyl 2-hydroxypropionate (ethyl lactate)
3-Methylbutyl acetate
Pentyl acetate
2-Methylpropyl propionate
Ethyl 2-methylbutanoate
Ethyl 3-methylbutanoate
Ethyl pentanoate
2-Ethoxyethyl acetate
Ethyl 4-oxopentanoate (ethyl levulinate)
Ethyl hexanoate (ethyl caproate)
Diethyl oxalate
2-Phenylethyl formate
Ethyl benzoate
Diethyl malonate
2-Phenylethyl acetate
Ethyl phenylacetate
3-Methylbutyl 3-methylbutanoate
Ethyl octanoate (ethyl caprylate)
Diethyl maleate
Diethyl succinate
Ethyl 3-phenylpropenoate (ethyl cinnamate)
Pentyl hexanoate (amyl caproate)
Ethyl nonanoate (ethyl pelargonate)

Table 5 (continued)
FLAVOR COMPOUNDS FOUND IN SHOYU

III. Esters (continued)

2-Phenylethyl butanoate
4-Formyl-2-methyloxyphenol acetate (vanillin acetate)
Ethyl 4-hydroxy-3-methoxybenzoate (ethyl vanillate)
Ethyl dodecanoate (ethyl laurate)
Ethyl tetradecanoate (ethyl myristate)
Ethyl hexadecanoate (ethyl palmitate)
Ethyl 9,12-octadecadienoate (ethyl linoleate)
Ethyl 9-octadecanoate (ethyl oleate)
Ethyl octadecanoate (ethyl stearate)

IV. Aldehydes (15)

Acetaldehyde
Propanal
2-Methylpropanal
Butanol
2-Methylbutanal
3-Methylbutanal
Pentanal
Hexanal
Benzaldehyde
2,3-Dihydro-4H-pyran-2-carbaldehyde
Phenylacetaldehyde
3-Phenyl-2-propenal (cinnamaldehyde)
2,5-Dimethyl-2,3-dihydro-4H-pyran-2-carbaldehyde
2-Methyl-3-phenyl-2-propenal (α-methylcinnamaldehyde)
4-Hydroxy-3-methoxybenzaldehyde (vanillin)

V. Acetals (5)

1,1-Diethyoxyethane
1,1-Diethoxy-2-methylpropane
1,1-Diethoxy-2-methylbutane
1,1-Diethoxy-3-methylbutane
1,1-Diethoxy-4-methyl-2-pentanol

VI. Ketones (17)

Acetone
2-Butanone
Hydroxyacetone (Acetol®)
2,3-Butanedione (biacetyl)
3-Hydroxy-2-butanone (acetoin)
4-Methyl-3-penten-2-one
4-Methyl-2-pentanone
2-Hexanone
2,3-Oentanedione
3-Hydroxy-2-pentanone
2-Hydroxy-3-methyl-2-cyclopenten-1-one (cyclotene)
5-Methyl-2-hexanone
2,3-Hexanedione
Acetophenone
3-Octanone
2,6-Dimethyl-4-heptanone
2-Methyl-3-octanone

VII. Acids (24)

Formic acid
Acetic acid
Propionic acid
(*E*)-2-butenoic acid (crotonic acid)
2-Methylpropanoic acid
Butanoic acid
2-Oxopropanoic acid (pyruvic acid)
2-Hydroxypropanoic acid (lactic acid)
2-Methyl-2-butenoic acid
2-Methylbutanoic acid
3-Methylbutanoic acid
Pentanoic acid
2-Oxobutanoic acid
4-Methylpentanoic acid
Hexanoic acid (caproic acid)
4-Oxopentanoic acid (levulinic acid)
Butanedioic acid (succinic acid)
Benzoic acid
Phenylacetic acid
Octanoic acid (caprylic acid)
Dodecanoic acid (lauric acid)
Hexadecanoic acid (palmitic acid)
9,12,-Octadecadienoic acid (linoleic acid)
9-Octadecacenoic acid (oleic acid)

Table 5 (continued) FLAVOR COMPOUNDS FOUND IN SHOYU

VIII. Phenols (16)

Phenol
1,2-Benzenediol (pyrocatechol)
4-Ethylphenol
2-Methoxyphenol (guaiacol)
4-(2-Hydroxyethyl)phenol (tyrosol)
4-Ethyl-1,3-benzenediol (4-ethylresorcinol)
4-Hydroxybenzoic acid
2-Methoxy-5-vinylphenol
4-Ethyl-2-methoxyphenol (4-ethylguaiacol)
2,6-Dimethoxyphenol
3,4-Dihydroxybenzoic acid (protocatechuic acid)
3-(4-Hydroxyphenyl)propenoic acid (*p*coumaric acid)
4-Hydroxy-3-methoxyacetophenone (acetovanillin)
4-Hydroxy-3-methoxybenzoic acid (vanillic acid)
3-(4-Hydroxy-3-methoxyphenyl)propenoic acid (ferulic acid)
4-Hydroxy-3,5-dimethoxybenzoic acid (syringic acid)

IX. Furans (16)

Furan
2-Methylfuran
2-Furfural
Furfuryl alcohol
Tetrahydrofurfuryl alcohol
1-(2-Furyl)-1-ethanone (2-furyl methyl ketone)
5-Methyl-2-furfural
1-(2-Tetrahydrofuryl)-1-ethanone (2-tetrahydrofuryl methyl ketone)
1-(2-Furyl)-1-propanone (ethyl 2-furyl ketone)
2-Furfuryl acetate
1-(3-Hydroxy-2-furyl)-1-ethanone (isomaltol)
5-Hydroxymethyl-2-furfural
1-(2,5-Dimethyl-3-furyl)-1-ethanone
Ethyl 2-furoate
3-Phenylfuran
2-Propenyl 2-furoate

X. Lactones (4)

4-Butanolide
2-Penten-4-olide (5-methyl-2(5H)-furanone)
2-Methyl-4-butanolide
4-Pentanolide

XI. Furanones (4)

2-Methyl-3-tetrahydrofuranone
4-Hydroxy-5-methyl-3(2H)-furanone
4-Hydroxy-2,5-dimethyl-3(2H)-furanone
4-Hydroxy-2(or 5)-ethyl-5(or 2)-methyl-3(2H)-furanone

XII. Pyrones (5)

3-Hydroxy-2-methyl-4H-pyran-4-one (maltol)
5-Hydroxy-2-methyl-4H-pyran-4-one
3-Methoxy-2-methyl-4H-pyran-4-one
3,5-Dihydroxy-2-methyl-4H-pyran-4-one
3,5-Dihydroxy-6-methyl-2,3-dihydro-4H-pyran-4-one

XIII. Pyrazines (25)

Pyrazine
2-Methylpyrazine
2,3-Dimethylpyrazine
2,5-Dimethylpyrazine
2,6-Dimethylpyrazine
2-Ethylpyrazine
5H-Cyclopenta(b)pyrazine
2-Methyl-6-vinylpyrazine
6,7-Dihydro-5H-cyclopenta(b)pyrazine
2,3,5-Trimethylpyrazine
2-Ethyl-5-methylpyrazine
2-Ethyl-6-methylpyrazine
2(or 3)-Methyl-5H-cyclopenta(b)pyrazine
6-Methyl-5H-cyclopenta(b)-pyrazine
7-Methyl-5H-cyclopenta(b)-pyrazine
2-Methyl-6,7-dihydro-5H-cyclopenta(b)-pyrazine
Tetramethylpyrazine
3-Ethyl-2,5-dimethylpyrazine
2,3-Diethylpyrazine
2,6-Diethylpyrazine
2(or 3),6(or 7)-Dimethyl-5H-cyclopenta(b)-pyrazine
2-Ethyl-6,7-dihydro-5H-cyclopenta(b)-pyrazine

Table 5 (continued)
FLAVOR COMPOUNDS FOUND IN SHOYU

XIII. Pyrazines (continued)

2-Ethyl-3,5,6-trimethylpyrazine
2,6-Diethyl-3-methylpyrazine
2,3,5-Trimethyl-6,7-dihydro-5H-cyclopenta(b)-pyrazine

XIV. Pyridines (7)

Pyridine
3-Methylpyridine
2,6-Dimethylpyridine
2-Ethylpyridine
2-Pyridylmethanol
3-Methoxypyridine
Ethyl-3-pyridinecarboxylate (ethyl nicotinate)

XV. Miscellaneous nitrogen compounds (6)

1-Methyl-2-pyrrolidinone
1-(-2-Pyrrolyl)-1-ethanone
Benzoxazole
1-(5-Methyl-2-pyrrolyl)-1-ethanone
1,5-Dimethyl-2-pyridone
2-Methylbenzoxazole

XVI. Sulfur compounds (11)

Methanethiol
Dimethyl sulfide
Ethanethiol
2-Propene-1-thiol
Dimethyl disulfide
3-(Methylthio)-1-propanol (methionol)
Phenylmethanethiol
Dimethyl trisulfide
2-Ethyl-3-methyl-1,3-oxathiane
3-(Methylthio)propyl acetate
1,1-Bis(methylthio)-2-methylpropane

XVII. Thiazoles (3)

2-Ethoxythiazole
2-Butoxythiazole
2-(Methylthio)benzothiazole

XVIII. Terpenes (3)

Borneol
Bornyl acetate
4-Methyl-2-(2-methyl-1-propenyl)-tetrahydropyran (*cis*-Rose oxide)

XIX. Miscellaneous compounds (2)

β-Methoxystyrene
1,5-Dimethoxynapthalene

milder by an increase in quantity of HEMF during manufacturing. HEMF is produced by yeasts (e.g., *Saccharomyces* or *Torulopsis*) during mash fermentation and its precursor may be formed during koji making.[21]

2. Oxidative Products of HEMF

Asao and Yokotsuka[22] found a weak acidic fraction of shoyu that had an intense and very good aroma. They isolated this aroma-producing compound from the weak acidic fraction and found its formula to be $C_7H_{12}O_5$. Further, they proposed the compound structure shown in Figure 4. Nunomura et al.[23] isolated oxidative products of HEMF (OX-HEMF) (Figure 5). The ^{13}C-NMR and IR spectra of OX-HEMF are pre-

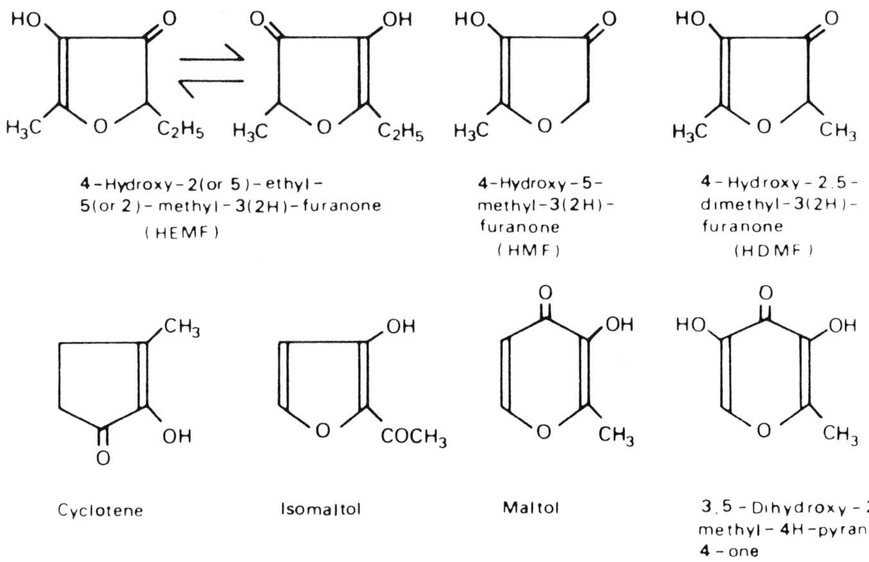

FIGURE 3. Caramel-like aroma compounds found in Japanese shoyu.

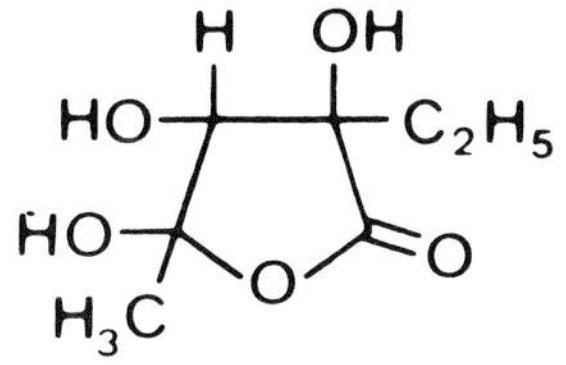

FIGURE 4. Proposed structure of flavor compound isolated by Asao and Yokotsuka.[22]

sented in Figures 6 and 7. Nunomura et al.[23] further showed that the compound isolated by Asao and Yokotsuka[22] was the same as OX-HEMF.

OX-HEMF is formed from HEMF under basic conditions. Autooxidation of HEMF gives α-keto acids, aldehydes, and carboxylic acids. However, HEMF appears to be stable in shoyu when stored under normal conditions. Various oxidation processes of HEMF are shown in Figure 8.[23]

3. Flavor Components of Koji, Matured Mash, and Shoyu Oil

Sasaki and Nunomura[24] obtained the flavor concentrates of koji by three methods: concentration of headspace gas aroma (top note aroma), carbon dioxide distillation of dichloromethane extract, and fractional extraction of dichloromethane extract. About 66 flavor components have been identified from the koji flavor concentrate by gas chromatography-mass spectrometry (Table 6). Phenylacetaldeyde and 1-octen-3-ol in the neutral fraction were responsible for the overall characteristic flavor of koji.

Matured mash and shoyu oil flavors were recently investigated by Nunomura et al.[25] The flavor concentrate from the matured mash was prepared by extraction with dichloromethane. About 43 components were identified from the flavor concentrate of matured mash (Table 7). The flavor fraction of shoyu oil was concentrated by means of ethanol vapor distillation under reduced pressure and 32 components from this fraction were identified (Table 8). Both the flavor concentrates from matured mash and

FIGURE 5. Structure of oxidation products of HEMF (OX-HEMF).[23]

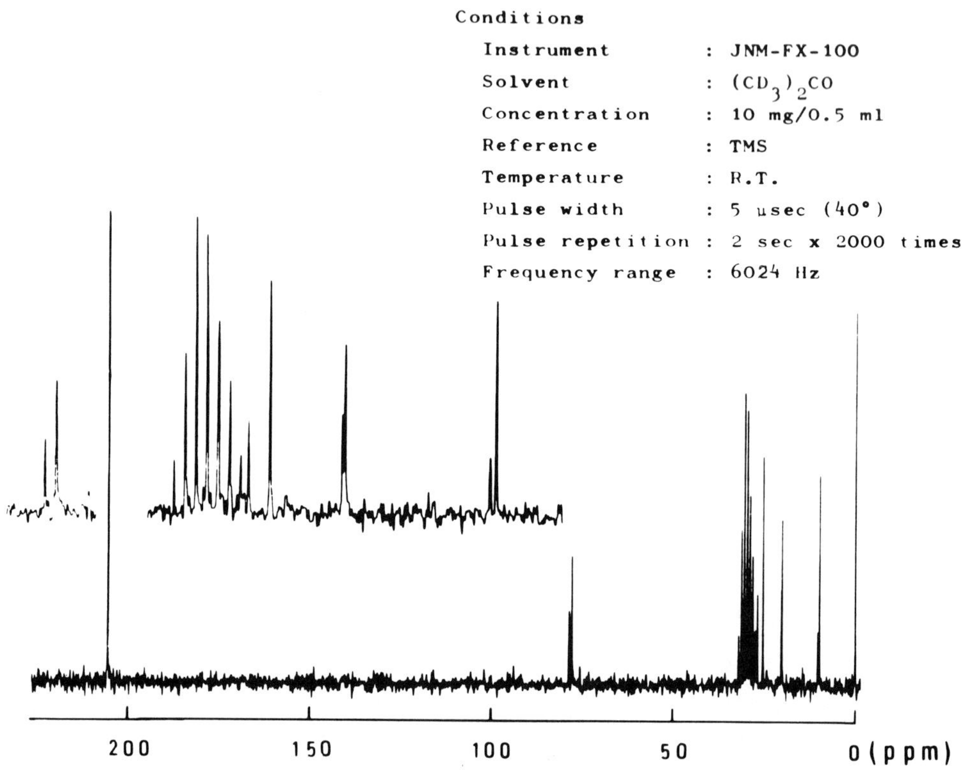

FIGURE 6. ^{13}C-NMR spectrum of OX-HEMF.

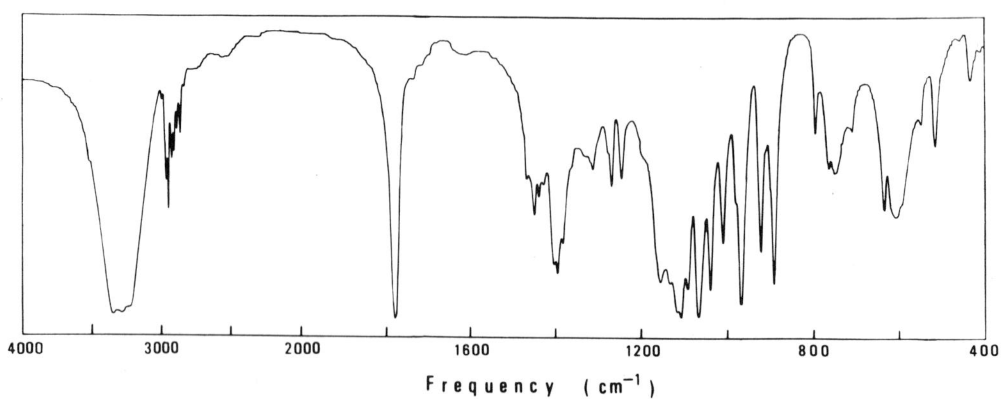

FIGURE 7. IR spectrum of OX-HEMF.

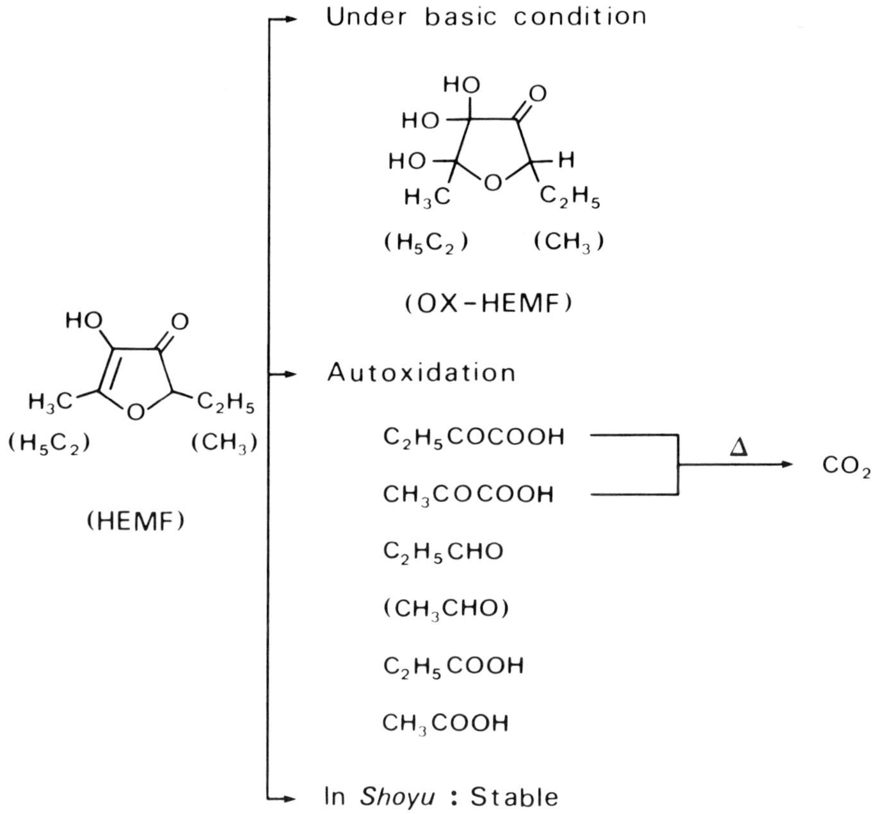

FIGURE 8. Oxidation products of HEMF.

shoyu oil contained appreciable amounts of many ethyl esters of fatty acids. These esters appear to contribute to the sweet aroma of matured mash.

Further, Nunomura et al.[25] prepared and analyzed flavor concentrates from the press cake. The ethyl esters of fatty acids are the major components in the flavor concentrate of press cake.

Gas chromatograms of the flavor concentrates of matured mash, press cake, and raw shoyu are presented in Figure 9. The peaks between 60 and 80 min retention time are due to ethyl esters of fatty acids (C_{14}-C_{20}). Apparently, the ethyl esters in matured mash remain in press cake and the lower-boiling compounds move to the liquid layer, raw shoyu. The most important component of the characteristic shoyu flavor, HEMF appears at 45 min retention time, and its peak is the largest in the raw shoyu gas chromatogram.

4. Quantitative Analysis of Flavor Compounds

Caution should be taken in the quantitative analysis of shoyu flavor components because they are volatile and delicate. Extraction of shoyu with methyl acetate[26] is one of the most suitable methods for preparation of flavor extracts for quantitative analysis (Figure 10). The recovery of individual components and coefficients of variation for reproducibility by this method are good. Individual components of the flavor extract can be quantified by gas chromatography. Typical quantitative data for individual flavor components of Japanese shoyu is presented in Table 9. Ethanol, lactic acid, glycerol, acetic acid, HMF, 2,3-Butanediol, isovaleraldehyde, and HEMF are the major flavor components in shoyu (Table 9).

Table 6
FLAVOR COMPONENTS FOUND IN SHOYU KOJI

I. Hydrocarbons (4)	
o-Xylene	*p*-Xylene
m-Xylene	*n*-Octane
II. Alcohols (11)	
Methanol	1-Pentanol
Ethanol	2-Methyl-1-penten-3-ol
2-Methyl-1-propanol	1-Hexanol
1-Butanol	3-Octanol
2-Methyl-1-butanol	1-Octen-3-ol
3-Methyl-1-butanol	
III. Esters (9)	
Methyl acetate	Ethyl 2-methylbutanoate
Ethyl acetate	Ethyl 3-methylbutanoate
Ethyl 2-methylpropionate	Ethyl hexadecanoate (ethyl palmitate)
Ethyl-2-hydroxypropionate (ethyl lactate)	Ethyl 9,12-octadecadienoate (ethyl linoleate)
3-Methylbutyl acetate	
IV. Aldehydes (9)	
Acetaldehyde	Benzaldehyde
Propanal	Phenylacetaldehyde
2-Methylpropanal	4-Hydroxy-3-methoxybenzaldehyde (vanillin)
Butanal	4-Hydroxy-3,5-dimethoxybenzaldehyde (syringaldehyde)
3-Methylbutanal	
V. Acetal (1)	
1,1-Dimethoxypropane	
VI. Ketones (8)	
Acetone	3-Hexanone
3-Hydroxy-2-butanone (acetoin)	2-Heptanone
4-Methyl-2-pentanone	3-Octanone
2-Hexanone	6-Methyl-2-heptanone
VII. Acids (6)	
Acetic acid	2-Methylbutanoic acid
2-Methylpropanoic acid	Pentanoic acid
Butanoic acid	Phenylacetic acid
VIII. Phenols (4)	
2-Methoxyphenol (guaiacol)	3-(4-Hydroxy-3-methoxyphenol)propenoic acid (ferulic acid)
2-Methoxy-5-vinylphenol	
2,6-Dimethoxyphenol	
IX. Furans (2)	
Furfuryl alcohol	2-Acetylfuran

Table 6 (continued)
FLAVOR COMPONENTS FOUND IN SHOYU KOJI

X. Lactone (1)

4-Butanolide

XI. Pyrones (3)

3-Hydroxy-2-methyl-4H-pyran-4-one (maltol)	3-Methoxy-2-methyl-4H-pyran-4-one
5-Hydroxy-2-methyl-4H-pyran-4-one	

XII. Pyrazines (3)

2,5-Dimethylpyrazine	2,3,5-Trimethylpyrazine
2,6-Dimethylpyrazine	

XIII. Miscellaneous nitrogen compounds (2)

1-Methyl-2-pyrrolidinone	Methyl nicotinate

XIV. Sulfur compounds (3)

Dimethyl disulfide	Bis(2-methylpropyl)disulfide
3-(Methylthio)propanal (methional)	

The head space gas of shoyu can also be quantitated for individual flavor components by gas chromatography.[27] The concentration of individual head space gas components for a typical shoyu are shown in Table 10. Ethanol and ethyl acetate are the major components in the head space of shoyu.

5. Change in Flavor Components During Pasteurization

Nunomura et al.[11] separated and quantitated the basic fraction from shoyu flavor concentrate for individual pyrazine compounds. The pyrazine compounds increase during pasteurization of shoyu (Table 11). These pyrazines appear to significantly contribute to the aroma of heated shoyu. Nunomura et al.[28] further studied the effects of heating or pasteurization on individual selected flavor components, such as furfuryl alcohol, 2-methylpropanol, and 3-methylbutanal. These flavor components increase linearly with heating time and exponentially with temperature. The relations are illustrated in Figures 11 to 14. Other flavor compounds such as propanal, dimethyl sulfide, ethylene sulfide, dimethyl disulfide, 3-(methylthio) propanal, HMF, HDMF, 3, 5-dihydroxy-6-methyl-2,3-dihydro-4H-pyran-4-one, ethyl 2-hydroxy propionate, 2-phenylethyl acetate, 3-hydroxy-2-butanone, and methyl 2-pyrrolyl ketone increase in concentration during heating. These compounds are considered to be produced by degradation of amino acids, sugars, HEMF, and other components.[28]

6. Flavor Components of Various Soy Sauces in the World

Nunomura and Sasaki[29] quantitatively estimated the flavor components of soy sauces manufactured in various countries of the world. They examined 39 samples of soy sauce from various countries (Table 12) for analysis of individual flavor components. The concentration of aldehydes (total), furfuryl alcohol, dimethyl sulfide, alcohols (total), ethyl formate, pentanoic acid, 3-methyl butanoic acid, and HEMF for these samples are presented in Figures 15 through 22. Aldehydes, furfuryl alcohols, and alcohols were detected in all samples (Figures 15 through 18). The concentration

Table 7
FLAVOR COMPONENTS FOUND IN MATURED MASH

I. Alcohols (9)	
Ethanol	L-2,3-Butanediol
2-Methyl-1-propanol	*meso*-2,3-Butanediol
1-Butanol	Furfuryl alcohol
2-Methyl-1-butanol	2-Phenylethanol
3-Methyl-1-butanol	
II. Esters (16)	
Ethyl acetate	Ethyl pentadecanoate
Ethyl lactate	Ethyl hexadecanoate (ethyl palmitate)
Ethyl octanoate	Ethyl heptadecanoate
Ethyl decanoate	Ethyl octadecanoate (ethyl stearate)
Ethyl benzoate	Ethyl 9-octadecanoate (ethyl oleate)
Ethyl phenylacetate	Ethyl 9,12-octadecadienoate (ethyl linoleate)
Ethyl dodecanoate (ethyl laurate)	Ethyl 9,12,15-octadecatrienoate (ethyl linolenate)
Ethyl tetradecanoate (ethyl myristate)	Ethyl eicosanoate (ethyl arachidate)
III. Acids (5)	
Acetic acid	Octadecanoic acid (stearic acid)
Butanoic acid	9,12-Octadecadienoic acid (linoleic acid)
3,4-Dimethoxyphenylacetic acid	
IV. Carbonyls (10)	
Propanal	3-Hydroxy-2-methyl-4H-pyran-4-one
3-Hydroxy-2-butanone (acetoin)	1-(2-Pyrrolyl)-1-ethanone
Hydroxyacetone (Acetol®)	4-Hydroxy-2,5-dimethyl-3(2H)-furanone
2-Furfural	4-Hydroxy-2(or 5)-ethyl-5(or 2)-methyl-3(2H)-furanone
4-Butanolide	
3-Methoxy-2-methyl-4H-pyran-4-one	
V. Phenols (3)	
4-Ethylguaiacol	2,6-Dimethoxyphenol
2-Methoxy-5-vinylphenol	

of dimethyl sulfide in sauces from Taiwan and the U.S. is higher than that from other countries (Figure 17). The alcohol content in the sauces indicates the extent of fermentation. A high concentration of ethyl formate was detected in sauces from Korea and Taiwan (Figure 19). The presence of high concentrations of dimethyl sulfide and ethyl formate in sauces from Taiwan, U.S., and Korea indicated that some sauces from these countries were chemical soy sauces. Formic acid is formed during hydrochloric acid hydrolysis of soybeans and subsequently ethyl formate is produced during fermentation. Pentanoic acid was also present in higher concentrations in the sauces from China, Taiwan, and U.S., further indicating that these sauces were chemical sauces (Figure 20). In general, chemical sauces contain appreciable amounts of dimethyl sulfide, formic acid, pentanoic acid, levulinic acid, and furfurols which give undesirable flavor to the sauces. 3-Methylbutanoic acid was detected in all sauces (Figure 21). This compound is produced by microorganisms during fermentation. The higher content of 3-methylbutanoic acid in sauces is preferred in Japan. HEMF was not detected in the

Table 8
FLAVOR COMPONENTS FOUND IN SHOYU OIL

I. Alcohols (10)	
Ethanol	1-Pentanol
1-Propanol	Furfuryl alcohol
2-Methyl-1-propanol	3-Methylthio-1-propanol (methionol)
1-Butanol	Benzyl alcohol
3-Methyl-1-butanol	2-Phenylethanol
II. Esters (13)	
Ethyl lactate	Ethyl hexadecanoate (ethyl palmitate)
Ethyl octanoate	Ethyl octadecanoate (ethyl stearate)
Ethyl benzoate	Ethyl 9-octadecenate (ethyl oleate)
Diethyl succinate	Ethyl 9,12-octadecadienoate (ethyl linoleate)
Ethyl phenylacetate	Ethyl 9,12,15-octadecatrienoate
Ethyl dodecanoate (ethyl laurate)	Ethyl eicosanoate (ethyl arachidate)
Ethyl tetradecanoate (ethyl myristate)	
III. Carbonyls (6)	
3-Hydroxy-2-butanone (acetoin)	3-Hydroxy-2-methyl-4H-pyran-4-one
1-(2-Furyl)-1-ethanone	4-Hydroxy-2(or 5)-ethyl-5(or 2)-methyl-3(2H)-furanone
3-Methoxy-2-methyl-4H-pyran-4-one	
1-(2-Pyrrolyl)-1-ethanone	
IV. Phenols (3)	
Guaiacol	4-Ethylphenol
4-Ethylguaiacol	

sauces except those from Japan and one Korean soy sauce (Figure 22). The high concentrations of HEMF in Japanese shoyu is still the most remarkable feature.

Soy sauces from different parts of the world can be classified into the following types, A, B, C, C′, and D on the basis of the concentrations of dimethyl sulfide and alcohols (Figure 23). Types A and B are chemical soy sauces, while C and D are mostly fermented soy sauces. Chemical soy sauces contain appreciable amounts of dimethyl sulfide. Soy sauce type D (Japanese shoyu) contains large amounts of HEMF. The types of soy sauces produced in each country are presented in Table 13.

C. Organoleptic Quality

Organoleptically preferable shoyu should have a good harmony of taste components (salty, acidic, sweet, bitter, and delicious) and aroma. The flavor of shoyu should always be checked in addition to ordinary analysis. If significant relationships between shoyu flavor components and sensory evaluation are found, the analysis of flavor components can be used for the control of manufacturing process. An accurate analysis of these components is necessary for this purpose.

Recently Mori et al.[30] studied the relationships between Japanese shoyu flavor components and sensory evaluation. They used five kinds of shoyu before pasteurization. All test samples were pasteurized. In the sensory test, odor preference of the test samples was evaluated. The rank order test was used for measuring differences among the samples. Thurstone's Case-V scaling procedure was applied to the test data, and the scale values quantified by the Case-V were used for the statistical analysis. Simultaneously, 27 flavor components were quantitatively analyzed by using the head space

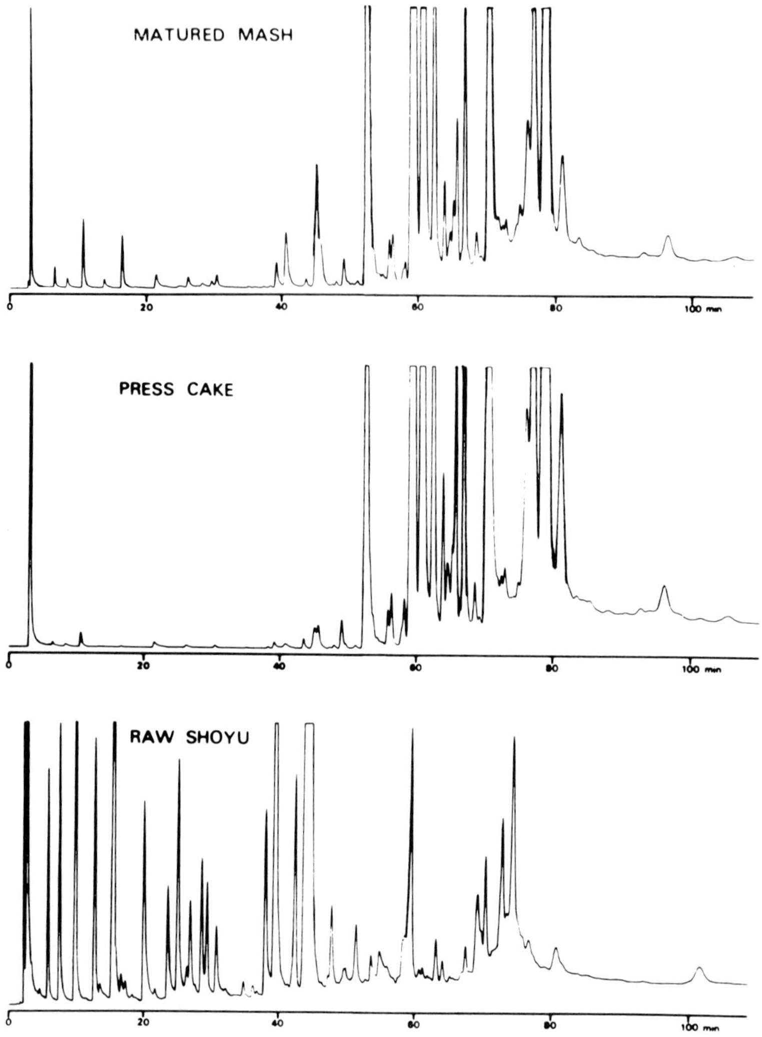

FIGURE 9. Gas chromatograms of flavor concentrates of matured mash, press cake, and raw shoyu.

gas and extraction methods. For statistical analysis, 8 models (linear, linear variant, logarithmic, hyperbolic, power, exponential, square root, and quadratic functions) were applied to combinations of data sets. Additionally, relations between the sum of concentrations of two components and the odor preference were also fitted with the 8 models. Then multiple regression analysis was used so that components with a statistically significant contribution could be distinguished. For estimating the best multiple regression equation, Allen's PSS (prediction sum of squares) and cross validation procedure for every combination of predictor variables were used.

As a result of this statistical analysis, three combinations of binary odor components accounting for the variation of sensory data were found. The combination of 4-ethyl-2-methoxyphenol (4-ethy-quaiacol, [4EG]) and 3-methylthio-1-propanol (methionol) was closely associated with the blending effect[31] and this combination accounted for about 28% of the variation of sensory data as determined by the following model:

Table 9
QUANTITATIVE ANALYSIS OF FLAVOR CONSTITUENTS IN JAPANESE SHOYU

Flavor component	Concentration (ppm)
Ethanol	31,501.10
Lactic acid	14,346.57
Glycerol	10,208.95
Acetic acid	2,107.74
4-Hydroxy-5-methyl-3(2H)-furanone (HMF)	256.36
2,3-Butanediol	238.59
Isovaleraldehyde	233.10
4-Hydroxy-2(or 5)-ethyl-5(or 2)-methyl-3(2H)-furanone (HEMF)	232.04
Methanol	62.37
Acetol®	24.60
Ethyl lactate	24.29
2,6-Dimethoxyphenol	16.21
Ethyl acetate	15.13
Isobutyraldehyde	14.64
Methyl acetate	13.84
Isobutyl alcohol	11.95
Furfuryl alcohol	11.93
Isoamyl alcohol	10.01
Acetoin	9.78
n-Butyl alcohol	8.69
4-Hydroxy-2,5-dimethyl-3(2H)-furanone (HDMF)	4.83
Acetaldehyde	4.63
2-Phenylethanol	4.28
n-Propyl alcohol	3.96
Acetone	3.88
Methionol	3.65
2-Acetylpyrrole	2.86
4-Ethylguaiacol	2.77
Ethyl formate	2.63
4-Butanolide	2.02
Methional	1.42
Dimethyl sulfide	0.04
4-Ethylphenol	trace

$$y = 8.937 + 22.612 \times \sqrt{\log (4EG + Methionol)} - 13.978 \times (4EG + Methionol) \qquad (1)$$

where "y" is the scale value for odor preference, and 4EG and Methionol refer to concentrations.

Next, in order to confirm the combination effect of 4EG and methionol which was distinguished by the statistical analysis, and experiment was carried out by adding these compounds (4EG and methionol) to shoyu.[32] After sensory evaluation and quantitative

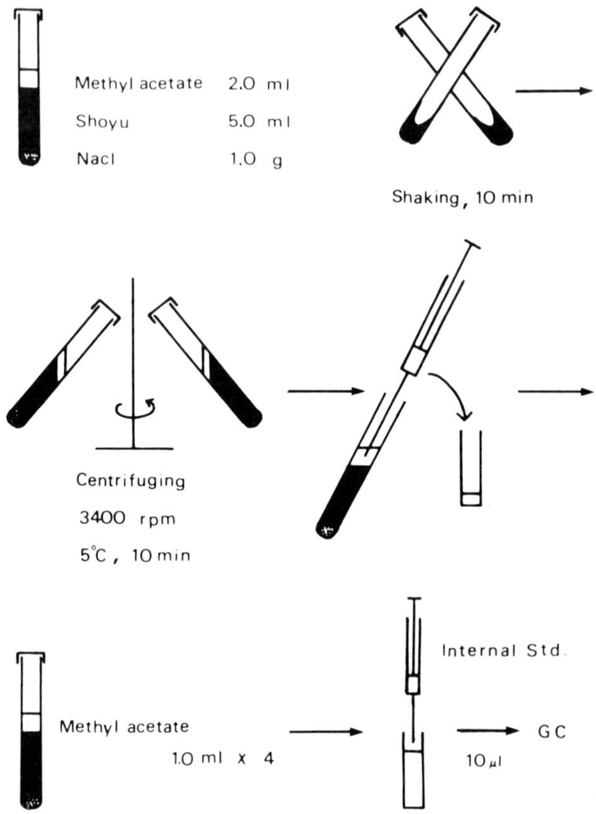

FIGURE 10. Procedure for direct extraction of shoyu with methyl acetate.

Table 10
QUANTITATIVE ANALYSIS OF COMPOUNDS OF HEADSPACE GAS FROM JAPANESE SHOYU

Compound	Concentration (ppm)
Ethanol	5605.18
Ethyl acetate	33.41
Methanol	9.45
Isovaleraldehyde	8.17
Isobutyraldehyde	6.38
Isobutyl alcohol	3.79
Acetaldehyde	3.76
Isoamyl alcohol	2.36
Acetone	2.09
Propionaldehyde	1.70
Ethyl formate	1.66
Dimethyl sulfide	0.88
n-Propyl alcohol	0.82
2,3-Pentanedione	0.76
n-Butyl alcohol	0.69
Dimethyl disulfide	0.03

Table 11
MAJOR PYRAZINE CONTENT FOR JAPANESE RAW AND HEATED SHOYU

Compound	Concentrations (mg/ℓ) Heated	Raw	Ratio of concentrations (Heated/Raw)
2-Methylpyrazine	0.075	0.024	3:1
Dimethylpyrazine	0.746	0.184	4:1
Ethyl-methylpyrazine	0.746	0.388	2:1
Trimethylpyrazine	0.050	0.040	1:1

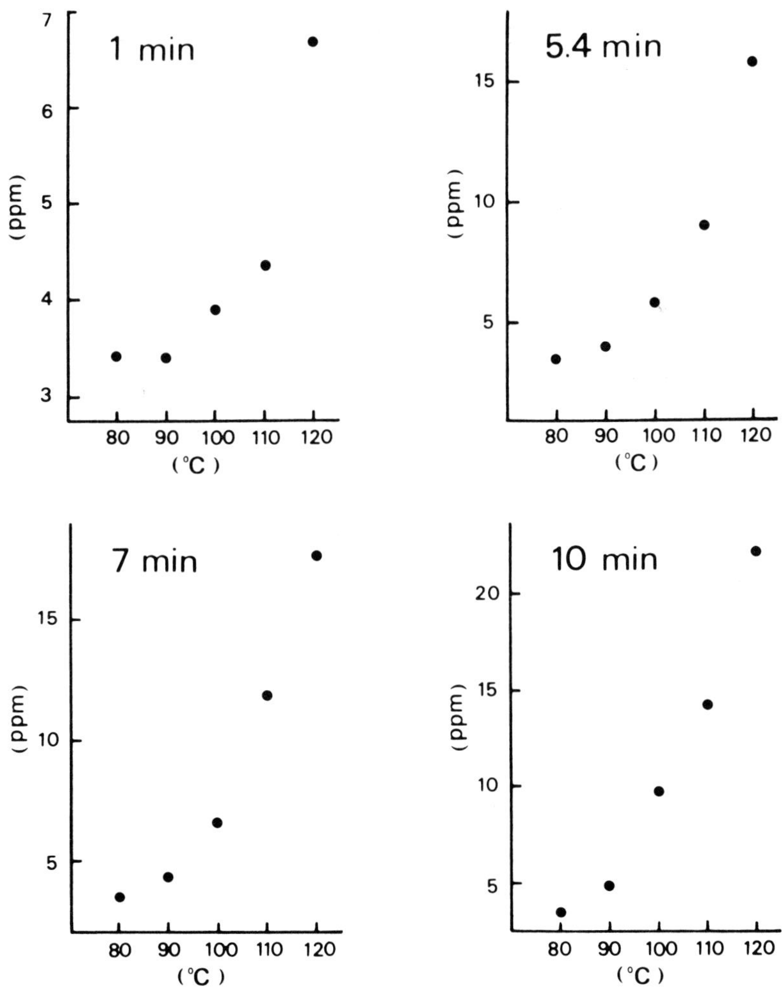

FIGURE 11. The effects of heating temperature on content of furfuryl alcohol in shoyu.

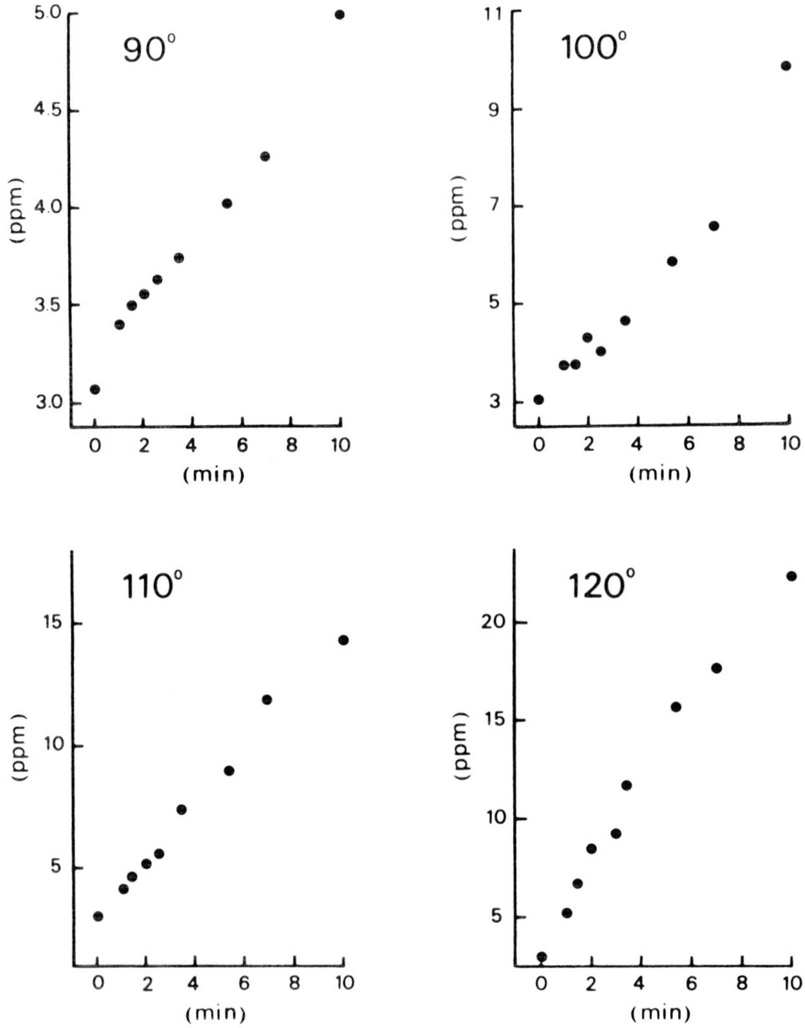

FIGURE 12. The effects of heating time on content of furfuryl alcohol in shoyu.

analysis of the flavor components, the statistical analysis (including multiple regression analysis) gave the following estimated equation.

$$y = 1.055 - 0.104 \times (4EG') + 0.006 \times (Met) - 0.103 \times (4EG')^2 - 0.045 \times (Met)^2 - 0.006 \times (4EG' \times Met)$$

y = the scale value of odor preference

$$4EG' = -0.234 + 2.188 \times \log(4EG)$$

$$Met = -4.106 + 1.091 \times \text{Methionol} \qquad (2)$$

The simplex design method was applied on odor preference surface generated from the multiple regression model (Equation 2), and the optimum concentrations of 4EG and methionol were found to be 0.8 ppm and 3.8 ppm, respectively.

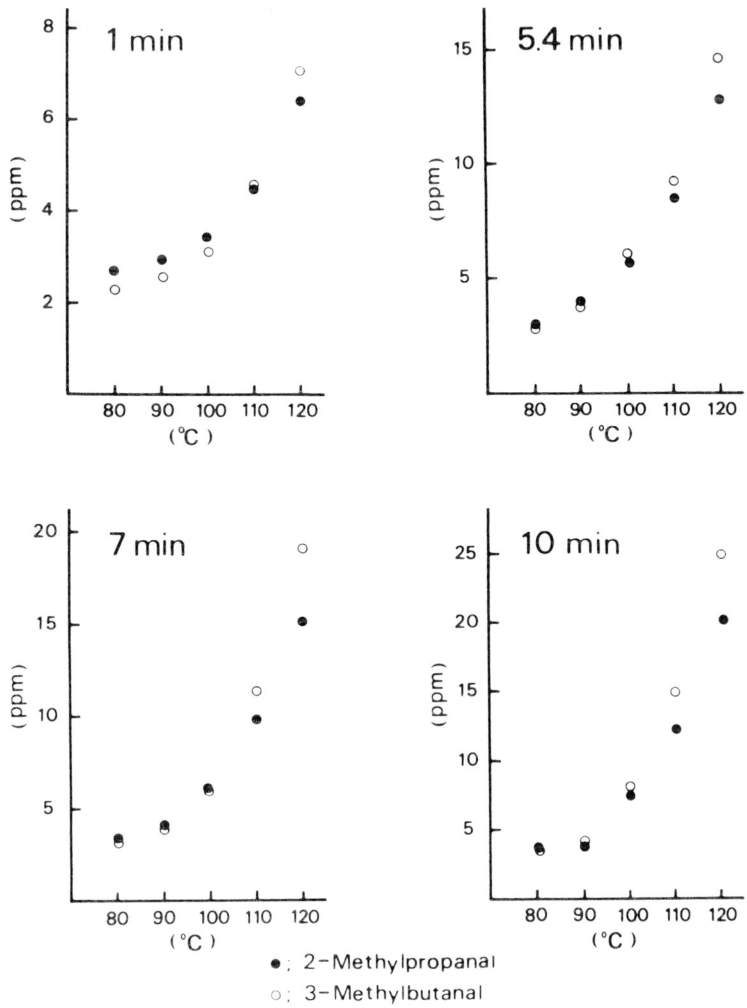

FIGURE 13. The effects of heating temperature on contents of 2-methyl propanal and 3-methyl butanal in headspace gas of shoyu.

4EG is formed during mash fermentation and aging by *Torulopsis*.[33] Both *Torulopsis* and *Saccharomyces* (which produces no 4EG) are reported to be important in Japanese shoyu manufacturing during mash making and aging. Methionol is produced by both these yeasts. The concentration of methionol can be used as an index for the extent of alcohol fermentation by yeasts.

V. SAFETY

Several researchers have reported on the safety of Japanese shoyu. The main concerns are the presence of mycotoxins in shoyu,[70] long term effects of consuming of shoyu, and mutagens in shoyu.

A. Mycotoxins

Secondary fungal metabolities that are toxic to humans or animal are referred to as "mycotoxins" and the disease initiated by these toxins is called "Mycotoxicosis".[71] The discovery of the aflatoxins[72,73] in the early 1960s resulted in an extensive reap-

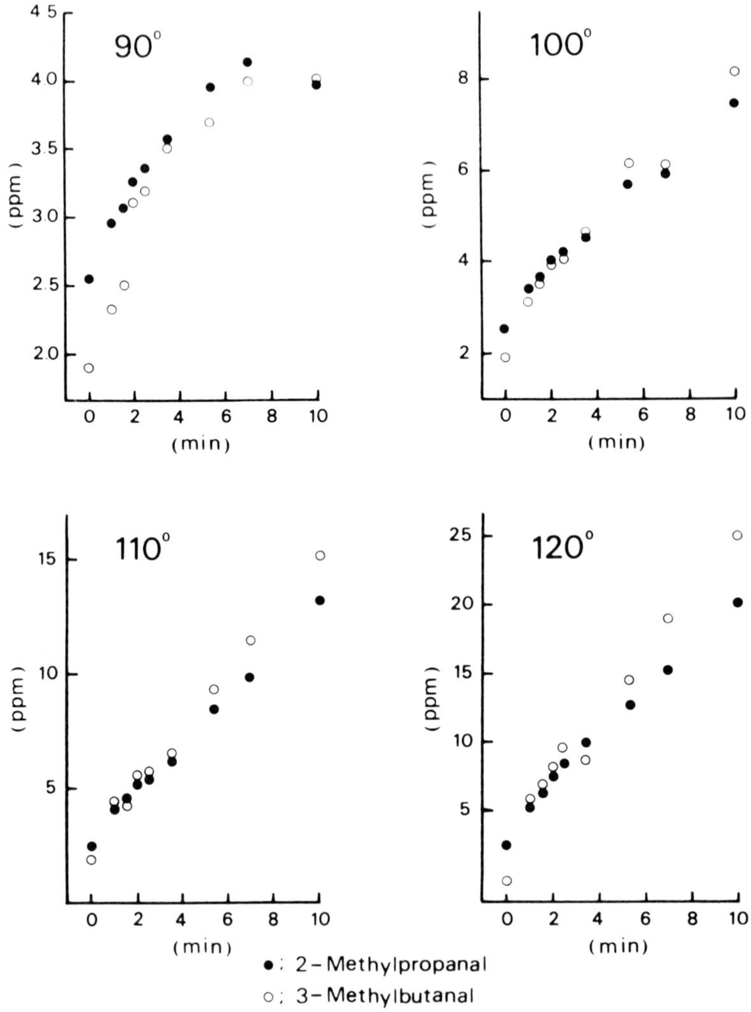

FIGURE 14. The effects of heating time of contents of 2-methyl propanal and 3-methyl butanal in headspace gas of shoyu.

Table 12
SOURCES OF SOY SAUCES USED FOR FLAVOR COMPONENT ANALYSIS

Countries	Number of samples used
People's Republic of China (China)	6
Singapore	5
Hong Kong	6
Korea	4
Republic of China (Taiwan)	4
U.S.	12
Japan	2
Total	39

Note: All of these soy sauces were prepared by using raw materials, soybeans, wheat, salt, and water.

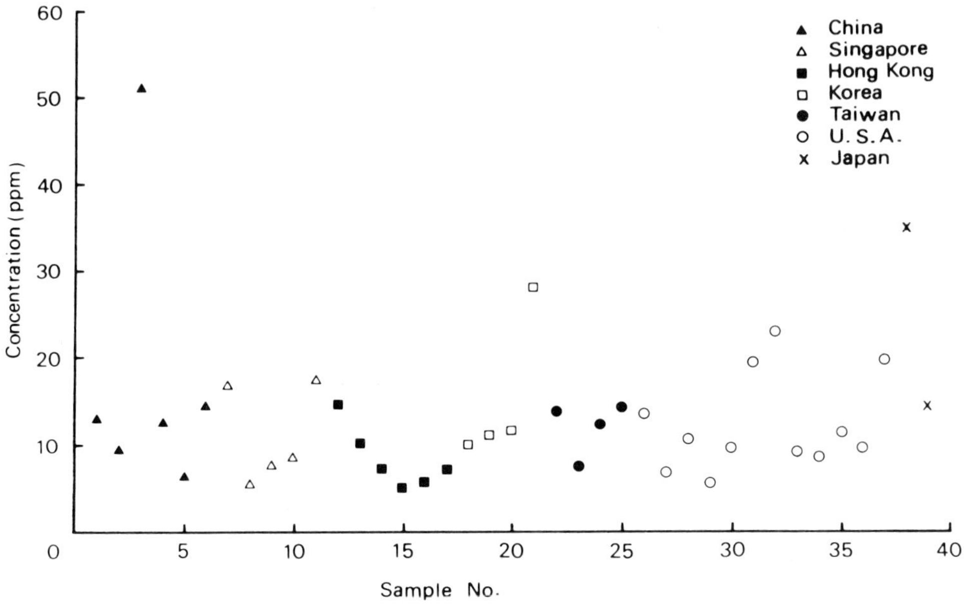

FIGURE 15. Concentration of aldehydes (propion-, isobutyr-, isovaler-) in various soy sauces.

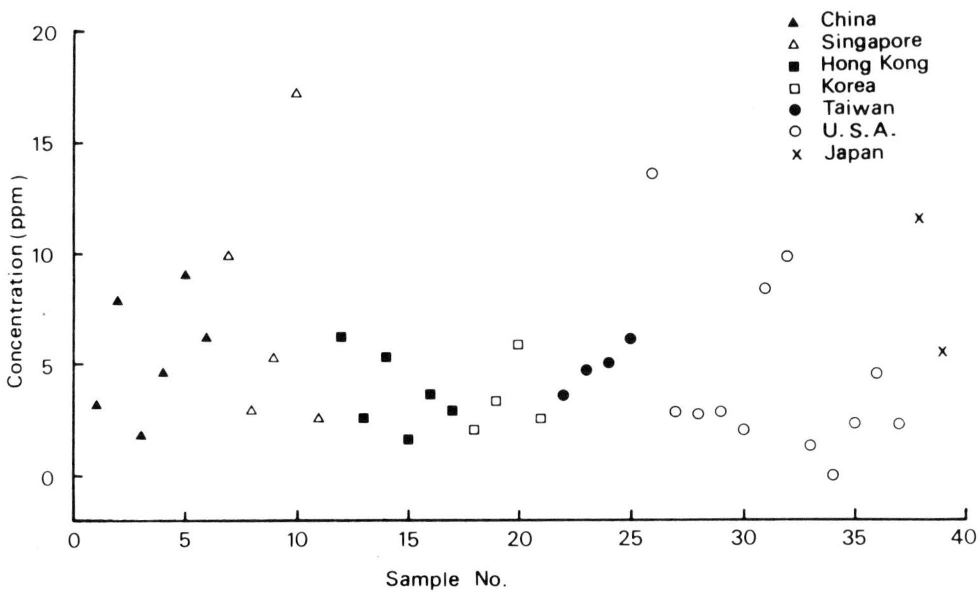

FIGURE 16. Concentration of furfuryl alcohol in various soy sauces.

praisal of the mycotoxin problem. The molds *Aspergillus flavus* and *Aspergillus paraciticus* are reported to produce aflatoxins.[74,75] Strains of *Aspergillus* (e.g., *A. oryzae* and *A. sojae*[76]) are commonly used in Japan in the preparation of shoyu and other fermented foods. These molds are closely related to the aflatoxin-producing molds. This relationship raises the question of whether the Japanese industrial molds produce aflatoxins under normal preparation conditions of fermented foods such as shoyu and miso. Several researchers[77-84] screened a large number of Japanese industrial mold strains

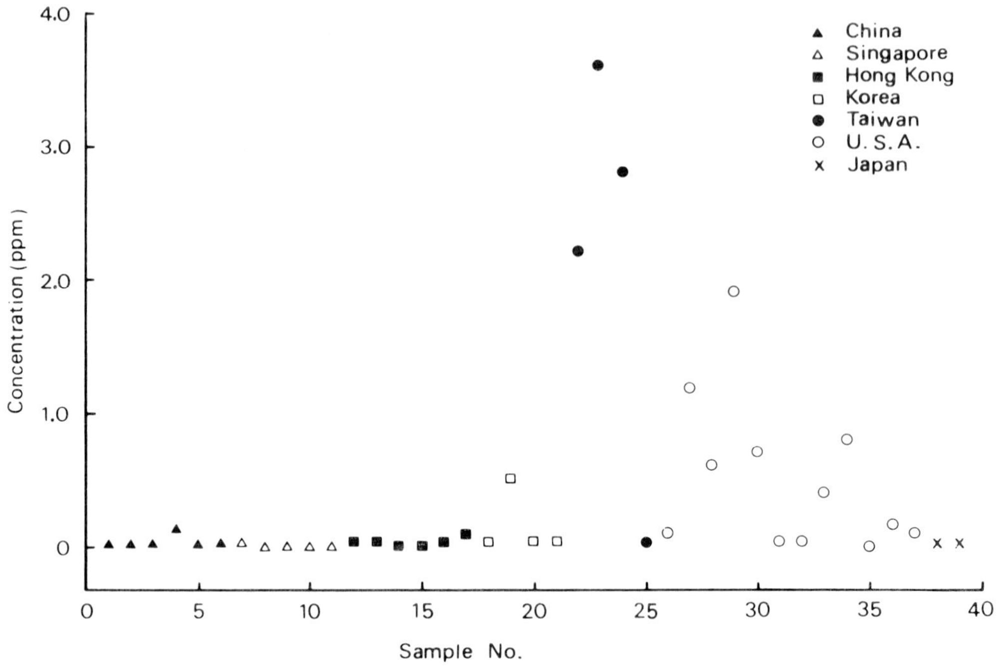

FIGURE 17. Concentration of dimethyl sulfide in various soy sauces.

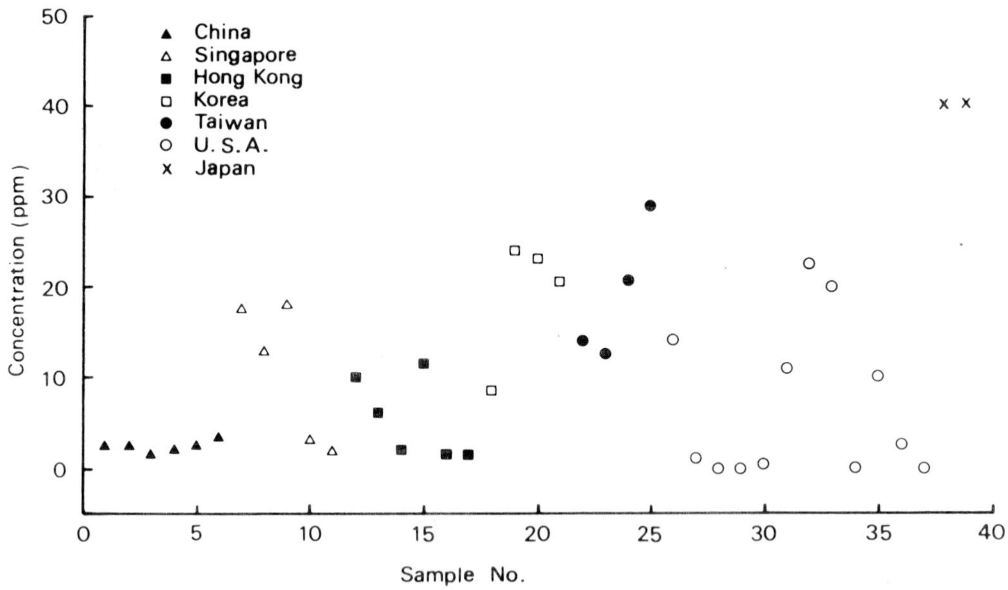

FIGURE 18. Concentration of alcohols (isobutyl-, n-butyl-, isoamyl-, methionol, 2-phenylethanol) in various soy sauces.

for their aflatoxin production. They found that none of these strains produced aflatoxins. However, the methods applied and the basis for conclusions by these researchers appear to have been limited to comparisons of thin layer chromatography (TLC) R_f values and the fluorescence spectra of aflatoxins and the compounds produced by the Japanese industrial molds.

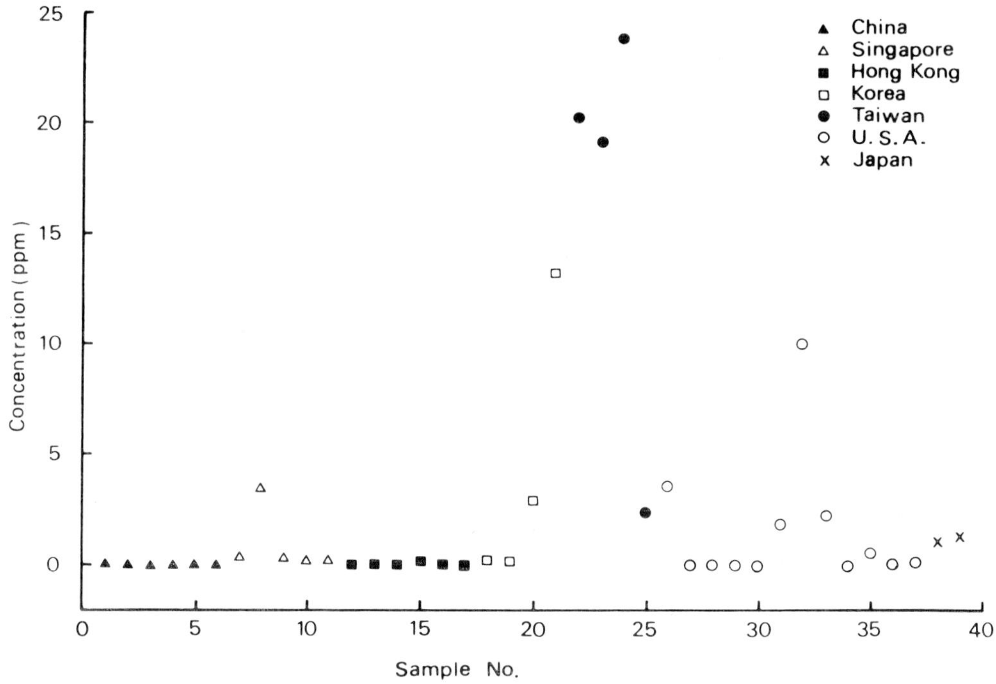

FIGURE 19. Concentration of ethyl formate in various soy sauces.

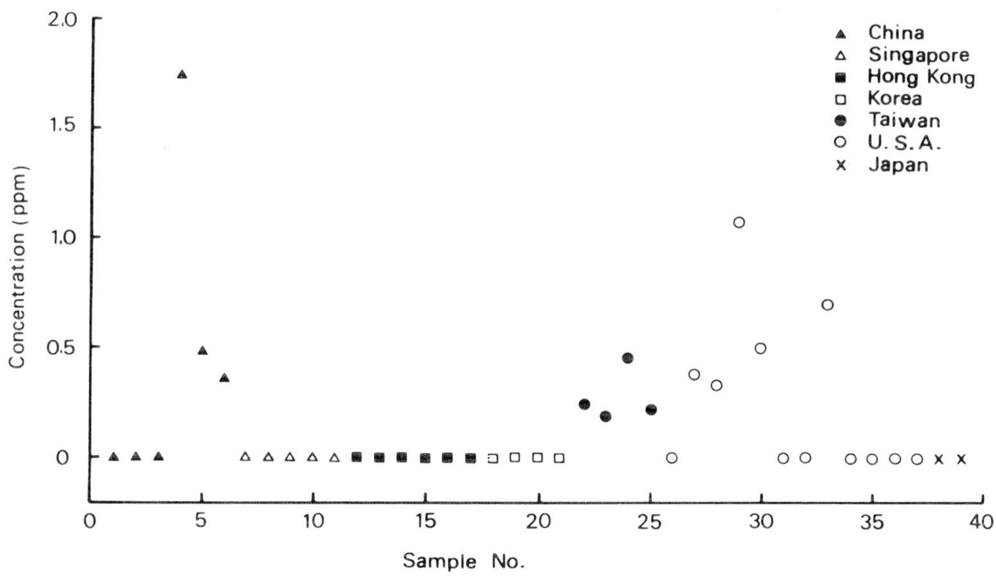

FIGURE 20. Concentration of pentanoic acid (n-valeric acid) in various soy sauces.

Yokotuska et al.[83] reported on extensive investigations on the toxin-producing capabilities of the Japanese industrial molds. About 73 mold strains (27 mold strains from shoyu, 16 from miso, 28 from alcoholic beverage, and 2 wild strains) were screened for production of fluorescent compounds. Fourteen strains produced compounds that were similar to the R_f values and bluish-violet fluorescence of aflatoxin

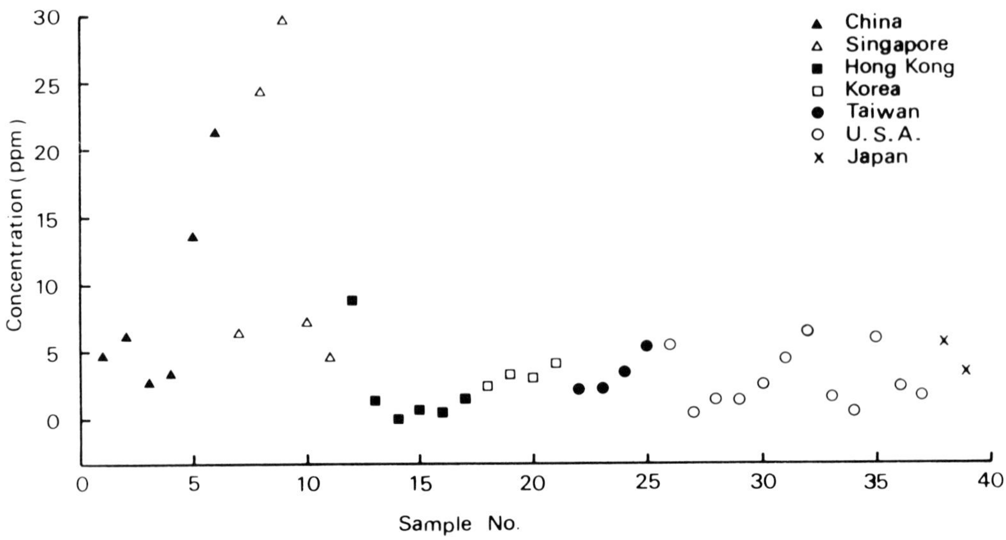

FIGURE 21. Concentration of 3-methyl butanoic acid (isovaleric acid) in various soy sauces.

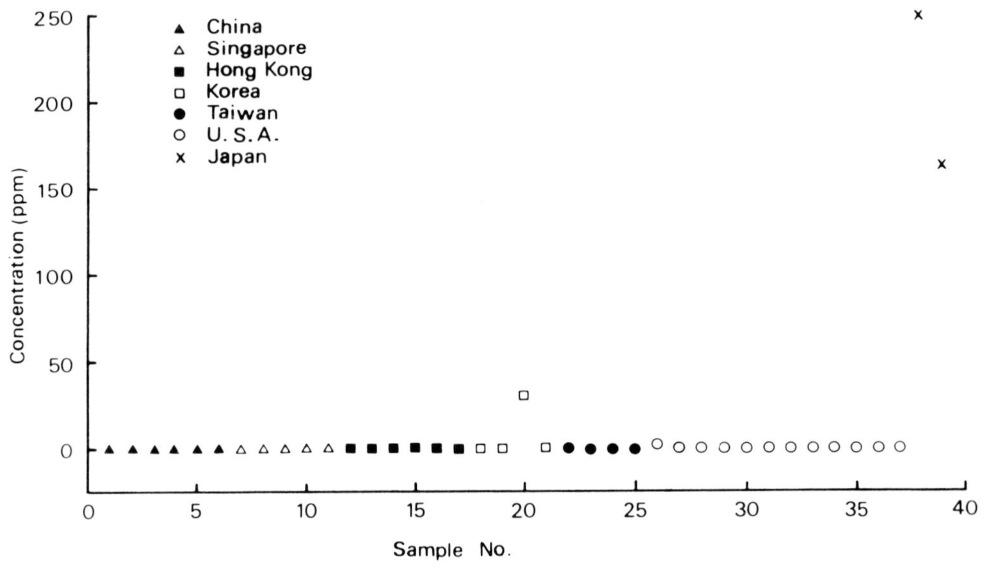

FIGURE 22. Concentration of 4-Hydroxy-2(or 5)-ethyl-5(or 2)-methyl-3(2H)-furanone (HEMF) in various soy sauces.

B_1. About eight strains produced compounds that were similar to those of aflatoxin G. The aflatoxin B-like compounds (B-0 to B-6) produced by one of the strains, *A. sojae* X-1, were separated on TLC using various solvent systems and determined by UV absorption spectra. It was concluded that these compounds may not be aflatoxins.[83-87] Some of the physico-chemical properties and structures of these compounds are present in Tables 14 and 15. All compounds are related to each other in structure and have a common pyrazine ring.

Yokotsuka et al.[90] and Sasaki et al.[91] isolated aflatoxin G-like compounds produced by *Aspergillus* strains from miso and other fermented foods. These aflatoxin G-like

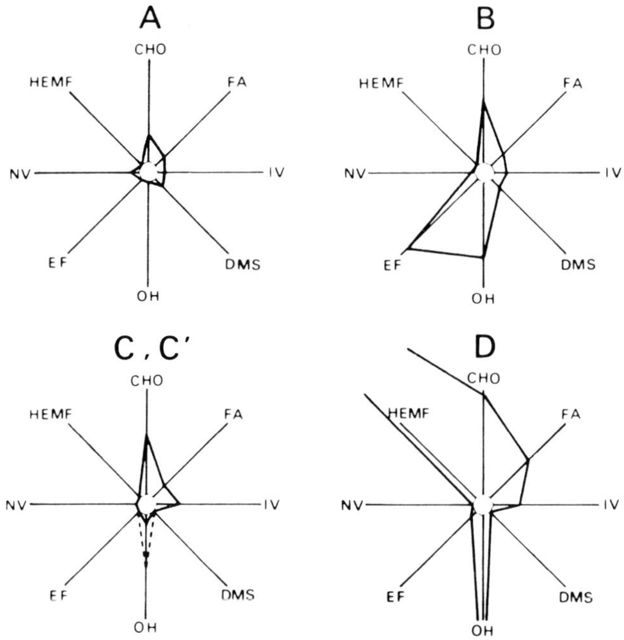

CHO : Aldehydes
FA : Furfuryl alcohol
IV : Isovaleric acid
DMS : Dimethyl sulfide
OH : Alcohols
EF : Ethyl formate
NV : n-Valeric acid
HEMF : HEMF

Type	DMS	Fermentation
A	+	−
B	+	+
C	−	(+)
C′	−	+
D	−	+ + (HEMF)

FIGURE 23. Classification of soy sauces based on dimethyl sulfide and alcohols.

Table 13
TYPES OF SOY SAUCE PRODUCED IN DIFFERENT COUNTRIES

Countires	Type	DMS[a]	Fermentation
China	C	−	(+)
Singapore	C	−	(+)
	C′	−	+
Hong Kong	C	−	(+)
Korea	B	+	+
	C′	−	+
Taiwan	B	+	+
U.S.	A	+	−
	B	+	+
	C′	−	+
Japan	D	−	++ (HEMF)

Note: See Figure 23 for basis of classification.

[a] Dimethylsulfide.

compounds have different physico-chemical properties than those of aflatoxin G_1. Sasaki and his co-workers[93,94] examined mold strains isolated from wheat flour, miso, and *A. sojae* X-1 for production of aflatoxin-like (fluorescent and nonfluorescent) compounds. The fluorescent pyrazine compounds (B-0 to B-6) produced by *A. sojae* X-1 were tested for toxicity in mice.[87,94] These compounds exhibited no acute toxicity in mice by intraperitoneally administered doses of 250 mg/kg and 500 mg/kg. Further, they isolated nonfluorescent pyrazine compounds (A-0 to A-5) produced by *A. sojae* X-1.[94] Some physico-chemical properties, structures, and toxicities of these compounds

Table 14
IDENTIFIED FLUORESCENT PYRAZINE COMPOUNDS OF *ASPERGILLUS SOJAE* X-1 (1)

Compd. No.		B-0	B-1	B-2	B-2'
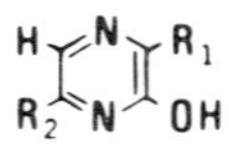	R_1	$-CH(CH_3)\ C_2H_5$	$-CH_2CH(CH_3)_2$	$-CH_2\ CH(CH_3)_2$	$-CH_2\ CH(CH_3)_2$
	R_2	$-CH(CH_3)\ C_2H_5$	$-CH(CH_3)\ C_2H_5$	$-CH_2\ CH(CH_3)_2$	$-CH(CH_3)_2$
Name of Compound		2-Hydroxy-3,6-di-sec-butyl-pyrazine	Deoxyaspergillic acid	Flavacol	2-Hydroxy-3-iso-butyl-6-isopropyl-pyrazine
Molecular formula		$C_{12}H_{20}O\ N_2$	$C_{12}H_{20}O\ N_2$	$C_{12}H_{20}O\ N_2$	$C_{11}H_{18}O\ N_2$
Elemental Analysis		C H N	C H N	C H N	C H N
	Found	69.1 9.3 13.3	70.1 9.6 13.4	69.4 9.4 13.8	68.4 9.3 14.3
	Calcd.	69.2 9.6 13.5	69.2 9.6 13.5	69.2 9.6 13.5	68.0 9.3 14.4
mp (°C)	Obtained	129	83	144	111
	Authentic	128	103 - 104	148.5 - 149	109
UV Spectrum λ_{max}^{EtOH} nm		229 , 325	229 , 326	230 , 326	231.5 , 329
Yield , mg / L (Name of medium)		69.9 / 260 (Czapek)	113 / 70 (Czapek)	140 / 225 (Czapek)	60 / 70 (Mayer)

Table 15
IDENTIFIED FLUORESCENT PYRAZINE COMPOUNDS OF *ASPERGILLUS SOJAE* X-1 (2)

Compd. No.		B-3	B-4	B-5	B-6	B-7
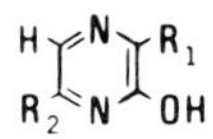	R_1	$-CH(CH_3)C_2H_5$	$-CH_2CH(CH_3)_2$	-	$-CH_2CH(CH_3)_2$	-
	R_2	$-C(CH_3)(OH)C_2H_5$	$-C(CH_3)(OH)C_2H_5$	-	$-C(OH)(CH_3)_2$	-
Name of Compound		2-Hydroxy-6-(1-hydroxy-1-methyl-propyl)-3-sec-butylpyrazine	Deoxyhydroxy-aspergillic acid	-	Deoxymuta-aspergillc acid	-
Molecular formula		$C_{12}H_{20}O_2N_2$	$C_{12}H_{20}O_2N_2$	$C_{12}H_{20}O_2N_2$	$C_{11}H_{18}O_2N_2$	$C_{11}H_{18}O_2N_2$
Elemental Alalysis		C H N	C H N	C H N	C H N	C H N
Found		64.1 8.8 12.4	64.4 8.9 12.4	64.5 8.9 12.5	63.0 8.3 13.1	63.0 8.6 13.0
Calcd.		64.3 9.0 12.5	64.3 8.9 12.5	64.3 8.9 12.5	62.9 8.6 13.3	62.9 8.6 13.3
mp (°C)	Obtained	120 - 120.5	107	153 - 154	133 - 134.5	158 - 159
	Authentic	120 - 120.5	105 - 106	-	135	-
UV Spectrum λ_{max}^{EtOH} nm		230 , 320	224 , 314	230 , 327	227 , 312	312
Yield , mg / L (Name of medium)		22 / 70 (Mayer)	80 / 111 (Mayer)	80 / 311 (Mayer)	80 / 26 (Mayer)	80 / 240 (Mayer)

are presented in Tables 16-18. Aspergillic acid and hydroxy-aspergillic acid are normal metabolities of *Aspergillus* molds. The toxicity of aspergillic acid and hydroxy-aspergillic acid was reported earlier in the literature. The toxicity of these compounds appears to be similar to that of aspergillic acid, LD_{50} approximately 100 mg/kg.

Aspergillic acid and its analogues (kojic acid, β-nitropropionic acid, and oxalic acid)[95,96] are produced by a number of industrial koji molds. Several strains of Japanese industrial molds have been screened for the production of aspergillic acid and its analogues.[97] Most strains (61 of 69) produced aspergillic acid and its analogues. Further, some selected mold strains were grown on a solid substrate composed of defatted soybeans and crushed wheat for production of aspergillic acid and its analogues. It was concluded that there is no hazard in human consumption of fermented foods such as shoyu contaminated with, or containing, aspergillic acid, kojic acid, β-nitropropionic acid, and oxalic acid.[97] Most of the industrial mold strains of *Aspergillus* do not produce other mycotoxins such as sterigmatocystin, ochratoxin, patulin, cyclopiazonic acid, and penicillic acid.[98]

B. Long Term Effects of Shoyu Consumption

MacDonald and Dueck[99] evaluated the long-term effects of feeding rats shoyu and its effect on gastric mucosa. The animals fed shoyu were smaller and healthier than the controls. Breast tumors developed in ten control rats, but none developed in those fed shoyu. It was concluded that shoyu did not appear to be carcinogen in rats. Yokotsuka[100] also reported on the acute and long term effects of feeding shoyu to mice and rats. He concluded that the acute toxicity of shoyu was accounted for by the toxicity of its sodium chloride content.

C. Mutagens

Mutagens from shoyu were first investigated by Lin et al.[101] using the Ames test. They studied mutagenicity by adding various levels of nitrite to shoyu. Mutagenic material was formed in shoyu when 2000 ppm of nitrite was present at pH 3.0. Addition of ascorbic acid to the shoyu prevented the formation of mutagenic products.

Nagahori et al.[102] observed 60 to 80% suppression of N-nitrosodimethylamine (a mutagenic product) formation, when a mixture of dimethylamine and nitrite was added to shoyu at pH 3.6. Wakabayashi et al.[103] reported that mutagenic materials are produced when shoyu and 2300 ppm nitric acid are mixed at pH 3.0 and left for 1 hr at 37°C. In addition, they isolated two precursors of the mutagen and identified these as (-)-(1S, 3S)-1-methyl- 1,2,3,4-tetrahydro-β-carboline-3-carboxylic acid [(-)-(1S,3S)-MTCA] and its stereoisomer (-)-(1R,3S)-MTCA. The mutagenicity of these compounds accounted for 16 to 61% of the total mutagenicity of the shoyu-nitrite mixture. Recently, tyramine was identified as a mutagen precursor in shoyu, becoming mutagenic after treatment with nitrite under acidic conditions. The mutagenic compound produced from tyramine was identified as 4-(2-aminoethyl)-6-diazo-2,4-cyclohexadienone.[104] It must be noted that in these experiments shoyu was treated with nitrite at a concentration of which is several times higher than that in the human mouth.[105]

Shibamoto[106] reported on mutagenicity of shoyu that was heated with 100, 500, 1000, and 2000 ppm sodium nitrite for 30 min at 80°C and pH 3.0. He concluded that mutagenic substances are formed in shoyu treated at a concentration of more than 2000 ppm of sodium nitrite and are not produced at a physiological concentration of sodium nitrite found in the human body. Judging from the published reports, it can be concluded that shoyu treated with nitrite at a concentration level detected in the human mouth does not induce mutagenicity. Most reviews supports the notion that shoyu is a safe food.[107-111]

Table 16
IDENTIFIED NON-FLUORESCENT PYRAZINE COMPOUNDS OF *ASPERGILLUS SOJAE* X-1 (1)

Compd. No.		A-0	A-1	A-2
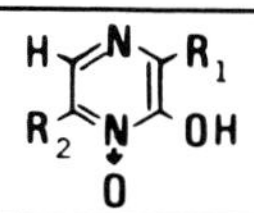	R_1	$-CH(CH_3)C_2H_5$	-	$-CH_3CH(CH_3)_2$
	R_2	$-CH(CH_3)C_2H_5$	-	$-CH(CH_2)C_2H_5$
Name of Compound		2-Hydroxy-3,6-di-sec-butylpyrazine-1-oxide	-	Aspergillic acid
Molecular formula		$C_{12}H_{20}O_2N_2$	-	$C_{12}H_{20}O_2N_2$
mp (°C)	Obtained (Cu-salt)[Zn-salt]	Oil(196)[-]	Oil(-)[-]	95 - 96(203.5)[229-230]
	Authentic(Cu-salt)[Zn-salt]	- (-)[-]	- (-)[-]	97 - 99(198-199)[-]
Yield , mg / 60 L (Mayer's medium)	Free form	136.8	192.4	-
	Cu-salt	186	-	129.5
	Zn-salt	-	-	508.3

Table 17
IDENTIFIED NON-FLUORESCENT PYRAZINE COMPOUNDS OF *ASPERGILLUS SOJAE* X-1 (2)

Compd. No.		A-3	A-4	A-5
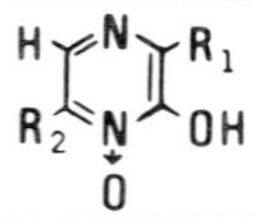	R_1	$-CH_2CH(CH_3)_2$	$-CH(CH_3)C_2H_5$	$-CH_2CH(CH_3)_2$
	R_2	$-CH(CH_3)_2$	$-C(CH_3)(OH)C_2H_5$	$-C(CH_3)(OH)C_2H_5$
Name of Compound		2-Hydroxy-3-iso-butyl-6-isopropyl-pyrazine-1-oxide	2-Hydroxy-6-(1-hydroxy-1-methyl-propyl)-3-sec-butylpyrazine-1-oxide	Hydroxyaspergillic acid
Molecular formula		$C_{11}H_{18}O_2N_2$	$C_{12}H_{20}O_3N_2$	$C_{12}H_{20}O_3N_2$
mp (°C)	Free form	94 - 95	120 - 121	148
	Cu-salt	215	120 - 120.5	232
Yield, mg / 60 L (Mayer's medium)	Free form	1257	359	-
	Cu-salt	2066	527	1276

Table 18
ACUTE INTRAPERITONEAL TOXICITY OF NON-FLUORESCENT PYRAZINE COMPOUNDS TO MICE

Compd. No.	Compound	Mortality (7 Days) Dose (mg/kg) 62.5	125	250
A-0	2-Hydroxy-3,6-di-sec-butylpyrazine 1-oxide	-	0 / 3	3 / 3
A-2	Aspergillic acid	LD_{50} 100 mg/kg (Value of reference)		
A-3	2-Hydroxy-3-isobutyl-6-isopropylpyrazine 1-oxide	0 / 3	3 / 3	-
A-4	2-Hydroxy-6(1-hydroxy-1-methylpropyl) 3-sec-butyl-pyrazine 1-oxide	0 / 3	3 / 3	-
A-5	Hydroxyaspergillic acid	LD_{50} 100 mg/kg (Value of reference)		
A-0 - A-5	Mixture	0 / 3	3 / 3	3 / 3 (LD_{50} 119 mg/kg)

Note: ddOM mice weighted about 20 g (↑).

VI. CONCLUSIONS

Soy sauce is now produced and consumed as an all-purpose seasoning in Japan, China, Southeast Asia, and U.S. It is used to enhance the flavor of meats, seafoods, and vegetables. In Japan, soy sauce is known as shoyu. It contains relatively high levels of salt, up to 19%. Shoyu has unique flavors that are formed during microbial fermentation. The high salt content of shoyu is one of the greatest obstacles in expanding its wide acceptance and consumption. On the other hand, the high salt content is necessary for controlled fermentation by yeasts, fermentation of nitrogenous compounds without putrefaction, and avoiding unwanted contaminants. The consumption of shoyu can only be increased if a reduction in salt is achieved without changing the desirable shoyu characteristics. Currently in the U.S., fermented shoyu accounts for more than 40% of the total distribution of soy sauce. About 30 years ago, most of the soy sauce consumed in the U.S. was a chemically prepared product. This indicates the wider acceptance of fermented shoyu in the Western Hemisphere.

Research is needed to further develop fermentation techniques and improve the manufacturing process for shoyu. Development of a multistage, continuous process using commercial enzymes and lower salt concentrations may reduce some of the production costs. Most of the production costs involved in shoyu are due to mash processing and filtration. Development of a rapid fermentation process by yeasts at lower salt concentrations would improve shoyu consumption patterns as well as production costs.

REFERENCES

1. Steinkraus, K. H., Ed., *Handbook of Indigenous Fermented Foods,* Marcel Dekker, New York, 1983, 433.
2. Ogawa, G. and Fujita, A., Recent progress in soy sauce production in Japan, in *Cereals for Food and Beverages,* Inglett, G. E. and Munck, L., Eds., Academic Press, New York, 1980, 381.
3. Beuchat, L. R., Fermented soybean foods, *Food Technol.,* 38, 64, 1984.
4. Abe, K., Yuasa, K., and Hashiba, H., Determination of neutral sugars in soy sauce by improved liquid chromatography using biochinconinate for dye reagent, *J. Agric. Food Chem.,* in press.
5. Tawara, Y., Analysis of soy sauce, *J. Pharm. Soc. Jpn. (Yakugaku Zasshi),* 61, 80, 1887.
6. Obata, Y. and Yamanishi, T., Studies on the flavors of soy. Part I. On the flavors which developed from hydrolyzed protein and amino acids, *J. Agric. Chem. Soc. (Jpn.),* 24, 226, 1950.
7. Yamanishi, T., Obata, Y., and Sano, M., Studies on the flavors of soy. V. Synthesis of sulfur-containing compounds, *J. Agric. Chem. Soc. (Jpn.),* 26, 125, 1952.
8. Yokotsuka, T., Studies on flavorous substances in soy. I. Isolation of flavorous substances from soy-cake, chiefly on the flavors of lower boiling point, *J. Agric. Chem. Soc. (Jpn.),* 23, 200, 1949.
9. Yokotsuka, T., Asao, Y., and Sakasai, T., Studies on flavorous substances in shoyu. XXVII. The production of 4-ethylguaiacol during shoyu fermentation, and its role for shoyu flavor, *J. Agric. Chem. Soc. (Jpn.),* 41, 442, 1967.
10. Nunomura, N., Sasaki, M., Asao, Y., and Yokotsuka, T., Identification of volatile components in shoyu (soy sauce) by gas chromatography-mass spectrometry, *Agric. Biol. Chem.,* 40, 485, 1976.
11. Nunomura, N., Sasaki, M., Asao, Y., and Yokotuska, T., Shoyu (soy sauce) volatile flavor components: basic fraction, *Agric. Biol. Chem.,* 42, 2123, 1978.
12. Nunomura, N., Sasaki, M., and Yokotuska, T., Isolation of 4-hydroxy-5-methyl-3(2H)-furanone, a flavor component in shoyu (soy sauce), *Agric. Biol. Chem.,* 43, 1361, 1979.
13. Nunomura, N., Sasaki, M., and Yokotsuka, T., Shoyu (soy sauce) flavor components: acidic fractions and the characteristic flavor component, *Agric. Biol. Chem.,* 44, 339, 1980.
14. Nunomura, N., Sasaki, M., and Yokotuska, T., Shoyu (soy sauce) flavor components: neutral fraction, *Agric. Biol. Chem.,* 48, 1753.

15. Nunomura, N., Sasaki, M., Asao, Y., and Yokotsuka, T., Isolation and identification of 4-hydroxy-2(or 5)-ethyl-5(or 2)-methyl-3(2H)-furanone, as a flavor component in shoyu (soy sauce), *Agric. Biol. Chem.*, 40, 491, 1976.
16. Rodin, J. O., Himel, C. H., Silverstein, R. M., Leeper, R. W., and Gortner, W. A., Volatile flavor and aroma components of pineapple. I. Isolation and tentative identification of 2,3-dimethyl-4-hydroxy-3(2H)-furanone, *J. Food Sci.*, 30, 280, 1965.
17. Shimizu, Y., Matsuto, S., Mizunuma, Y., and Okada, I., Studies on the flavors of roast barley (mugicha). VI. Separation and identification of 5-hydroxymaltol, maltol, 5-methylcyclopent -2-en-2-ol-1-one and other compounds, *Agric. Biol. Chem.*, 34, 845, 1970.
18. Jurch, G. P., Jr., and Tatum, J. H., Degradation of D-glucose with acetic acid and methylamine, *Carbohydr. Res.*, 15, 233, 1970.
19. Re, L., Maurer, B., and Ohloff, G., Ein Einfacher Zugang zu 4-hydroxy-2,5-dimethyl-3(2H)-furanon (Furaneol), einem Aromabestandteil von Ananas und Erdbeere, *Helv. Chim. Acta*, 56, 1882, 1973.
20. Ohloff, G., Importance of minor components in flavors and fragrances, *Perfumer & Flavorist*, 3(1), 11, 1978.
21. Nunomura, N. and Sasaki, M., unpublished data, 1985.
22. Asao, Y. and Yokotsuka, T., Studies on flavorous substances in soy sauce. XVIII. Isolation of $C_7H_{12}O_5$ compound in soy sauce, and it structure (1), *J. Agric. Chem. Soc. (Jpn.)*, 35, 831, 1961.
23. Nunomura, N., Sasaki, M., Asao, Y., and Yokotsuka, T., On the oxidation products of the most important flavor component, HEMF in shoyu, in *Abstr. Pap. Annu. Meet. Agric. Chem. Soc. Jpn.*, 448, 1977.
24. Sasaki, M., and Nunomura, N., Flavor components in shoyu (soy sauce) koji, in *Abstr. Pap. Annu. Meet. Agric. Chem. Soc. Jpn.*, 237, 1983.
25. Nunomura, N., Sasaki, M., and Yokotsuka, T., Flavor components in matured mash of shoyu (soy sauce), in *Abstr. Pap. Annu. Meet. Agric. Chem. Soc. Jpn.*, 282, 1980.
26. Sasaki, M., Nunomura, N., and Yokotsuka, T., Quantitative analysis of flavor components in shoyu (soy sauce), in *Abstr. Pap. Annu. Meet. Agric. Chem. Soc. Jpn.*, 282, 1980.
27. Sasaki, M. and Nunomura, N., Constituents of top note aroma of shoyu (soy sauce), *J. Chem. Soc. (Jpn.) (Pure Chem. Sect.)*, 102, 736, 1981.
28. Nunomura, N., Mori, S., Sasaki, M., and Motai, H., Heating effects on flavor components in shoyu (soy sauce), in *Abstr. Pap. Annu. Meet. Agric. Chem. Soc. Jpn.*, 237, 1983.
29. Nunomura, N. and Sasaki, M., Comparison in flavor compounds between Japanese shoyu and the soy sauces of other countries, in *Abstr. Pap. Annu. Meet. Agric. Chem. Soc. Jpn.*, 129, 1984.
30. Mori, S., Nunomura, N. and Sasaki, M., Correlation between sensory preference and compounds of Japanese soy sauce flavor, in *Abstr. Pap. Annu. Meet. Agric. Chem. Soc. Jpn.*, 36, 1982.
31. Tanaka, T., Saito, N., Nakajima, T., and Yokotsuka, T., Studies on sensory evaluation of soy sauce. VII. Statistical research on the effect of blending, *J. Ferment. Technol.*, 47, 237, 1969.
32. Mori, S., Nunomura, N. and Sasaki, M., Special combination of 4-ethyl-2-methoxyphenol and 3-(methylthio)-1-propanol to enhance Japanese soy sauce flavor, in *Abstr. Pap. Annu. Meet. Agric. Chem. Soc. Jpn.*, 236, 1983.
33. Asao, Y., Sakasai, T., and Yokotsuka, T., Studies on the flavorous substances in shoyu (soy sauce). XXVI. Flavorous components produced by yeast fermentation (2), *J. Agric. Chem. Soc. (Jpn.)*, 41, 434, 1967.
34. Ishizu, H., Aroma components of soy sauce, *Cooking Sci. Jpn. (Chori Kagaku)*, 2, 156, 1969.
35. Goto, S., Constituents and composition of aroma concentrate from soy sauce. I. On neutral fraction, *Eiyo to Shokuryo Jpn.*, 26, 135, 1973.
36. Yokotsuka, T. and Asao, Y., Studies on flavorous substances in soy sauce. XIX. The sturcture of $C_7H_{12}O_5$ compound (2) and α-diketone compounds in soy sauce, *J. Agric Chem. Soc. (Jpn.)*, 35, 837, 1961.
37. Ikeda, K., Kawaguchi, M. and Kodama, S., *J. Chem. Soc. (Jpn.) (Pure Chem. Sect.)*, 43, 956, 1922.
38. Kihara, K., Studies on flavorous substances in soy sauce. I. Volatile ingredients; II. Nonvolatile ingredients; III. Comparison between soy sauce and chemical hydrolazate of soybean, *J. Chem. Soc. (Jpn.) (Ind. Chem. Sect.)*, 43, 876, 1940.
39. Yokotsuka, T., Studies on flavorous substances in soy. IX. On high boiling point's fraction. (4). Isolation of phenolic flavorous substances. XII. Flavorous substances in raw soy, *J. Agric. Chem. Soc. (Jpn.)*, 27, 276, 1953.
40. Yokotsuka, T., Studies on flavorous substances in shoyu. XIV. Flavorous substances in heated shoyu (3), *J. Agric. Chem. Soc. (Jpn.)*, 30, 66, 1956.
41. Yokotsuka, T., Studies on flavorous substances in soy. X. Flavors with high boiling fraction (5) precursors of phenolic flavorous substances, *J. Agric. Chem. Soc. (Jpn.)*, 27, 334, 1953.
42. Kosuge, R., Zenda, H., Tsuji, K., Yamamoto, T., and Narita, H., Studies on flavor components of foodstuffs. I. Distribution of tetramethylpyrazine in fermented foodstuffs, *Agric. Biol. Chem.*, 35, 639, 1971.

43. Akabori, S., Studies on the flavorous ingredients of soy sauce. I. *J. Chem. Soc. (Jpn.) (Pure Chem. Sect.)*, 57, 828, 1936.
44. Akabori, S. and Kaneko, T., Studies on the flavorous ingredients of soy sauce. II. Isolation and synthesis of a sulfur-containing compound from soy sauce, *J. Chem. Soc. (Jpn.) (Pure Chem. Sect.)*, 57, 832, 1936.
45. Asao, Y. and Yokotsuka, T., Studies on flavorous substances in soy sauce. XVI. Flavorous substances in raw soy sauce (2), *J. Agric. Chem. Soc. (Jpn.)*, 32, 617, 1958.
46. Asao, Y. and Yokotsuka, T., Studies on the flavorous substances in soy sauce. XX. α-diketone compound in soy sauce (2), *J. Agric. Chem. Soc. (Jpn.)*, 35, 1211, 1961.
47. Asao, Y. and Yokotsuka, T., Studies on the flavorous substances in soy sauce. XXIII. α-diketone compound in soy sauce (3), *J. Agric. Chem. Soc. (Jpn.)*, 37, 563, 1936.
48. Fukai, T., The relation between soy sauce fermentation and fats and oils. I. On the ethyl esters of fatty acids and free fatty acids in soy sauce oil, *J. Agric. Chem. Soc. (Jpn.)*, 5, 485, 1929.
49. Ishizu, H., Gas chromatography of chemical components in soy sauce. II. Variation of odor on heating, *Mukogawajoshidikiyo Jpn.*, 11, S-113, 1962.
50. Matsumoto, K., *Rep. Govt. Inst. Brewing Jpn.*, 85, 28, 1921.
51. Morimoto, S. and Murakami, M., *Shointandaikiyo Jpn.*, 7, 69, 1966.
52. Morimoto, S. and Murakami, M., Studies on a quick ripening process for soy sauce. V. A new quick ripening process based on alcohol fermentation. A qualitative analysis of soy sauce flavor components using gas-liquid chromatography, *J. Ferment. Technol.*, 44, 467, 1966.
53. Morimoto, S. and Murakami, M., Studies on fermentation products from aldehyde by microoganisms. I. The fermentative production of furfurly alcohol from furfurol by yeasts (1), *J. Ferment. Technol.*, 45, 467, 1967.
54. Nakazima, M., Imaki, T., and Takei, S., Studies on the flavorous ingredients of soy sauce. I. On the lower boiling compounds and acidic compounds in soy sauce aroma and flavor, *J. Chem. Soc. (Jpn.), (Pure Chem. Sect.)*, 70, 40, 1949.
55. Shoji, K., *Proc. Imp. Acad. (Jpn.)*, 3, 324, 1927.
56. Shoji, K. and Onuki, M., Studies on fermentation products by polarography. VI. On the soy flavor, *Sci. Pap. Inst. Phys. Chem. Res.*, 11, 277, 1932.
57. Shoji, K., Flavor of soy sauce. II and III, *Sci. Papers Inst. Phys. Chem. Res.*, 15, 166, 1936.
58. Taira, T., Distribution of 2, 3-butylene-glycol in the fermented products in Japan, *J. Agric. Chem. Soc. (Jpn.)*, 2, 734, 1926.
59. Taira, T., Studies on the flavorous compounds in soy sauce. I, *J. Soc. Brewing Jpn.*, 21, (11), 31, 1926.
60. Tomiyasu, Y., Distrubution of acetoin and 2,3-butyleneglycol in the fermented products, *J. Agric. Chem. Soc. (Jpn.)*, 3, 758, 1927.
61. Yamada, M., On the origins of some aldehydes in the fermentation products. II, *J. Agric. Chem. Soc. (Jpn.)*, 4, 1, 1928.
62. Yamada, K. and Goan, M., Studies on soy sauce flavor by gas chromatography. I. Research of flavor components by headspace gas analysis, *Seasoning Sci. (Chomi Kagaku)*, 16(2), 1, 1969.
63. Yokotsuka, T., Studies on flavorous substances in soy. II. Flavors resulting from heating. (1) islolation of and acetal, $C_{10}H_{22}O_3$, *J. Agric. Chem. Soc. (Jpn.)*, 24, 335, 1950.
64. Yokotsuka, T., Studies on flavours substances in soy. VII. Flavors with high boiling points. (2) Amyl alcohol and amyl acetate, *J. Agric. Chem. Soc. (Jpn.)*, 25, 401, 1951.
65. Yokotsuka, T., Studies on flavorous substances in soy. VI. Flavors with high boiling points. (1) Palmitic acid ethyl estre and linolic acid ethyl ester, *J. Agric. Chem. Soc. (Jpn.)*, 25, 206, 1951.
66. Yokotsuka, T., Studies on flavorous substances in soy. VIII. Flavors with high boiling points. (3) Benzaldehyde and benzoic acid ethyl ester, *J. Agric. Chem. Soc. (Jpn.)*, 25, 446, 1951.
67. Yokotsuka, T., Studies on flavorous substances in soy. XI. Flavors of lower boiling fractions. (2) *J. Agric. Chem. Soc. (Jpn.)*, 27, 359, 1953.
68. Yokotsuka, T., Studies on flavorous substances in soy. XIII. On the yeast-static substances in brewed soy and their increase in the heated soy, *J. Agric. Chem. Soc. (Jpn.)*, 28, 114, 1954.
69. Yukaya, M., Transformation of tyrosine in the aging of soy sauce mash, *J. Chem. Soc. (Jpn.)*, 38, 358, 1917.
70. Sarkisov, A. Ch., Mycotoxicosis, in *Govt. Edit. Agr. Lit.*, Moscow, 1954, 216.
71. Forgacs, J. and Carll, W. T., Mycotoxicoses, in *Adv. Vet. Sci.*, Vol. 7, Brandly, C. A. and Jungherr, E. L., Eds., 1962, 272.
72. Van der Zijden, K. J., Boldingh, J., Barett, C. B., Ord, W. O., and Philp, J., Isolation of crystalline form of a toxin responsible for turkey X disease, *Nature*, 195, 1060, 1962.
73. Asao, T., Buch, G., Abdel-Karder, M. M., Change, S. B., Wick, E. L. and Wogan, G. N., Structure of alfatoxin B_1 and G_1, *J. Am. Chem. Soc.*, 89, 6745, 1965.
74. Nesbitt, B. E., O'Kelly, J., Sargent, K. and Sheriden, A., Toxic metabolities of *Aspergillus flavus*, *Nature*, 195, 1962.

75. Hesseltine, C. W., Sorenson, W. G., and Smith, M., Taxonomic studies of the aflatoxin-producing strains in the *Aspergillus flavus* group, *Mycologia*, 62, 123, 1970.
76. Raper, K. B. and Fennell, D. I., *The Genus Aspergillus*, Waverly Press, Baltimore, 1965, 357.
77. Aibara, K. and Miyaki, K., Qualitative and quantitative analysis of aflatoxin, in *Proc. Annu. Meet. Agric. Chem. Soc. (Jpn.)*, 86, 1965.
78. Manabe, M., Matsuura, S., and Nakano, M., Studies on the fluorescent compounds in fermented foods, part 1. Chloroform soluble fluoresecnt compounds produced by molds, *J. Food Sci. Technol.*, 15, 341, 1968.
79. Murakami, H., Takase, S. and Kuwabara, K., Non-productivity of aflatoxin by Japanese industrial strains of *Aspergillus*. II. Production of fluorescent substances in rice koji and their identification by absorption spectrum, *J. Gen. Appl. Microbiol.*, 13, 97, 1963.
80. Masuda, Y., Mori, K., and Kuratsune, M., Studies on aflatoxins in *Proc. Annu. Meet. Jpn. Cancer.*, 167, 1965.
81. Hesseltine, C. W., Shotwell, O. L., Ellis, J. J., and Stubblefield, R. D., Aflatoxin formation by *Aspergillus flavus, Bacteriol. Rev.*, 30, 795, 1966.
82. Kinosita, R., Ishiko., T., Sugiyama, S., Seto, T., Igarasi, S., and Goetz, I. E., Mycotoxins in fermented food, *Cancer Res.*, 28, 2296, 1968.
83. Yokotsuka, T., Sasaki, M., Kikuchi, T., Asao, Y., and Nobuhara, A., Studies on the compounds produced by molds. I. Fluorescent compounds produced by Japanese industrial molds, *J. Agric. Chem. Soc.(Jpn.)*, 41, 32, 1967.
84. Dunn, G., Newbold, G. T., and Spring, F. S., Synthesis of flavacol, a metabolic product of *Aspergillus flavus, J. Chem. Soc.*, 2586, 1949.
85. Ducher, J. D., Aspergillic acid: an antibiotic substance produced by *Aspergillus flavus* (3), the structure of hydroxyaspergillic acid, *J. Biol. Chem.*, 232, 785, 1958.
86. Sasaki, M., Kikuchi, T., Asao, T., and Yokotsuka, T., Studies on the compounds produced by molds. II. Fluorescent compounds produced by Japanese industrial molds (2). *J. Agric. Chem. Soc. (Jpn.)*, 41, 154, 1967.
87. Sasaki, M., Asao, Y., and Yokotsuka, T., Studies on the compounds produced by molds. III. Fluorescent compounds produced by Japanese industrial molds (3), *J. Agric. Chem. Soc. (Jpn.)*, 42, 288, 1968.
88. Sasaki, M., Kaneko, Y., Oshita, K., Takamatuse, H., Asao, Y., and Yokostuka, T., Studies on the compounds produced by molds. IV. Isolation isocoumarin compounds, *J. Agric. Chem. Soc. (Jpn.)*, 34, 1296, 1970.
89. Nishikawa, E., Biochemistry of molds (2). One metabolic compounds by *Aspergillus melleus* Yukawa. I, *J. Agric. Chem. Soc. (Jpn.)*, 9, 772, 1933.
90. Yokotsuka, T., Kikuchi, T., Sasaki, M., and Oshita, K., Aflatoxin-G like compounds with green fluorescence produced by Japanese industrial molds (Note), *J. Agric. Chem. Soc. (Jpn.)*, 42, 581, 1968.
91. Sasaki, M., Takamatsu, H., Oshita, K., and Yokotsuka, T., Studies on the compounds produced by molds. VIII. Isolation of lumichrome from the culture filtrate of *Aspergillus oniki* 1784 (Note), *J. Agric. Chem. Soc. (Jpn.)*, 48, 569, 1974.
92. Karrer, P., Salomon, H., Schopp, K., Schlitter, E., and Fritzche, H., Lumichrome, a new irradiation product of lactoflavin, *Helv. Chim. Acta*, 17, 1010, 1934.
93. Sasaki, M., Oshita, K., and Yokotuska, T., Studies on the compounds produced by molds. IX. Studies on the fluorescent compounds produced by *Aspergillus flavus* link ex Fr(WF-3-8) and *Aspergillus flavus* link ex Fr(68-SA-R-7), *J. Agric. Chem. Soc. (Jpn.)*, 49, 57, 1975.
94. Sasaki, M., Asao, Y., and Yokotsuka, T., Studies on the compounds produced by molds. V. Isolation of non-fluorescent pyrazine compounds (2). *J. Agric. Chem. Soc. (Jpn.)*, 42, 351, 1965.
95. Yabuta, T., The constitution of kojic acid, a γ-pyrone derivative formed by *Aspergillus oryzae* from carbohydrates, *J. Chem. Soc. (Jpn.)*, 45, 575, 1924.
96. Bush, M. T., Touster, O., and Brockman, J. E., The production of β-nitropropionic acid by a strain of *Aspergillus flavus, J. Biol. Chem.*, 188, 685, 1951.
97. Yokotsuka, M., Oshita, K., Kikuchi, E., Sasaki, M., and Asao, Y., Studies on the compounds produced by molds. VI. Aspergillic acid, kojic acid, β-nitropropionic acid and oxalic acid in shoyu-koji, *J. Agric. Chem. Soc. (Jpn.)*, 43, 189, 1969.
98. Yokotsuka, T. and Sasaki, M., Risks of mycotoxin, in *Fermented Foods in Advances in Biotechnology, Vol. 2 Fuel, Chemicals, Foods, and Waste Treatment*, Moo-Young, M. and Robinson, C. W., Eds., Pergamon Press, Oxford, 1981, 461.
99. McDonald, W. C. and Duek, J. W., Long-term effect of shoyu (Japanese soy sauce) on the gastric mucosa of the rat, *J. Nat. Cancer Inst.*, 56, 1143, 1976.
100. Yokotsuka, T., Shoyu presentation at SCOGS (Select Committee of GRAS Substances) Hearing, Hilton Hotel, Bethesda, Md., July 26, 1977.

101. Lin, J. Y., Wang, H. I., and Yeh, Y. C., The mutagencity of soybean sauce, *Food Cosmet. Toxicol.*, 17, 329, 1979.
102. Naghori, T., Motai, H., and Okuhara, A., On the substances in shoyu to surpress the nitrosation of dimethylamine, *J. Jpn. Soc. Food Nutr.*, 33, 151, 1980.
103. Wakabayashi, K., Ochiai, M., Saito, H., Tsuda, M., Suwa, Y., Nagao, M., and Sugimura, T., Presence of 1-methyl-1,2,3,4-tetrahydro-β-carboline-3-carboxylic acid, a precursor of a mutagenic nitroso compound, in soy sauce, *Proc. Natl. Acad. Sci.*, 80, 2912, 1983.
104. Ochiai, M., Watanabe, K., Nagao, M., and Sugimura, T., Tyramine is a major mutagen percursor in soy sauce, being convertible to a mutagen by nitrite, *Gann.*, 75, 1, 1984.
105. Walters, C. L. and Smith, P. L. R., The effect of water-borne nitrate on salivary nitrite, *Food. Cosmet. Toxicol.*, 19, 297, 1981.
106. Shibamoto, T., Possible mutagenic constituents in nitrite-treated soy sauce, *Food. Chem. Toxic.*, 21, 745, 1983.
107. Yokotsuka, T., Sasaki, M., Kikuchi, T., Asao, Y., and Nobuhara, A., Production of fluorescent compounds other than aflatoxins by Japanese industrial toxins, in *Biochemistry of Some Food-borne Microbial Toxins*, Mateles, R. I. and Wogan, G. BN., Eds., MIT Press, Cambridge, 1967, 131.
108. Yokotsuka, T., Asao, Y., Sasaki, M., and Oshita, K.,Pyrazine compounds produced by molds, in *Proc. First U.S.-Japan Conf. Toxic Micro-organisms.* U.S. Department of Interior and UJNR Panel on Toxic Microorganisms, Washington, D.C., 1970, 133.
109. Yokotsuka, T., Recent advances in shoyu research, in *The Quality of Foods and Beverages, Vol. 2, Chemistry and Technology,* Charalambous, G. and Inglett, G., Eds., Academic Press, New York, 1981, 171.
110. Hesseltine, C. W., Fermented foods in the Orient with emphasis on soy sauce. *J. Jpn. Soy Sauce Res. Inst.* 10, 60, 1984.
111. Yokotsuka, T. and Sasaki, M., Risks of mycotoxin in fermented foods, (unpublished data) 1984.

Chapter 3

MISO

H. Ebine

TABLE OF CONTENTS

I. INTRODUCTION

A. History

Miso is a commonly used fermented soybean food in Japan, China, and other parts of the Orient. It is referred to as "chiang" in China, "jang" in Korea, "miso" in Japan, "tao-tjo" in Indonesia and Thailand, and "tao-tsi" in the Philippines. The prototype of miso was introduced to Japan from Korea and China at least 1300 years ago. Misho (the original type of miso) was first cited as "show (jang)" and "shi" in Taihouryou in Japan in 702. Although the production method for miso was not described, it probably resembled that of jang in Korea. The process for miso preparation includes pounding and mashing cooked soybeans, shaping it into balls, wrapping the balls in rice straw, hanging the wrapped material under the eves, and mixing the dried balls with salt and water. This process was often seen at farm houses in prewar Japan. A modified production process uses a motor-driven extruder for ball-making, a koji mold starter mixed with toasted barley flour, and a koji fermenter. This modified method for making soybean miso is widely used in the prefectures of Aichi, Mie, and Gifu, located in the central part of Japan.

The miso (as introduced in Japan) has been changed by employing rice koji (an essential for the production of sake), resulting in the development of characteristic and unique miso. Originally miso was made and consumed at Buddhist monasteries. Buddhists and envoys introduced miso to Japan from China. They produced fermented foods such as kinzanji miso after returning home. The main characteristics of their process was to make koji from cooked whole soybeans or dehulled whole soybeans mixed with barley or wheat. This method was not developed for making miso except kinzanji miso, but was employed widely for shoyu-making.

The industrial production of rice miso, the most popular variety being salty rice miso, was started about 1645 in Sendai. Soybean miso was developed in 1625 in Aichi Prefecture. Thereafter, the number of miso factories in Japan increased to about 5000 toward the end of World War II (1945). Since the introduction of miso, many modifications have been made in the raw materials and the production methods used to give a large number of varieties of miso.

B. Consumption Patterns and Production

Miso is used as an ingredient in soups. Soups are prepared by dissolving a lump of miso and condiments in hot water to which previously cooked vegetables, tofu, fried tofu, mushrooms, meat, fish, and other suitable ingredients are added. Sometimes miso is also used for preparing neri-miso (sweet simmered miso), a side dish prepared by cooking with vegetables and other ingredients as mentioned above, with or without sugar and other condiments. It is also used as a salty meat-like seasoning much as shoyu (soy sauce). Miso is often used as a base for pickling with meat, egg, fish, shellfish, and vegetables.

Daily per capita consumption was 28 to 30 g during the years from 1951 to 1959. From that time until 1982, it gradually decreased to the level of 17 g. This amount of miso is enough for making one bowl of miso soup.

Miso can be classified into three major types on the basis of the raw materials (Table 1): rice miso made from rice, soybeans, and salt; barley miso from barley, soybeans, and salt; and soybean miso from soybeans and salt. Rice miso and barley miso types are further classified into sweet, medium salty, and salty groups by taste. Each of these groups is further divided into white, light yellow, and red by color of the product. According to the Munsell Color Book, all the colors of miso are in the hue from YR to R. The amount of rice miso, barley miso, and soybean miso (including mixed miso), was, respectively, 451,880, 69,220, and 57,200 tons produced in Japan from 180,000

Table 1
VARIETIES OF MISO

Varieties	Taste	Color	Principal areas of production in Japan
Rice miso	Sweet	White	Kyoto, Kagawa, Hiroshima
		Red	Tokyo
	Medium salty	Light yellow	Shizuoka
	Salty	Light yellow	Nagano
		Red	Tohoku province, Hokkaido, Niigata, Kanto province, Hokuriku province, Sanin province
Barley miso	Sweet	Light yellow	Kyushu province, Chugoku province, Ehime
	Salty	Red	Kyushu province, Kanto province
Soybean miso	Salty	Red	Aichi, Mie, Gifu

tons of soybeans, 600 tons of defatted soybeans, 104,000 tons of rice, 25,000 tons of barley, and 70,000 tons of salt in 1982. In addition, the amount of home-made miso in Japan can be roughly estimated at 72,000 tons in 1980.

Salty rice miso is the most typical variety produced in Japan. Barley miso types are produced principally in Kyushu Province, and prefectures of Ehime, Yamaguchi, and Hiroshima. A small amount of barley miso is also produced in the districts located in the northern part of Tokyo. Soybean miso types are produced and consumed principally in the three prefectures Aichi, Mie, and Gifu located in the central part of Japan.

The cost for one kg of packaged miso ranges from 300 to 1200 yen. The wide range in the cost of miso is in part due to the cost of raw materials, labor, and duration of fermentation (from two weeks for sweet miso, to three years for some varieties of soybean miso and salty rice miso).

II. PREPARATION

A. Raw Materials

1. Soybeans

Yellow soybeans, rich in protein, are generally used. Japanese domestic soybeans are more suitable for making rice or barley miso types than soybeans imported from China, U.S., or Canada. The soybean varieties with high water-absorbing capacity, high carbohydrate and low calcium content are used as substrates for miso. Soybean grits made from hard soybeans have improved water absorbing capacity, therefore are preferred for miso preparation. Soybean grits have a preferable consistency to those of cooked soybeans.[1-3] Defatted soybeans are not suitable for miso preparation, since they give poor color and consistency to the final product.[4]

2. Rice

Nonglutinous polished rice is used for making rice koji. Rice containing unripened kernels, different shapes, damaged kernels, or foreign materials (including insects, ticks, sand, and soil) are not suitable. The soft type or *Oryza sativa* var. *japonica* is preferred over *Oryza sativa* var. *indica* (hard type), for cooking and making rice koji and miso. The hard type of rice is difficult to cook by ordinary cooking method. The soft types of rice are available in Japan, Taiwan, Korea, Italy, Spain, and the U.S.

3. Barley

Barley, including naked barley produced in western Japan, is used for making barley koji. Barley with thin coats and a brilliant yellow color is preferred for barley miso.

Table 2
VARIETIES OF MISO AND ASSOCIATED SUBSTRATES FOR KOJI PREPARATION, SHAPES OF KOJI, AND MICROORGANISMS USED FOR KOJI PREPARATION

Miso name	Substrate for koji preparation	Shape of koji	Microorganisms	Place
Soybean miso	Soybeans	Ball or cylinder	*Aspergillus oryzae* *Aspergillus sojae*	Central Japan
Doenjang	Soybeans	Ball or cube	*A. oryzae* *A. sojae*	Korea
Kochujang	Soybeans	Ball or cube	*A. oryzae* *A. sojae*	Korea
Hama-natto	Soybeans	Granular	*A. oryzae*	Japan
Tauco	Soybeans and tapioca	Granular	*A. oryzae* and *Rhizopus oligosporus*	West Java, Indonesia
Kinzanji miso	Soybeans and barley or wheat	Granular	*A. oryzae*	Japan
Rice miso	Rice	Granular	*A. oryzae*	Japan
Barley miso	Barley	Granular	*A. oryzae*	West Japan

4. Salt and Water

Food grade salt with more than 95% purity is generally employed. Salt containing less than 1 ppm of iron is used for preparation of miso, to avoid the browning effect by iron. The presence of ferric iron in water accelerates the browning reaction of miso during fermentation and after ripening. Tap water is used for washing, soaking, cooking of raw materials, and for mixing cooked soybeans, koji, and salts. When hard water containing calcium is used for soaking or cooking of soybeans, the cooked soybeans become harder than soybeans cooked with soft water.

B. Types of Koji

The basic process in manufacturing miso consists of two steps: (1) the koji-making process, and (2) the brine-fermentation process. Koji is prepared from carbohydrate-containing materials such as rice, barley, or soybeans, and is inoculated with a mold culture. After fermentation, the resulting koji is mixed with salt, cooked whole soybeans, and water. The mixture is inoculated with yeast and lactic acid bacteria to produce miso. Varieties of miso and the associated substrates for koji preparation, the shapes of koji, and the microorganisms involved, are all presented in Table 2. Bo[5] reported that koji had been used for manufacturing "chiang" (a type of soybean paste) 4000 years ago. The prototype soybean pastes of today apparently originated in China at least 2000 years ago. The substrate for koji preparation and shape of koji, as well as size of koji, affect the manner of koji making and the properties of finished koji. Ball-shaped soybean (miso-dama) kojis are employed especially at farmhouses in Japan and widely in Korea. The preparation method for miso-dama at factories has been improved by employing a specially designed extruder and pure-cultured starter of *Aspergillus oryzae* for koji cultivation (Table 2). However, the conventional method seen at farmhouses naturally ferments the miso-dama wrapped in rice straws. This method prevents serious contamination by aerobic *Bacillus subtilis* inside the ball and is adequate for the development of lactic acid bacteria for natural fermentation.

Another method of koji preparation is for granular koji. This koji is prepared from cooked soybeans. In this method a large surface area favorable for the growth of mold provides aerobic conditions which favors production of essential enzymes. Granular soybean koji generally has a large amount of the essential enzymes for fermentation

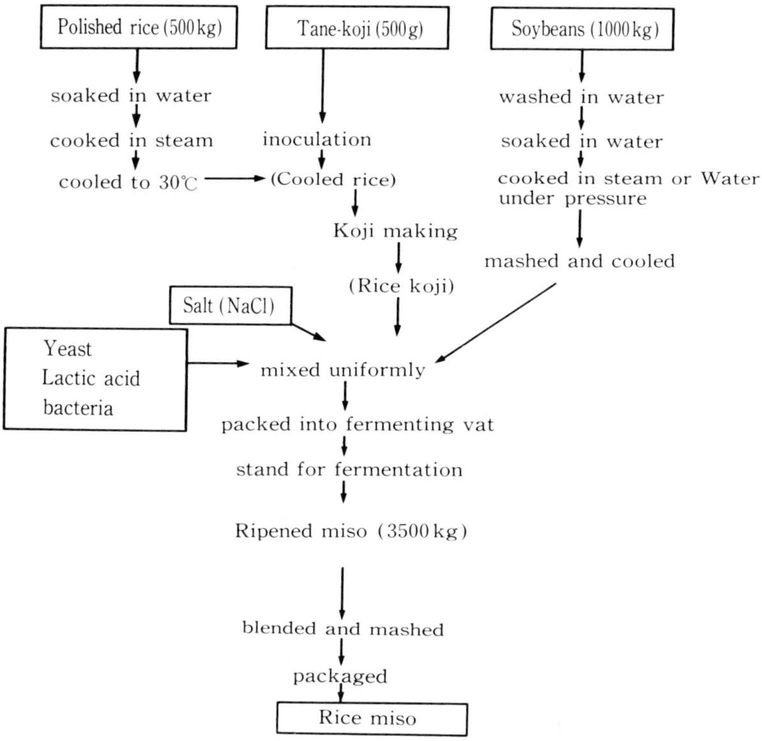

FIGURE 1. Flow chart of the manufacture of salty rice miso.

providing a shorter fermentation period and an increase in the level of hydrolysis of the higher molecular weight constituents. However, granular soybean fermentation is often accompanied by a danger of contamination with *Bacillus subtilis* due to the higher moisture level of the cooked soybeans. To reduce the moisture level on the surface, a small amount of flour from parched barley or wheat is added to provide adequate conditions for the mold growth.

A third method is the preparation of granular koji with rice or barley, specifically done in miso factories in Japan. This method was developed in the districts of Japan where sake (rice wine) is produced using rice koji. Granular rice or barley is suitable for preparing a clean koji.

Another method of koji preparation is to use a mixture of cooked soybeans and barley or wheat. This method results in koji with high enzyme activity, especially proteolytic enzymes. It is widely used for soy sauce preparation, where koji with higher proteolytic enzymes is preferred.

C. Commercial Preparation of Miso

There are many varieties of miso which differ from each other depending on the manufacturing process. Rice miso is the most popular type. It is produced throughout Japan and accounts for about 80% of the total miso production. The manufacturing method of salty rice miso is presented in Figure 1, and soybean miso production is presented in Figure 2. Figure 3 shows typical soybean, rice, and barley miso.

1. Rice Miso

a. Koji Preparation

Polished and cleaned rice is first washed and soaked in water overnight at room

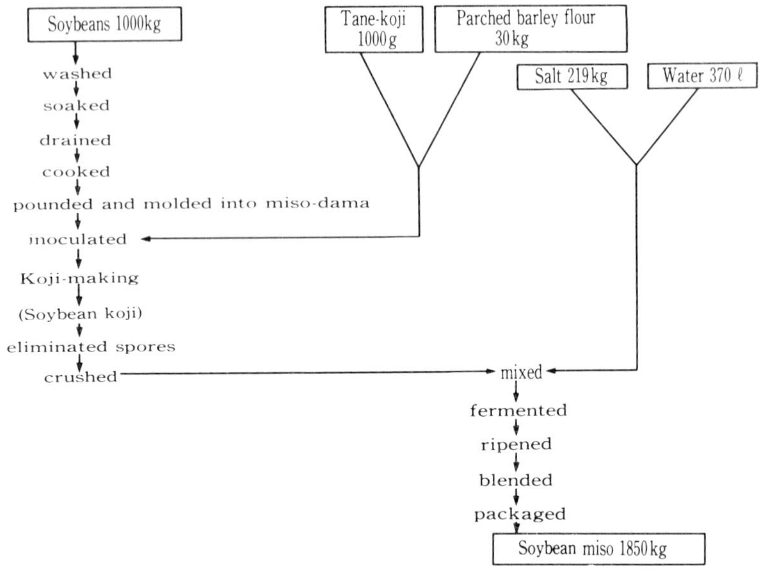

FIGURE 2. Flow chart of the manufacture of soybean miso.

FIGURE 3. Typical varieties of miso.

temperature. The soak water is then discarded and the water-absorbed rice is cooked in an open cooker for 40 min with steam. The cooked rice is cooled to 35°C and inoculated or sprinkled with tane-koji (spores of *Aspergillus oryzae*) and mixed well. One gram of tane-koji containing 10^9 or more of spores inoculates about 1 kg of cooked rice. The inoculated rice is incubated for about 15 hr in a container at 30°C and 96% relative humidity. The spores start to germinate after the first 5 hr of incubation and eventually develop all over the rice, raising the temperature gradually to 36°C or more. In the koji tray method, the mold which developed on the rice is mixed and spread on many trays to a depth of 4 cm. The trays containing young rice koji are stacked one upon the other. The young rice koji is allowed to develop for up to 40 hr. The koji tray method was used by industry until the koji fermenter was introduced in 1955. Now it is seen only on the village level at factories or farmhouses. At present, industrial preparation of koji is carried out in a koji fermenter equipped with an air conditioner to maintain temperature (30°C) and relative humidity (96%), and a mechanical stirrer to break the sponge-like lump during cultivation. The mold develops vigorously, raising the temperature of the rice. The completed koji is taken out after 40 to 48 hr of fermentation and mixed with salt to stop further growth of mold.

b. Cooking of Soybeans

Mechanically cleaned soybeans are washed and soaked in water overnight at room temperature. The soaking time can be reduced by elevating the temperature of the soaking water. Absorption of water to double the original weight of the beans is necessary to soften and smooth the soybeans. There are two ways to cook soybeans, one is with steam, the other is boiling the soybeans in water. To cook with steam, the soaked soybeans are first drained, transferred into a cooker, and then cooked for 20 to 30 min under pressure at 115°C. Generally, batch-type cookers are widely used in factories. Continuous cookers with rotary bulbs and belt conveyors are used in the large scale factories. In the boiling method, soybeans are cooked with four parts water without presoaking. Cooking in water is desirous, since soft, light cooked soybeans are obtained. This results in a light colored miso. Cooking in water has the disadvantage of increasing solids loss (20 to 30%) from soybeans. This causes an additional problem in waste treatment.[6] The desired hardness of cooked soybeans varies from 0.3 kg (weight necessary to break a cooked soybean) for salty red miso, to 0.5 kg for other salty miso.

c. Mixing and Mashing

Cooked soybeans are mixed with salted rice koji and a water inoculum of yeast and lactic acid bacteria. The mixed inoculated soybeans are roughly mashed by passing them through a motor-driven chopper equipped with a cutting plate with many 5 mm perforations. The cooked soybeans are sometimes chopped first and mixed with salted rice koji and a water inoculum. Homogeneity of the mashed mixture is important for normal fermentation. The chopped, homogenous mixture is then packed into either wooden fermentation vats (167 cm deep and 217 cm in diameter), steel vats coated with epoxy resins, stainless steel vats, or glass-lined resin vats.

Recently, large cylindrical fermenters capable of holding 60 to 100 tons of miso have been used at large-scale factories. Small stainless steel vats with a one ton capacity are used in place of larger vats (4 to 6 ton capacity) in small factories.

After packing tightly, the surface of the miso in the vats is covered with a resin sheet to prevent contamination. The sheet is pressed with weights equivalent to 5 to 10% of the miso. The miso mixture in the vats is allowed to ferment for up to 6 months or more at an ambient temperature. Presently, the miso filled vats are put in a fermenting chamber, where the temperature and relative humidity (RH) can be controlled automatically. The temperature is maintained at 30°C and 60 to 70% RH. The fermentation period for ripening varies from 10 to 14 days for sweet miso, to 3 to 4 months for salty miso. During fermentation, miso is transferred from the original vat to another vat at least twice to improve homogeneity of the mixture and accelerate the fermentation by yeast.

Fully fermented miso is removed from the fermenter, blended if necessary, and mashed by passing it through a chopper having a steel plate cutter with perforations of 1 to 2 mm. During mashing, miso is pasteurized by passing it through a steam injector which is attached to the chopper. Finally, the finished miso is packaged in plastic film bags or plastic cubic containers that hold 500 to 1000 g of product. Packing raw fermented miso into containers results in swelling during distribution and marketing because of the presence of viable yeast cells and subsequent refermentation. In order to prevent swelling, the containers containing raw miso are generally pasteurized in hot water (90°C), with a specially designed pasteurizer,[7] or by adding 2% ethyl alcohol to the raw fermented packed miso containers.

2. Soybean Miso

Soybean miso is prepared from soybeans, salt, and small amounts of parched barley flour (Figure 2). Soybeans are washed and soaked in water to a 1.5- to 1.6-fold weight increase in the soybeans. The soaked soybeans are then cooked at atmospheric pressure for 30 to 40 min, then under a pressure of 0.7 kg/cm² for 90 to 120 min. The cooked soybeans are cooled to 60° C and made into miso-dama (cooked soybean ball) by passing the cooked soybeans through a simple extruder specially designed for miso-dama preparation. The size of the miso-dama varies from 13 to 66 mm in diameter. Small miso-dama gives a large surface area for the development of aerobic mold, *Aspergillus oryzae.* Large miso-dama provides anaerobic conditions in their interior for the development of anaerobic lactic acid bacteria. The latter ferment carbohydrates into lactic acid which is necessary for protection from microbial infections and for formation of flavor. The fermentation period can be reduced by the use of small miso-dama, but larger miso-dama give improved color and flavor to the miso.

The miso-dama is cooled to 30°C and mixed with parched barley flour containing tane-koji (spores of *A. oryzae*). The amount of parched barley flour employed is usually equivalent to 2 to 30% of the dried soybeans, but often it is increased when the moisture content of miso-dama is high. The well-mixed mixture is fermented at 27 to 35°C in a koji fermenter similar to that for rice koji preparation. One ton of soybeans yields about 1200 to 1300 kg of miso-dama koji. The fermented miso-dama koji is pressed and cut into 10 mm-thick pieces by passing it through a roller. The cut pieces of miso-dama koji are mixed with salt and water or brine. About 200 kg of salt and 410 kg of water are mixed with 1200 to 1300 kg of miso-dama koji to obtain the final mixture (1770 to 1870 kg). The mixture is packed into vats similar to that of rice miso preparation and is fermented for about 6 months when started in the spring, and one year when started in autumn (Figure 3). When large miso-dama is used, it requires 2 years or more for fermentation.

D. Microorganisms Involved

Although many varieties of microorganisms have been isolated and identified from miso, the principal and indispensable microorganisms in miso fermentation are koji molds, salt tolerant yeasts, and lactic acid bacteria.

1. Koji Molds

Two species of *Aspergillus,* namely *A. oryzae* and *A. sojae,* are widely used as the inoculum for making tane-koji (mold starter). Koji is a Japanese given name to the mold-cultured steamed cereals or beans. The major role of koji is to supply various types of enzymes (amylolytic, proteolytic,and lipolytic) which hydrolyze the constituents of the substrates. The low molecular weight compounds resulting from the action of these enzymes are involved in the development of flavors and act as nutrients for the microorganisms during miso ripening. Koji-molds produce minor nutrients including vitamins (thiamin, riboflavin, vitamin B_6, and pantothenic acid) and other growth factors.[8]

2. Bacteria

Streptococcus faecalis and *Streptococcus faecium* are the dominant weak salt-tolerant lactic acid bacteria. They grow well during koji preparation and produce lactic acid which prevents the growth of bacillus and clostridium. Since they are not tolerant to the salt concentration of miso, they disappear after the mixing of koji with brine.

Pediococcus halophilus and *Pediococcus acidi lactici* are the most dominant lactic acid types. They play an important role in the fermentation of miso. These organisms can survive in 23% brine. Since the acid tolerance of these bacteria is comparatively weak, they cannot survive below pH 5.0. The role of these bacteria in miso fermentation is to produce organic acids from carbohydrates, which lowers the pH of the fermenting miso and accelerates the growth of salt-tolerant yeasts.[9] *P. acidi lactici* is more acid tolerant and less salt tolerant than *P. halophilus. P. acidi lactici* can grow under low levels of salt (less than 10 percent) and produces lactic acid until the pH comes down to 4.0. It is recognized as a causative microorganism of acidification of miso of low salt.[10-12]

Micrococcus epidermidis, Micrococcus varians, Micrococcus congromeratus, and *Micrococcus caseolyticus* are dominant aerobic bacteria and grow well on rice koji producing small amounts of acid. Their numbers decrease sharply after mixing the koji with brine. *Bacillus subtilis* is one of the harmful microorganisms of fermented foods, and is considered as harmful when it grows to more than 10^9/g in miso. Generally, it is propagated in koji and survives as spores and vegetative cells in miso.

Clostridium toanum, Clostridium butyricum, Clostridium tertium, Clostridium multifermentus, and *Clostridium perfringens* are heat-resistant, spore-forming bacteria that are not desired in fermented foods. These are often isolated from off-flavored koji. Since the salt tolerance of these bacteria is low, they cannot grow in salty miso, but survive as spores which can germinate under low salt conditions. They often cause swelling in packed sweet miso by production of carbon dioxide and hydrogen.[13-15]

3. Yeasts

Saccharomyces rouxii, Torulopsis versatilis, and *Torulopsis etchellsii* are the salt-tolerant yeasts involved in the miso fermentation. *S. rouxii* grows in a high salt medium (18% sodium chloride) of pH 4.0 to 5.0 and is the main yeast involved in the fermentaiton of miso. It ferments glucose and maltose, but not galactose, sucrose, or lactose. *S. rouxii* produces high yields of glycerol, up to 50% of the fermented glucose under aerobic conditions and high sodium chloride medium. *T. versatilis* and *T. etchellsii* have been isolated from miso fermentation medium. They grow well and ferment in a 21% sodium chloride medium. Further, they produce ethanol from sugars and phenolic compounds such as 4-ethylguaiacol, 4-ethylphenol, and 2-phenylethanol.[16,17] These compounds give a characteristic flavor to the fermented miso.

Saccharomyces rouxii var. *halomembranis* nov. var., *Pichia* spp. and *Hansenula* spp. are the aerobic film yeasts, which often produce films on substrate surfaces resulting in off-flavor of the products.

III. COMPOSITION

A. Chemical Composition

The approximate composition of various types of miso is presented in Table 3. The moisture content in miso types ranges from 42.6 to 45.7%. Moisture often includes ethylalcohol (1 to 2%) and other volatile compounds. Sweet and barley miso have higher amounts of carbohydrates compared to others. This is due to a short fermentation period. Glucose (10 to 13%) is the predominant sugar in rice miso followed by isomaltose, fructose, galactose, galacturonic acid, stachyose, mannose, melibiose, arabinose, and xylose.[20] The miso types are poor sources of fiber and vitamins, and contain less than 3.5% of fiber. Appreciable amounts of calcium and phosphorus are present in various types of miso.

Protein in the substrates is rapidly digested at early stages of fermentation. Protein solubility degree (the ratio of water soluble nitrogen to total nitrogen, %) is approxi-

Table 3
CHEMICAL COMPOSITION OF VARIOUS TYPES OF MISO[18,19,44]

Component	Sweet miso	Salty light yellow miso	Salty red miso	Barley miso	Soybean miso
Moisture (%)	42.6	45.4	45.7	44.0	44.9
Protein (%)	9.7	12.5	13.1	9.7	17.2
Fat (%)	3.0	6.0	5.5	4.3	10.5
Carbohydrate (%)	36.7	19.4	19.1	28.3	11.3
Crude fiber (%)	1.2	2.5	2.0	1.7	3.2
Ash (%)	6.8	14.2	14.6	12.0	12.9
Salt (%)	6.1	12.4	13.0	10.7	10.9
Calcium (mg/100g)	80.0	100.0	130.0	80.0	150.0
Phosphorus (mg/100g)	130.0	170.0	200.0	120.0	250.0
Iron (mg/100g)	3.4	4.0	4.3	3.0	6.8
Thiamin (mg/100g)	0.05	0.03	0.03	0.04	0.04
Riboflavin (mg/100g)	0.1	0.1	0.1	0.1	0.12
Niacin (mg/100g)	1.5	1.5	1.5	1.5	1.2
Vitamin B12 (μg/100g)	—	0.17	—	—	—

Table 4
AMINO ACID CONTENT (mg/g miso) OF VARIOUS TYPES OF MISO[22,23]

	Edo miso[a]		Sendai miso[b]		Hacho miso[c]	
Amino acid	Total	Free	Total	Free	Total	Free
Arginine	8.8	3.5	8.7	3.7	6.8	0.5
Lysine	8.6	2.4	7.4	2.8	9.5	4.5
Histidine	1.5	0.3	1.4	0.3	3.3	0.9
Glycine	6.4	1.4	7.4	2.3	12.6	4.7
Valine	7.5	2.0	8.1	2.2	11.2	5.0
Leucine	11.7	3.2	10.2	4.1	22.8	5.0
Isoleucine	11.5	3.6	11.3	3.6	18.2	7.3
Methionine	2.3	1.2	1.2	0.9	3.7	1.8
Serine	8.6	2.9	6.0	2.0	11.2	5.7
Threonine	4.3	1.4	4.4	1.6	6.8	2.6
Phenylalanine	5.5	1.7	5.9	2.1	6.0	3.3
Tyrosine	4.3	1.5	3.4	1.5	7.1	2.7
Tryptophan	0.9	0.3	1.0	0.2	2.0	0.3
Proline	6.4	2.7	5.8	2.4	10.7	6.4
Aspartic acid	9.6	2.3	5.6	1.8	16.4	4.6
Glutamic acid	20.0	4.9	25.6	6.9	35.2	9.9

[a] Represents a variety of sweet rice miso.
[b] Represents a variety of salty rice miso.
[c] Represents a variety of soybean miso.

mately 60%, whereas protein degrading degree (the ratio of formol nitrogen to total nitrogen, %) is 25%.[21] The amino acid content of sweet rice, salty rice, and soybean miso is presented in Table 4. Glutamic acid is the major amino acid in miso. This gives meat-like flavor to the fermented product, miso. It is derived from glutamine, by action of the enzyme glutaminase which is produced by the mold.[24] Methionine is converted to methinol during yeast fermentation. This gives a strong characteristic aroma to the fully ripened miso.[25,26] The level of fat in miso varies from 3.0 to 10.5% depend-

Table 5
PHYSICAL CHARACTERISTICS OF MISO[7]

Density (g/cm^3)	Specific heat (cal/g °C)	Thermal conductivity (cal/cm min °C)	Apparent viscosity (g/cm sec)10^2
1.22—1.24	0.71—0.79	0.061—0.07	2.70—16.60

Table 6
COLOR OF VARIOUS TYPES OF MISO
(C.I.E. COLOR NOTATION)[30]

	White miso	Salty light yellow miso	Salty miso	Barley miso	Soybean miso
Y (%)	33.19	26.76	10.33	12.78	2.57
x	0.416	0.435	0.473	0.466	0.483
y	0.420	0.426	0.418	0.429	0.403

ing on the raw materials and their ratio in the ingredients. According to Kiuchi,[27] the fatty acid composition of most miso types is similar to that of soybeans, but the linolic acid content varies remarkably among soybean miso types. During fermentation, hydrolysis of lipids is evident by an increase in the free fatty acids and a gradual decrease of linolic acid. Lipids in rice miso are also hydrolyzed gradually during fermentation to produce glycerol and free fatty acids including linolic, oleic, palmitic, and linolenic acids. These acids can be esterified with ethanol resulting in the formation of characteristic aromas in miso. Esterification of free fatty acids occurs only when the ethanol level is over 0.1%.[28]

Lactic, acetic, citric, pyroglutamic, succinic, malic, and pyruvic are the major organic acids produced in miso. The content of lactic acid is 50, 40 to 150, 100 to 200, and 100 to 400 mg/100g in white, salty light yellow, salty red, and soybean miso, respectively. During fermentation, citric and malic acids decrease while others increase.

Ethanol and its esters of organic acids are the most important aroma components in rice and barley misos. Isobutyl, butyl, and isoamylalcohols are produced by yeasts from amino acids during fermentation. They also give characteristic aroma to well-fermented rice or barley miso. Vanillic acid and ferulic acid are the major components of the characteristic aroma of barley miso.[29] Furthermore, ferulic acid is changed to 4-ethylguaiacol (a strong aroma compound) in miso that is fully fermented by *Torulopsis*.[16]

B. Physical Properties

Physical properties of miso, including density, specific heat, thermal conductivity, apparent viscosity and color, are shown in Tables 5 and 6. Large variation was found in the apparent viscosity of miso. (Table 5). A small variation was reported in density, specific heat, and thermal conductivity of miso. White miso has a high Y(%) value compared to other miso types (Table 6). This is due to use of different ingredients in preparation of various miso types. Soybean miso has lowest Y(%) value. The variation between miso types in x, y values appeared to be small.

C. Organoleptic Qualities

Miso is a salty paste and has a meat-like flavor. Its texture resembles that of slightly hard peanut butter. Each type of miso has its own characteristic flavor and aroma. White miso is rich in sweetness and has a light aroma with less meat-like favor. Salty

Table 7
FATTY ACID CONTENT OF MISO[a]

Miso type	Palmitic acid	Stearic acid	Oleic acid	Linoleic acid	Linolenic acid
Salty light yellow rice miso	13.4	3.2	25.2	52.2	6.9
Salty red rice miso	11.5	2.8	22.6	54.6	8.3
Barley miso	14.1	2.8	20.1	55.2	7.1
Soybean miso	9.7	4.1	20.0	58.1	9.3

[a] Expressed as percent of the total fatty acids.

light yellow rice miso is salty with meat-like flavor and a fermented aroma. Salty red rice miso has a matured saltiness with meat-like flavor and deep fermented aroma. Barley misos, including mellow and salty or light yellow and red varieties have a characteristic aroma which originates from the phenolic compounds.[29] Soybean miso has a matured saltiness with deep meat-like flavor and deep, heavy soy aroma.

D. Other Characteristics

Miso has a strong antioxidant activity because of the presence of antioxidants.[30,31] Also, miso has a high buffer capacity which makes it possible to maintain a constant pH value even when miso is mixed with other foods.[32] The insoluble proteins in miso absorb odors such as those of fish, shellfish, and wild animals. Because of these properties, miso can be used as an ingredient for making soup or as a seasoning in variety of foods including meats, seafoods, and vegetables.

IV. NUTRITIONAL QUALITY

Water-soluble components (other than sodium chloride) account for about 30% of salty rice miso. These water-soluble components include free amino acids, peptides, free sugars, and organic acids which are all highly digestible.[33-35]

The amino acid pattern of miso resembles that of soybeans. According to Obata et al.,[36] the biological value of nonsalty miso and low-salt miso is 73.2 and 70.7, respectively, compared to casein which has a biological value of 80.1.

The pattern of fatty acids in lipids resembles that of soybeans. The amount of unsaturated fatty acids including linolic and linolenic acids is over 60% in various types of miso (Table 7). The peroxide value of the fat and oil in fully ripened rice miso is low (0.04 to 0.06 meq/kg) indicating the presence of a strong antioxidant activity.[37]

The content of sodium, potassium, and calcium is comparatively high in various types of miso. The high content of sodium has often been criticized when considering current recommendations concerning salt intake. According to Yamori et al.,[38] however, the Na/K ratio plays a very important role in the control of blood pressure. The ratio (Na/K) of salty miso is 10. This ratio can be reduced to 5 to 6 by preparing a miso soup using potassium rich foods such as potatoes, leafy vegetables, and seaweed.[39] Because of the high calcium content, miso (consumed as miso soup) supplies about 15% of calcium intake for the inhabitants of Tohoku province.[40]

The thiamin, riboflavin, and niacin content of various types of miso is presented in Table 3. Thirty to fifty percent of the thiamin in raw soybeans is lost during soaking and cooking.[41] Riboflavin is synthesized by *A. oryzae* during koji making and remains stable in miso.[42] The riboflavin content can be increased by using a mutant mold strain.[43] Vitamin B-12 content produced by bacteria in salty light yellow miso is 0.17

FIGURE 4. Kinzanji miso packaged in a plastic transparent container.

μg/100 g.[44] The folic acid content of white miso, salty rice miso, barley miso, and soybean miso, respectively, is 13.4 to 17.5, 7.6 to 17.5, 7.3 to 11.4, and 15.8 to 17.7 μg/100 g.[45]

Phytic acid is an antinutrient that is present in soybeans and cereals. It easily reacts and forms stable complexes with nutritionally important minerals such as zinc, calcium, iron, and copper resulting in their insolubilization and reduced bioavailability.[45] Phytic acid, however, is hydrolyzed by mold phytase during the fermentation.[47,48]

According to Hirayama,[49] daily intake of miso soup reduces the mortality rates for gastric cancer in humans. This study was carried out on a large scale involving 122,261 males and 142,857 females aged 40 or more in 29 health center districts of Japan between 1966 and 1978. Okazaki et al.[50] found a significant antimutagenic effect with the lipid fraction of miso.

V. TOXICOLOGY

In one study, no aflatoxin-producing mold strains were found among 238 mold strains collected from factories where koji is prepared for miso and its related products. Similarly, no aflatoxin was detected in 28 rice kojis from miso factories, 108 commercial misos, and 30 homemade misos.[51] Absence of mycotoxins in fermented foods has been recently described by Wang and Hesseltine.[52]

VI. MISO-LIKE PRODUCTS

A. Kinzanji Miso

Kinzanji miso (Figure 4) is said to have been introduced from China to Yuasa, Wakayama Prefecture, Japan, by the famous Buddist priest, Kakushin (AD 1255)[53] Kinzanji miso is principally produced in the prefectures of Shizuoka, Wakayama, Hiroshima, and in Shikoku Province. Similar products, namely *Hishiho* in Chiba Prefecture and *Oname* in Saitama Prefecture, are produced. Kinzanji miso is not used as an ingredient of miso soup but is used as a side dish. The approximate annual production of kinzanji miso in Japan is 10,000 tons.

The preparation method for kinzanji miso varies from district to district and from factory to factory. The traditional and typical product is made from wheat, soybeans, salt, and several varieties of vegetables including salted eggplant, East Indian lotus, oriental pickling melon, and perilla seeds. The manufacturing process for kinzanji miso is presented in Figure 5.

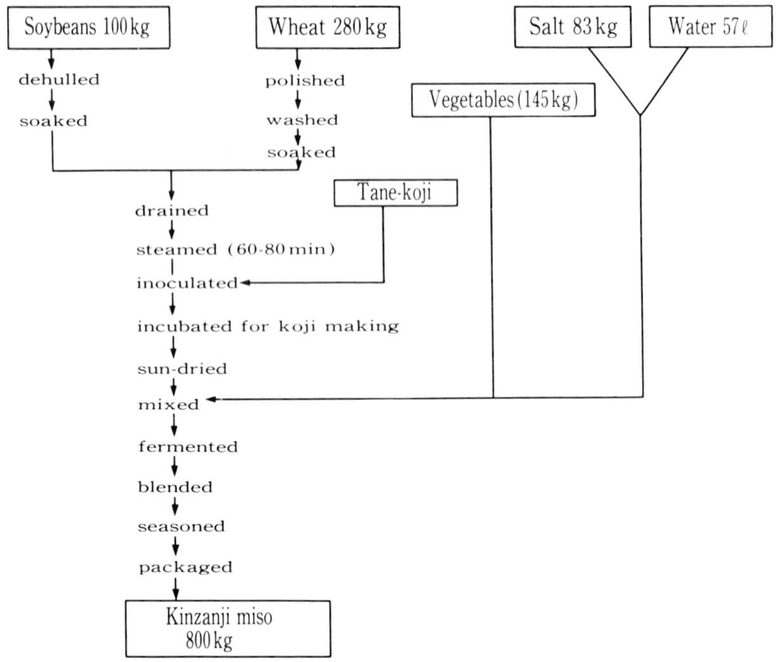

FIGURE 5. Flow chart of the manufacture of Kinzanji miso.

Dehulled soybeans are slightly roasted and soaked in water. In addition, polished wheat is washed and soaked. The soaked soybeans and wheat are drained and cooked in steam for making koji, following the method for barley koji. The finished koji is partially sun-dried for two days. The partially dried koji is mixed with brine and salted vegetables and fermented in vats for about 6 months. The ripened miso is often blended and mixed with sugar and other suitable condiments and packaged in small plastic cups (300 to 500 mℓ) for marketing.

The chemical composition of kinzanji miso is shown in Table 8. The physico-chemical and organoleptic properties of kinzanji miso are similar to those of barley miso.

B. Hama-Natto (Tera-Natto)

Hama-natto consists of granular, salty fermented soybean nuggets (Figure 6). It is believed to have been introduced to Japan directly from China by Buddhists. In China, Hama-natto is described as "Shin" in the old literature and "Chimin Yaoshu" in AD 450. Hama-natto is also known as "Tera-natto" and "Daitokuji-natto" because it was first prepared at temples with these names. Hama-natto is produced at Hamamatsu in Shizuoka Prefecture. It is mainly consumed as a side dish. The annual production of Hama-natto is less than 1000 tons and is declining.

The preparation method for Hama-natto varies from place to place. The cleaned soybeans are washed and soaked in water for 3 to 4 hr at room temperature and drained (Figure 7). The drained soybeans are cooked for 5 to 6 hr in a cooker, and cooled to 30°C. The cooked soybeans are inoculated with a mold starter (tane-koji in parched barley or wheat flour) and mixed well. The mixture is spread over a koji tray and fermented for 40 to 48 hours at 27 to 30°C. The finished koji is sun-dried to a moisture content of 22%. The partially dried soybean koji is mixed with brine (Baumé 15°) and fermented in vats for 8 months. The fully ripened soybean koji is taken out from the fermenting vat and sun-dried on a mat. Ginger pickled in Tamari (a variety of shoyu) is mixed with the dried soybean koji at a level of 7%. A small amount of Japanese pepper is also added, if necessary.

Table 8
CHEMICAL COMPOSITION OF HAMA-NATTO AND KINZANJI MISO

Component	Hama-natto	Kinzanji miso
Moisture (%)	24.4	34.3
Protein (%)	18.6	6.9
Fat (%)	8.1	3.2
Carbohydrate (%)	29.2	47.3
Crude fiber (%)	2.9	1.5
Ash (%)	16.8	6.8
Salt (%)	14.2	5.8
Calcium (mg/100g)	110.0	40.0
Phosphorus (mg/100g)	330.0	130.0
Iron (mg/100g)	5.9	1.7
Sodium (mg/100g)	5600.0	2300.0
Potassium (mg/100g)	1000.0	280.0
Thiamin (mg/100g)	0.04	0.12
Riboflavin (mg/100g)	0.35	0.18
Niacin (mg/100g)	4.1	2.3

FIGURE 6. Hama-natto packaged in a bamboo sheath.

The chemical composition of Hama-natto is presented in Table 8. It has a high sodium content. There are some studies on reducing the salt level to 7%. Kiuchi, et al.[56] investigated the changes in lipid content that occur during processing and fermentation of Hama-natto. The triglyceride content decreased from 99.0% to 25.1%, whereas the free fatty acid content increased from 0 to 70.8% during fermentation. According to Kon,[57] many free acids are detected in Hama-natto and their pattern is similar to that of soybean miso.

Asano and Saito[58] isolated from Hama-natto koji two strains of *A. oryzae* that had strong proteolytic and amylolytic activity. Kon and Ito[59] isolated *Micrococcus* spp., *Streptococcus* spp., and *Pediococcus* spp. from the fermented Hama-natto. These microorganisms may play an important role in the formation of characteristic flavors. There were no toxicological problems detected from Hama-natto due to the presence of other microorganisms.

C. Doenjang (Soybean Paste) and Kochujang (Red Pepper Soybean Paste)

Fermented soybean products including *Jang* and *Kanjang* were recorded in Korea in AD 680.[59] Doenjang and Kochujang are prepared widely in Korean homes. Doenjang

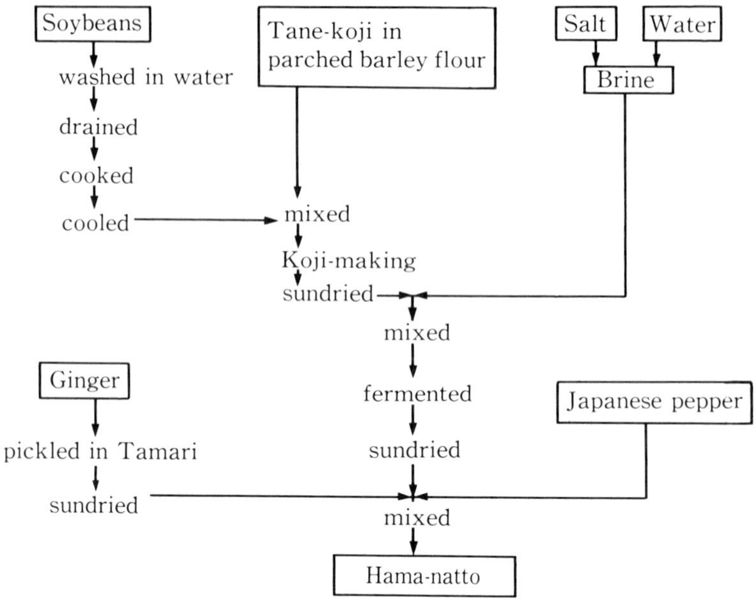

FIGURE 7. Flow chart of the manufacture of Hama-natto.

is used untreated with the rice or as a base for the soup or stew. Kochujang is also used untreated with rice or as an ingredient of various stews. The daily per capita consumption of Doenjang and Kochujang in 1976 was 15 g and 10 g, respectively. According to Mheen et al.,[60] the amount of Doenjang produced was 268,757 tons in 1979, of which 20% was commercial product. The total production of Kochujang in 1979 was 130,195 tons, of which 25% represented commercial product. The commercial products are milder in taste compared to home-made products. People prefer the sharp and delicate flavored traditional products.

The preparation method for Doenjang is shown in Figure 8. Washed and soaked soybeans are cooked in water for 5 to 6 hours, pounded and molded into balls of 0.5 to 1.0 kg called "Meju", which corresponds to miso-dama in Japan. Meju balls are wrapped in rice straw, hung under rafters or eves for 1 to 3 months until the meju is covered with a white mold. The mold covered ball is broken into pieces and sun-dried. The dried meju is mixed well with brine (20% salt) in earthenware containers of 4 to 40 ℓ capacity. The mixture is fermented for six months in the summer. At the end of fermentation, the liquid part that accumulates in the container is filtered off to make Kanjang (a soy sauce), and the remaining semi-solid portion in the container is allowed to ferment further for making Doenjang.

The preparation method for Kochujang varies widely depending on the place and availability of raw materials. A mixture of ground meju (8%), steamed cereal flour (mainly rice with or without wheat flour) (12%), salt (10%), and water (33%) is fermented for 2 to 3 months.

Meju is produced industrially by inoculating *A. oryzae* for use in individual households. The presence of *Mucor* spp., *Pencillium* spp., *Rhizopus* spp., *A. oryzae, B. subtilis, B. pumilis, Rhodotorula flava, Torulopsis datilla,* and many other microorganisms in meju has been reported.[61] However, the typical microorganisms involved in meju fermentation are *A. oryzae, A. sojae,* and *B. subtilis.*

Doenjang is light grayish-brown and slightly chunky with a salty, meat-like aroma and taste. Kochujang (Figure 9) is a bright red paste, and has a smooth texture with an appearance of tomato ketchup. It is softer than Japanese miso and has a hot flavor.

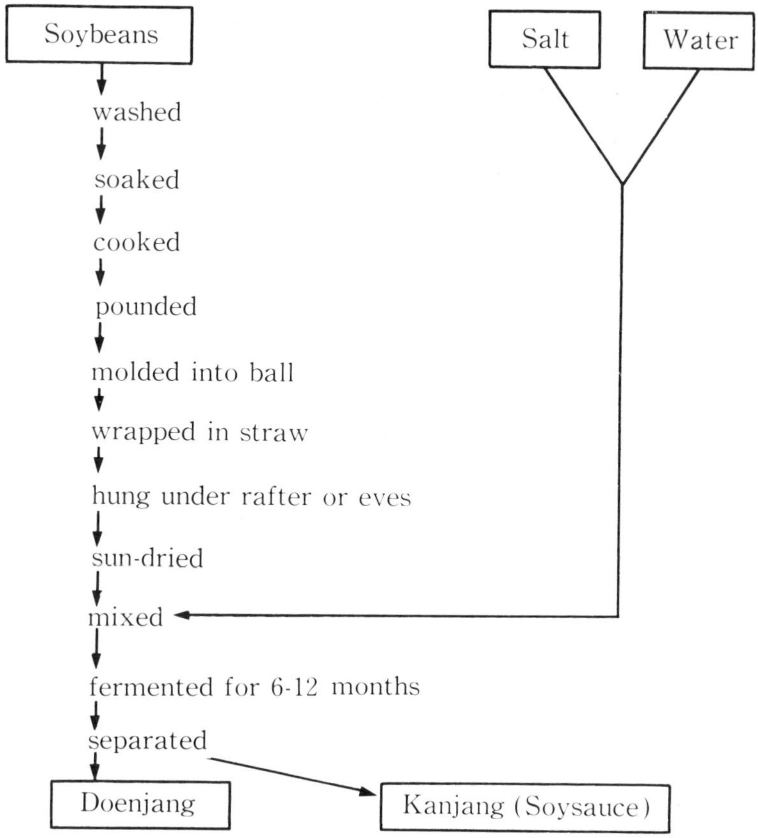

FIGURE 8. Flow chart of the manufacture of Doenjang.

FIGURE 9. Kochujang packaged in a small glass container in Japan.

Kochujang is the sole spiced, fermented soybean food in Korea. Its place as characteristic seasoning in foods will probably not change. The chemical composition of Doenjang and Kochujang is presented in Table 9.

D. Tauco

Tauco appears to have been introduced from China, because in Chinese "tau" and "co" mean soybeans and chiang respectively. Tauco is mainly produced in Indonesia, around the town of Cianjur in west Java. It is used as an ingredient of soups and

Table 9
CHEMICAL COMPOSITION OF DOENJANG AND KOCHUJANG[62,63]

Component	Doenjang	Kochujang
Moisture (%)	51.0—58.3	47.7—55.1
Protein (%)	12.0	8.9
Fat (%)	4.1—9.9	2.5—9.5
Carbohydrate (%)	10.7	25.9
Reducing sugars (%)	0.1—0.2	7.6
Crude fiber (%)	2.6—3.7	—
Ash (%)	17.9	15.9—21.0
Salt (%)	14.9—17.8	13.5—16.6
Calcium (mg/100g)	122.0	126.0
Phosphorus (mg/100g)	141.0	72.0
Thiamin (mg/100g)	0.04	0.35
Riboflavin (mg/100g)	0.20	0.35
Niacin (mg/100g)	—	1.2—1.5
Ascorbic acid (mg/100g)	—	10.0
β-Carotene (μg/100g)	—	210.0

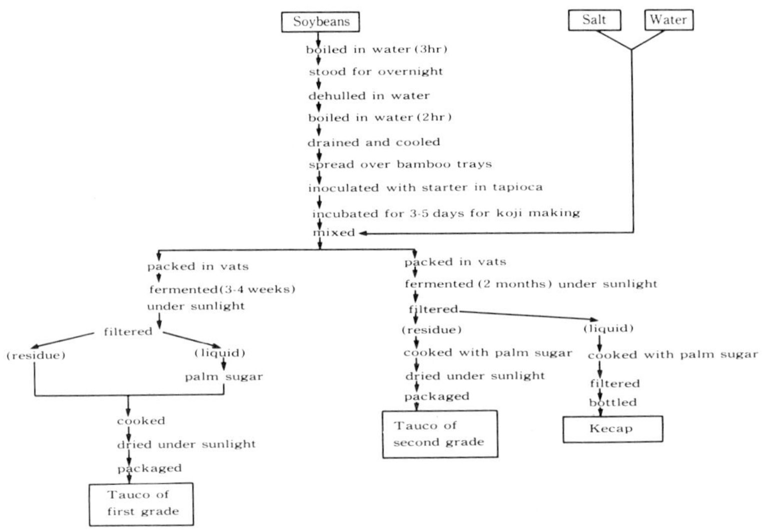

FIGURE 10. Flow chart of the manufacture of Tauco.

seasonings. In 1977 it was reported that a 300 mℓ bottle of tauco costs about $0.48 and a package of semisolid (like soybean miso) costs $0.12.[63] Although the manufacturing method varies from factory to factory, tauco is produced basically through a two-step fermentation process using yellow soybeans (Figure 10). The first step is koji making, followed by a second fermentation in brine (20% salt) for 2 months in vats. At the end of the fermentation period, the brine is drained through a filter cloth, palm sugar (25%) is added, the mixture is cooked and ripened (Kecap). Semi-solid tauco (Figure 11) is prepared by mixing the filtrate with the residue and palm sugar, cooking the mixture and sun-drying it into semisolid form.

The microorganisms involved in koji making are *Rhizopus oligosporus, R. oryzae,* and *A. oryzae.* According to Saono, et al.,[64] in Bogor and Cianjur areas, 11 out of 18 tauco makers use *Rhizopus* spp., while the remaining 7 utilize *A. oryzae. Aspergillus*

FIGURE 11. Semi-solid type Tauco which is similar in appearance to Japanese soybean miso.

sp. and *Rhizopus* spp., respectively, grow under lower and higher humidities. Tauco prepared with *A. oryzae* contains more digestible proteins than that made with *Rhizopus* spp. From the fermented mash, microorganisms such as *Lactobacillus delbrückii, Hansenula* spp., and *Saccharomyces rouxii* have been isolated and identified.

VII. CONCLUSIONS

Miso is a salty, smooth-to-chunky paste with a meat-like flavor. It is used as an ingredient in soups and as a seasoning for meats, seafoods, and vegetables. Miso production in Japan decreased from 590,000 tons in 1973 to 572,000 tons in 1983. The decrease in miso production may be due to: (1) changes in food patterns, i.e., from traditional Japanese to western foods, and (2) concerns about the high sodium chloride content in miso. According to Shurtleff and Aoyagi,[65] the total miso consumption increased in the U.S. from 1122 tons in 1975 to 2349 tons in 1982. If the salt level could be lowered, some of the problems related to sodium would be lessened and consumption probably would increase. Some methods have been studied to prevent an abnormal fermentation when producing a low-salt miso. These have included the use of ethylalcohol (up to 3%) in place of salt, the use of a pure culture of yeast at a level of 10^6 cells/g miso, and a reduction in moisture to 43%. Though these methods prevented an abnormal fermentation, the flavor balance was not satisfactory because of low saltiness. Therefore, further studies should be conducted to find new flavoring materials to replace salt. Freeze-dehydration has been used to make dehydrated miso, which can then be employed as an ingredient for instant miso soup. Dehydration causes some changes in flavor and taste. Further improvements in this area are necessary. Development of new food dressings such as barbecue sauces and salad dressings from miso should be investigated.[66]

REFERENCES

1. Ebine, H., Matsushita, Z., Sasaki, H., Evaluation of U.S. soybeans as raw material for making miso, *Rep. Natl. Food Res. Inst.*, No. 25, Yatabe, Ibaraki, Japan, 1971, 126.
2. Shibasaki, K. and Hesseltine, C. W., Miso. I. Preparation of soybeans for fermentation, *J. Biochem. Microbiol. Technol. Eng.*, 3, 161, 1961.
3. Smith, A. K., Hesseltine, C. W., and Shibasaki, K., U.S. Patent, 2,967,108.
4. Ebine, H., Ito, K., and Nakano, M., On the defatted soybean miso, *Rep. Natl. Food Res. Inst.*, No. 10, Yatabe, Ibaraki, Japan, 1955, 133.
5. Bo, Thi-An, History and production of Chiang and soy sauce, *J. Brew. Soc. (Jpn.)*, 77, 366, 1982.
6. Ebine, H., Integrated research on waste reclamation, in Bioconversion of Organic Residues for Rural Communities, papers presented at the Conference on the State of Art of Bioconversion of Organic Residues for Communities, Institute of Nutrition of Central America and Panama, Guatemala City, Nov. 13-15, 1978, 122.
7. Inoue, M., Takagi, S., and Murata, W., Pasteurizer and seasoning for miso, *Miso no Kagaku to Gijutsu*, 148, 3, 1966.
8. Fukui, S., Tani, Y., and Kishibe, T., Changes of the amount of several B-vitamins in the koji, *J. Ferment. Technol. (Jpn.)*, 33, 239, 1955.
9. Kambe, C., Iwasa, T., and Sasaki, T., Production of organic acids by *Pediococcus halophilus* in soy sauce fermentation, *Nippon Nogei Kagaku Kaishi*, 52, 329, 1978.
10. Ito, K. and Ebine, H., Studies on lactic acid bacteria in miso. IV. On *Pediococcus* group, *Rep. Natl. Food Res. Inst.*, No. 19, Yatabe, Ibaraki, Japan, 1965, 89.
11. Ito, K. and Ebine, H., Studies on lactic bacteria in miso. II. Investigation on their distribution in various types of miso, *Rep. Natl. Food Res. Inst.*, No. 18, Yatabe, Ibaraki, Japan, 1964, 9.
12. Ito, K. and Ebine, H., On the acidification by lactic acid bacteria, *Rep. Natl. Food Res. Inst.*, No. 19, Yatabe, Ibaraki, Japan, 1965, 109.
13. Ito, K. and Ebine, H., Swelling caused by *Clostridium*, *Rep. Natl. Food Res. Inst.*, No. 22, Yatabe, Ibaraki, Japan, 1967, 46.
14. Ito, K. and Ebine, H., Swelling caused by *Clostridium*, *Rep. Natl. Food Res. Inst.*, No. 22, Yatabe, Ibaraki, Japan, 1967, 54.
15. Ito, K. and Ebine, H., Salt-tolerance of *Clostridium*, *Rep. Natl. Food Res. Inst.*, No. 22, Yatabe, Ibaraki, Japan, 1967, 61.
16. Asao, Y. and Yokotsuka, T., Studies on flavorous substances in soy sauce. XVII. Guaiyacyl components, *Nippon Nogei Kagaku Kaishi*, 32, 622, 1961.
17. Yamamoto, A., Sasaki, K., and Saruno, R., The flavors of sake. IV. Separation and identification of ferulic acid, vanillic acid and vanilline, *Nippon Nogei Kagaku Kaishi*, 35, 715, 1961.
18. Ebine, H. and Ito, K., Miso, in *Chemical Compounds of Fermented and Brewed Products*, Sato, S., Ed., The Brewing Society of Japan, Tokyo, 1977, 335.
19. Resources Council, Science and Technology Agency, Japan, *Standard Tables of Food Composition in Japan*, 4th rev. ed., Toyko, Japan, 1982.
20. Hondo, T. and Mochizuki, T., On the free sugar of miso, *Nippon Shokuhin Kogyo Gakkaishi*, 26, 461, 1979.
21. Mochizuki, T., Yasuhira, H., Hondo, S., Ouchi, I., Rokugawa, K., and Itoga, K., Studies on the changes of several components during miso making, *Ferment. Technol. Today*, 1972, 663.
22. Tamura, G., Kirimura, J., Hara, H., and Sugimura, K., The microbial determination of amino acid in miso, *Nippon Nogei Kagaku Kaishi*, 26, 483, 1952.
23. Sugimura, K., Taira, H., Ebisawa, H., and Sakurai, Y., Variation of total and free amino acids content in soybean miso processing, *J. Jpn. Soc. Food and Nutr.*, 14, 414, 1962.
24. Fujinami, H., Moshizuki, T., Sagawa, I., and Mori, M., Changes of glutamic acid, glutamine and pyroglutamic acid during fermentation of miso at different temperature, *J. Brew. Soc. (Jpn.)*, 78, 466, 1983.
25. Ebine, H., Nakano, M., and Kuroha, I., On the manufacture of enriched miso. VI. Effect of added methionine on the quality of miso, *Rep. Natl. Food Res. Inst.*, No. 23, Yatabe, Ibaraki. Japan, 1968, 5.
26. Kakezawa, M. and Ebine, H., unpublished data, 1983.
27. Kiuchi, K., Ohta, T., and Ebine, H., Accumulation of diglycerides and monoglycerides and decrease of unsaturated free fatty acids in miso, *Nippon Shokuhin Kogyo Gakkaishi*, 23, 455, 1976.
28. Yasuhira, H., Yonetani, I., and Mochizuki, T., On the aroma of miso, *Rep. Shishu Miso Inst.*, No. 11, Nagano, Japan, 1970, 10.
29. Kuribayashi, Y., Phenolic compounds in the extract of barley koji, *Nippon Shokuhin Kogyo Gakkaishi*, 14, 549, 1967.
30. Munesawa, T., Kosugi, N., and Sagara, T., On miso pasteurizer, *Miso no Kagaku to Gijutsu*, 148, 8, 1976.

31. Watanabe, T., Effect of miso pickles on the oxidation of foods, *Miso no Kagaku to Gijutsu,* 31, 1, 1983.
32. Yamaguchi, N., Yokoo, Y., and Fujimaki, M., Antioxidative activities of miso and soy sauce on linoleic acid, *Nippon Shokuhin Kogyo Gakkaishi,* 26, 71, 1979.
33. Ito, K., Sakurai, E., and Inoue, T., Immunochemical and electrophoretic studies on the soluble proteins of mame miso, *Nippon Shokuhin Kogyo Gakkaishi,* 29, 147, 1982.
34. Ito, K., Immunobiochemical studies on the soluble proteins of rice miso, *Nippon Shokuhin Kogyo Gikkaishi,* 30, 38, 1983.
35. Ito, K., Immunobiochemical studies on the soluble proteins of rice miso during maturing process, *Nippon Shokuhin Kogyo Gakkaishi,* 30, 43, 1983.
36. Obata, Y., Matuno, N., and Tamura, U., Biological value of tofu and miso, *Rep. Natl. Nutr. Inst.,* No. 20, Tokyo, Japan, 1932, 20.
37. Ebine, H., unpublished data, 1975.
38. Yamori, M., Horie, R., Nara, Y., Kihara, M., and Toda, M., Studies on stroke prevention in animal models and their supportable epidomiological evidence, in *Cerebrovascular Disease, New Trends in Surgical and Medical Aspects,* Banett, H., Paloletti, P., Flamm, E., and Bramabiolla, G., Eds., Elsevier North-Holland Biomedical Press, Amsterdam, 1981, 47.
39. Honda, M., Miyanoshita, H., Osawa, Y., and Mochizuki, T., Determination of sodium and potassium in miso soup, *Miso no Kagaku to Gijutsu,* 330, 29, 1981.
40. Shimada, A., Situation of miso soup in food life, *Miso no Kagaku to Gijutsu,* 331, 2, 1981.
41. Sakurai, Y., Components of miso IV. Acids of miso, *Nippon Nogei Kagaku Kaishi,* 10, 486, 1934.
42. Ebine, H., Nakajima, M., and Nakano, M., Studies on enriched miso. IV. Enrichment of vitamins and calcium, *Rep. Natl. Food Res. Inst.,* No. 10, Yatabe, Ibaraki, Japan, 1955, 155.
43. Mogi, M., Nakajima, S., Iguchi, N., and Yosida, Y., Studies on fortified bean-paste (miso), *J. Ferment. Technol. (Jpn.),* 29, 302, 1951.
44. Takahashi, J., Some resources of vitamin B-12, *J. Food Nutr.,* 8, 25, 1955.
45. Miyamoto, T., Murata, K., and Kawamura, M., Folic acid content of some fermented soybean products and vegetables, *Vitamin* (Jpn.), 47, 233, 1973.
46. Maga, J. A., Phytate: Its chemistry, occurrence, food interaction, nutritional significance, and method of analysis, *J. Agri. Food Chem.,* 30, 1, 1982.
47. Sakamoto, M. and Iida, S., Studies on the effect of phytin hydrolysate on the growth of sake yeast, *J. Ferment. Technol. (Jpn.),* 37, 11, 1959.
48. Wang, H. L., Swain, E. W., and Hesseltine, C. W., Phytase of molds used in oriental food fermentation, *J. Food Sci.,* 45, 1262, 1980.
49. Hirayama, T., Does daily intake of soybean paste reduce gastric cancer risk?, presented at the Annual Convention of Japanese Cancer Association, Sapporo, October 4 to 8, 1981, 5.
50. Okazaki, H., Kano, U., and Kimura, S., Anti-mutagenicity of food component, presented in the 37th Convention, Japanese Society of Nutrition and Food Science, Osaka, May 8 to 10, 1983, 10.
51. Manabe, M., Matsuua, S., and Nakano, M., Studies on the fluorescent compounds in fermented foods. I. Chloroform-soluble fluorescent compounds, *Nippon Shokuhin Kogyo Gakkaishi,* 15, 341, 1968.
52. Wang, H. L. and Hesseltine, C. W., Oriental fermented foods, in *Industrial Microbiology,* 4th ed., AVI Publishing, Westport, Conn., 1982, 492.
53. Kawamura, W., *Miso Enkaku Shi (History of Miso),* Zenkoku Miso Kogyo Kydo Jumiai Rengokai, Tokyo, 1958, 423.
54. Nishiyama, B. and Kumashiro, Y., *Chimin Yaoshu,* Asian Keizai Suppan, Tokyo, 1975.
55. Shih, S. A., *An Agricultural Encyclopedia of the 6th Century,* Science Press, Peking, 1962, 107.
56. Kiuchi, K., Ohta, T., Ito, K., Takabayashi, H., and Ebine, H., Studies on lipids of natto, *J. Agr. Food Chem.,* 24, 404, 1976.
57. Kon, M., Studies on hama-natto. I. Amino acids, *Rep. Shizuoka University,* No. 3, Shizuoka, Japan, 1970, 185.
58. Asano, K. and Saito, Y., Studies on the molds for Daitokuji-natto manufacturing, *J. Ferment. Technol. (Jpn.),* 32, 360, 1954.
59. Kon, M. and Ito, K., Studies on hama-natto, *Rep. Natl. Food Res. Inst.,* No. 30, Yatabe, Ibaraki, Japan, 1975, 232.
60. Mheen, T. I., Kown, T. W., and Lee, C. H., Traditional fermented foods in Korea, in *Traditional Food Fermentation as Industrial Resources in ASCA Countries,* Saono, S., Winarno, F. G. and Karjadi, D., Eds., Medan, Indonesia, 1982, 63.
61. Steinkraus, K. H., Ed., *Handbook of Indigenous Fermented Foods,* Marcel Dekker, New York, 1983, 479.
62. Office of Rural Development, *Food Composition Table,* Suwon, Korea, 1977.
63. Steinkraus, K. H., Ed., *Handbook of Indigenous Fermented Foods,* Marcel Dekker, New York, 1983, 482.

64. Saono, J. K. D., Baba, T. and Matsuyama, A., Problems to be assessed for further development of traditional food fermentation in Indonesia, in *Traditional Food Fermentation as Industrial Resources in ASCA Countries,* Saono, S., Winarno, F. G. and Karjadi, D., Eds., Medan, Indonesia, 1982.
65. Shurtleff, W. and Aoyagi, A., *The Book of Miso,* 2nd ed., Ten Speed Press, Berkeley, California, 1983, 240.
66. Hesseltine, C. W., New Ways to use Miso, *Miso no Kagaku to Gijutsu,* 263, 5, 1976.

Chapter 4

SUFU

Y.-C. Su

TABLE OF CONTENTS

I. INTRODUCTION

Sufu (tau-zu), a traditional Chinese food, is a highly flavored, soft cream cheese-type product. It is prepared by overgrowing soybean curd (tofu) with a mold of the genus *Actinomucor, Mucor,* or *Rhizopus* and further fermenting the mold-overgrown curd in a salt-brine/rice-wine mixture.[1,2] During fermentation, the mold and rice wine mixture imparts additional flavor to the product. Sufu, as seen in the market, is usually comprised of red, pale yellow, or white colored blocks (2 to 4 cm square and 1 to 2 cm thick) (Figure 1). Pale yellow or white sufu are untreated, whereas red sufu is colored with hung-chu (red koji), which is prepared by growing the mold *Monascus anka* or *Monascus purpureus* on cooked rice.[1-4] Other additives are frequently incorporated into the brine solution, either to give additional color or flavor to various types of sufu. Because of numerous dialects used in China and the difficulties of phonetic rendering from Chinese to English, the following synonyms for sufu are found: tou-fu-ju, fu-ju, fu-su, fu-ru, tou-fu-ru, fu-yu, and foo-yue. Sufu is also known as to-fu-zu in Mandarin, tau-zu or tao-hu-yi in Taiwan,[1-4] and nyu-fu or fu-nyu in Japan.[5,6] Sufu is the name that first appeared in the literuare.[7] Literally, sufu means "molded milk" and tosufu means "molded bean milk". In the west, sufu is referred to as Chinese cheese or soybean cheese.[8,9]

The substrate for sufu is soybean curd, which is prepared by soaking soybeans overnight in water, grinding them with added water to make a slurry, heating the slurry to a boil, and finally filtering the slurry through a cheese cloth to obtain an aqueous extract (soybean milk). Crude sea salt, calcium salts, or magnesium salts are added to the hot filtered soybean milk to precipitate proteins and other materials as curd. The precipitated soybean milk is filtered and pressed to separate the whey and soybean curd.

Manufacture of soybean curd began during the era of the Han dynasty. The "Pen Ts' ao" or Chinese Materia Medica of 1596, compiled by Li Shi-Chin, indicated that soybean curd was invented by Liu An (179 BC to 122 BC), King of Wainan.[1,10] It is not known when sufu production began. The food encyclopedia written by Wang Su-Hsiung (1861) of the Ch'ing dynasty described the food as follows: "Hardened tofu is difficult to digest and not healthful for children, elderly persons or ill persons. Sufu prepared from tofu is better because it is aged and very good for patients."[1,2]

Sufu is produced both commercially and domestically in China and Taiwan. In Taiwan, the annual production of sufu was 12,000 tons in 1982. Annual consumption per person was 660 g.[1,11] Sufu is consumed directly as a condiment or is cooked with vegetables or meats. Either way, sufu adds zest to the bland taste of the rice-vegetable diet. Because of sufu's creamy texture, it may be suitable for use as a cracker spread or as an ingredient for dips and dressings in the western world.[12]

II. METHOD OF PREPARATION

A. Raw Materials, Microorganisms Involved, and Preparation of Inoculum

A selected quality of soybean are generally used for making sufu. Other ingredients such as crude sea salt, sodium chloride, red koji (cooked rice overgrown with *M. anka* or *M. purpureus*), rice wine, jaggery, red pepper, and other spices are required in the preparation of sufu.

The process of making sufu had been considered a natural phenomenon until 1920, when a microorganism believed to be responsible for sufu fermentation was isolated and identified.[7] Wai[13,14] did some research on microorganisms found in sufu, using the samples collected from the fermentation chambers of factories in Shaoshing and Soochow, China. He successfully isolated a mono-mucor from the fermenting sufu sam-

FIGURE 1. Sufu product.

ples. Almost 40 years later, Wai[15] reinvestigated the sufu fermentation and developed a pure-culture fermentation. A variety of fungi are involved in the sufu fermentation. They are *Rhizopus chinensis* var. *chungyuen, Actinomucor elegans, Mucor hiemalis, Mucor silvaticus,* and *Mucor praini.*[10,15,17] The fungi, namely *Rhizopus chinensis* var. *chungyuen,* are most commonly present in commercial sufu.[10] The type of mold used in sufu fermentation should have certain qualities[12] such as:

1. Producing enzyme systems with high proteolytic and lipolytic activity, to act upon the rich protein and lipid substrates of tofu.
2. Produce white or yellowish white color mycelium mat.
3. Texture of the mold mycelial mat should be dense and tenacious so that the film formed on the surface of the pehtze will serve as an envelope to protect the finished sufu from distortion in the shape.
4. The mold growth does not produce any disagreeable odor, astringent taste, or mycotoxins

A pure culture inoculum of *Actinomucor, Mucor,* or *Rhizopus* spp. is prepared by growing the organism on filter paper impregnated with a culture medium. The culture medium consists of 30 g sucrose, 3 g $NaNO_3$, 1 g K_2HOP_4, 0.5 g KCl, 0.5 g $MgSO_4$ $7H_20$, 0.1 g $FeSO_4$ $7H_20$, 50 g potato, 2 g sodium glutamate and 1 ℓ water.[1] This medium is used to moisten the circular sheet of filter paper, which has been placed in a petri dish and steam sterilized for 40 min. The sterilized filter paper is inoculated with spores from agar slant cultures. For best growth of mycelia, controlled temperature, good aeration, and a relative humidity of 80% are necessary.[1] Wai[10] developed a similar method for pure culturing of fungi on filter paper impregnated with a modified Czapek-Dox medium. The fully grown filter paper cultures can be used for inoculation of new batches of tofu.

B. Preparation of Sufu

The preparation method for sufu consists of five steps: (1) preparation of tofu, (2) preparation of pehtze ("pehtze" is freshly prepared soybean curd grown with the mold, but not yet processed and aged to become sufu), (3) salting, (4) processing, and (5) aging. The flow sheet for preparation of sufu is shown in Figure 2.

1. Preparation of Tofu

Selected quality soybeans are washed, soaked overnight in water, and ground with added water to a milky slurry in a stone mill. A water to bean ratio of 10:1 is common. The ground slurry is boiled or steamed for about 20 min and filtered through cheese cloth to separate soy milk. A suitable quantity (about 2.5 to 3.5% of dry weight of soybeans) of coagulant such as calcium sulfate, magnesium sulfate or sea salt is added to the hot soy milk. Generally, 20% higher weight coagulant is used to produce tofu for sufu preparation than that used for regular tofu. After the addition of coagulant to the soy milk, the mixture is agitated vigorously to break the curd into small pieces. The agitated mixture is set aside for 10 min to complete the coagulation process. The coagulated material is referred to as curd or tofu. Tofu is separated from the whey by filtering and pressing the coagulated mixture. Pressing is done in a cloth-lined wooden box to form firm blocks of desired size.[9] The firm blocks of tofu are cut into small cubes (2 × 4 × 4 cm) weighing approximately 18 g (Figure 3).[4] Tofu contains 9% water-insoluble proteins, 0.3% water-soluble nitrogen compounds (as protein), 4% lipids, and 83% water.[9] Tofu used for preparation of sufu is generally pressed harder than ordinary tofu.

2. Preparation of Pehtze

Pehtze (Figure 3) is soybean curd (tofu) overgrown with greyish hair-like mycelium of molds of the genus *Actinomucor* spp., *Mucor* spp., or *Rhizopus* spp.[1,2] These fungi are normal contaminants in rice straw. Traditionally, inoculation is performed by placing tofu cubes on the rice straw, but this method does not yield a high quality pehtze or sufu because of undesirable contaminating microorganisms.[1] To avoid contamination and bacterial spoilage in traditional sufu preparation, tofu cubes are exposed to bright sunlight for several hours prior to inoculation with mold. Heat from the sun dehydrates the surface of the tofu cubes, making them less susceptible to bacterial spoilage. However, using this initial treatment can result in the loss of quality and flavor characteristics of sufu. Wai[10] developed an improved method for production of sufu. In this method, tofu cubes cut into blocks measuring 2 × 4 × 4 cm are mounted on bamboo chopsticks or aluminum rods (2 mm diameter) and surface dried in a hot air oven at 100°C for about 15 min. Hesseltine[16] proposed a modification of this heat treatment procedure. In his procedure, tofu cubes are first immersed in an acidic saline solution (prepared by dissolving 6 g NaCl and 2.5 g citric acid in 100 mℓ water) which will prevent the growth of putrefactive bacteria while allowing mold growth. The

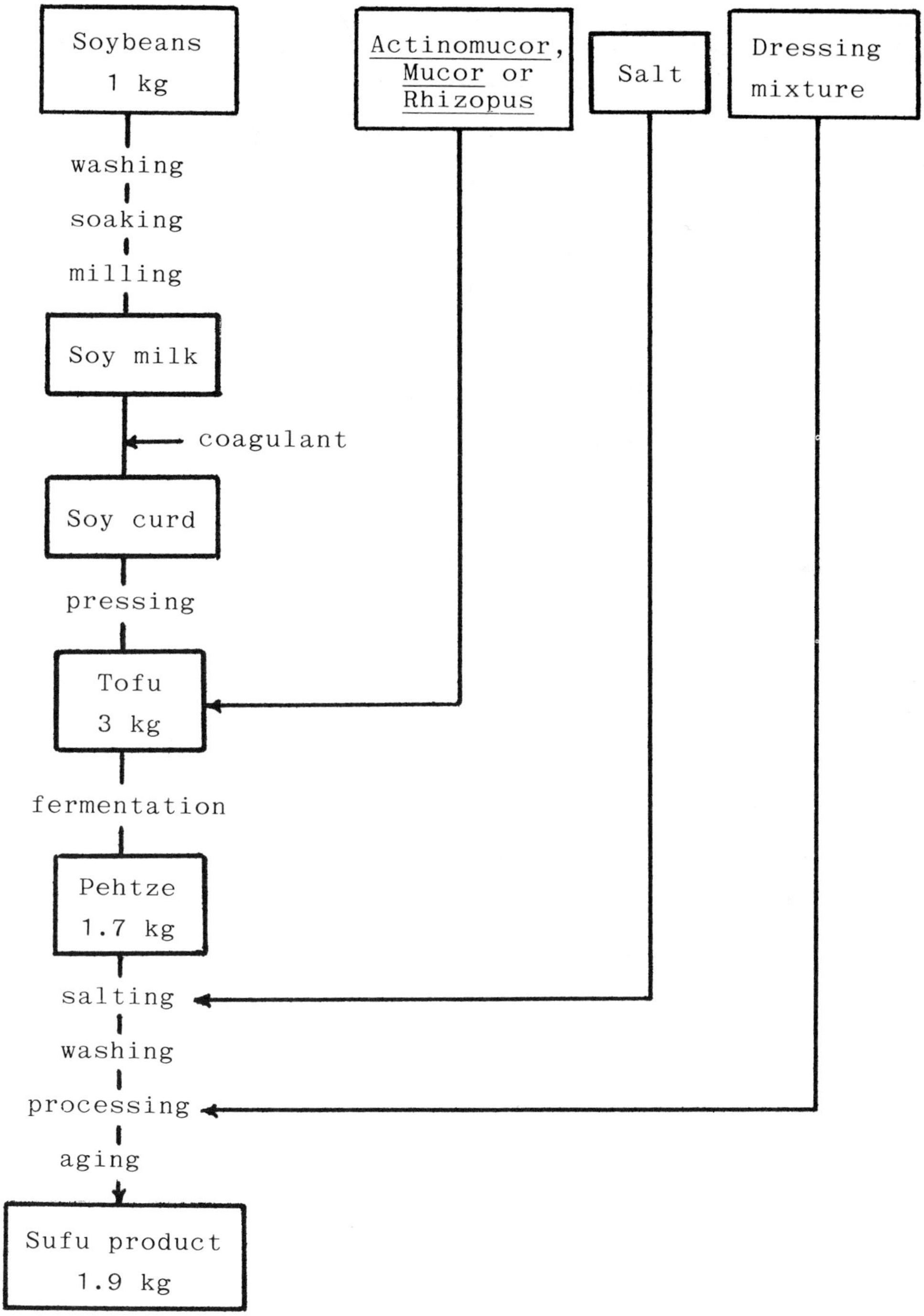

FIGURE 2. Flow sheet for the preparation of sufu.[3]

treated cubes are placed on a tray having an opening at the bottom for air circulation during hot air treatment. Lactic acid can be substituted for citric acid in the saline solution. The dip in an acidic saline solution is considered the most important step for pretreatment of raw tofu. The trays containing acid-dipped tofu cubes or blocks are dried in a hot air oven.

FIGURE 3. Tofu and pehtze.

a. Inoculation and Incubation

The hot tofu cubes or blocks are cooled to 20°C in a refrigerator. A pure culture inoculum of *Actinomucor* spp., *Mucor* spp., or *Rhizopus* spp. grown on the filter paper impregnated with culture medium is inoculated onto the surface of the tofu blocks. The inoculated tofu blocks are put in a tray for incubation. If the tofu blocks adhere to the bottom of the tray, mold mycelia may not grow well. Rice straw or sedges are placed between bottom of the tray and tofu blocks to avoid the adherence of tofu blocks to the tray.[17] The trays containing inoculated tofu blocks are placed in an incubator at 20°C or 12°C for incubation depending upon the pure culture inoculum of mold used. In a scaled up commercial method, twenty or more trays containing inoculated tofu blocks are stacked one upon another. The room temperature is maintained at 10 to 20°C. After 3 to 7 days of incubation, white fungal mycelium can be seen on the surface of tofu blocks and at that time they are taken out and immediately salted.[1,2]

Different times and temperatures are used for sufu fermentation depending upon the mold culture. *Rhizopus chinensis* var. *chungyuen* grows well in 7 days at 12°C compared to *Mucor hiemalis* and *Mucor silvaticus,* which grow in 3 days at 20°C. *Rhizopus chinesis* var. *chungyuen* is one of the common molds in commercial sufu.[10] However, *M. hiemalis* No. 28 and *M. silvaticus* No. 508 produce a more dense mycelial mat and tenacious film on the surface of the tofu blocks than does R. *chinensis* var. *chungyuen.* Because of this, *Mucor* sp. is more preferred in preparation of sufu (Figure 4). However, from a safety standpoint, low temperature incubation is more desirable for the inhibition of bacterial growth. *Actinomucor elegans* can also be used in the sufu preparation.[15] This mold grows rapidly on tofu blocks in 2 days at 20°C. Also, *A. elegans* is used commercially in the preparation of pehtze.

Lin[17] isolated *Mucor hiemalis, M. silvaticus, M. praini,* and *Rhizopus chinensis* from pehtze and rice staw. He reported that incubation time and temperature varied from 3 to 7 days and 12 to 25°C, respectively, for these molds. Recently Lin et al.[18] found that the mycelium of *Mucor* spp. spread densely over the surface of tofu and grew outward in the first 2 days of incubation (Figure 5). However, the mold mycelium grows very little inward from the surface cracks of tofu. After 2 days, a 0.1 cm thick fungal mat is formed on top of the tofu block and the mycelium penetrates to a depth of about 0.18 cm (Figure 6). In contrast, Wang[19] reported that mold growth is limited to the surface of the tofu blocks in sufu fermentation.

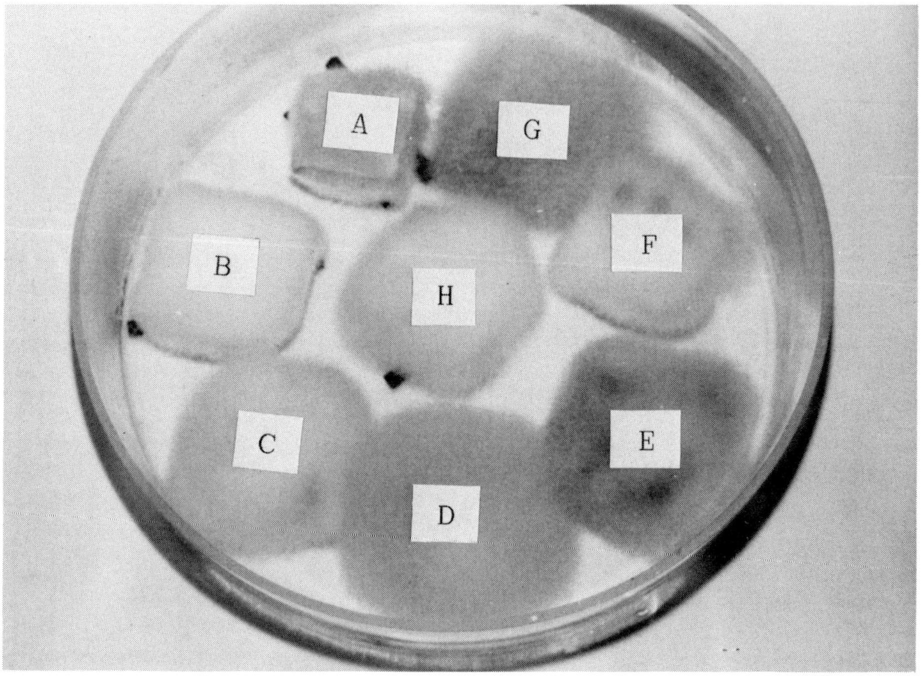

FIGURE 4. Tofu freshly grown with a fungus: (A) *Aspergillus niger*, (B) *Mucor hiemalis*, (C) *Mucor silvaticus*, (D) *Actinomucor elegans*, (E) *Rhizopus delemar*, (F) *Mucor pusillus*, (G) *Rhizopus chinesis* var. *chungyuen*, and (H) *Mucor praini*.

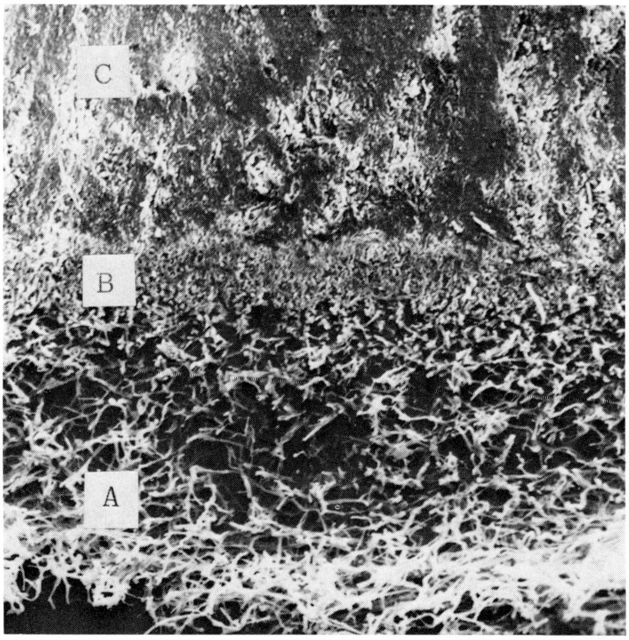

FIGURE 5. The structure of the soybean curd surface with mycelium (magnification × 60): (A) Mycelial mat, (B) Layer of mycelium and soybean curd, and (C) Soybean curd protein.

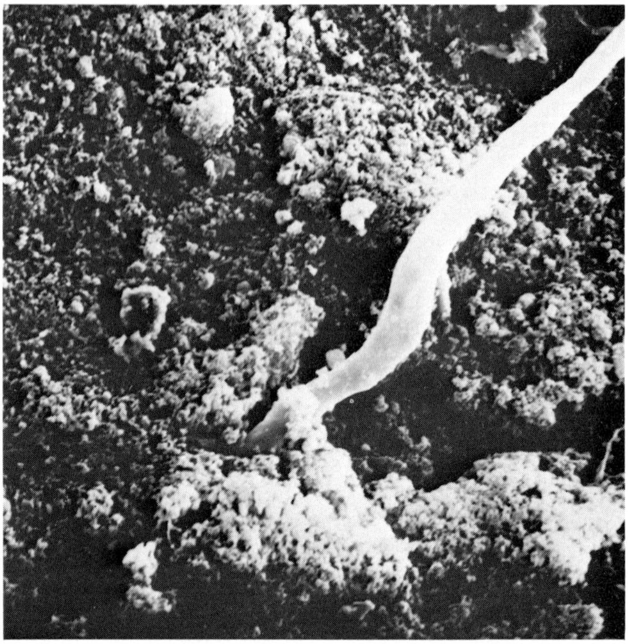

FIGURE 6. Mycelial growth into the interior of soybean curd. (Magnification × 1000.)

3. Salting

Pehtze is transferred to large earthenware jars (having a volume of 7hℓ) for further curing. It is put in layers in the jars. Each layer of pehtze is sprinkled with a layer of salt and kept for 3 to 4 days. During this period, the pehtze absorbs most of the salt. The salted pehtze is removed from the jars, washed with water, and transferred to another jar for further processing.[4]

Freshly prepared pehtze can be salted in a number of other ways. In one conventional method, pehtze is immersed in an alcoholic saline solution consisting of 12% NaCl and 10% ethanol (generally rice wine or distilled liquor is used). Pehtze immersed in alcoholic saline solution can be sold as such.[1] Wai[10] attempted to age and preserve pehtze by dipping the product in various solutions:

1. 0.1% Benzoic acid + 0.05% ethyl acetate
2. 2% Acetic acid + 0.05% ethyl acetate
3. 1% Propionic acid + 0.05% ethyl acetate
4. 0.01% Benzaldehyde + 0.05% ethyl acetate
5. 1% Orange peel oil + 0.05% ethyl acetate
6. 1% Acetic acid + 2% NaCl and Kao-Liang wine to make up 5% alcohol content
7. Kao-Liang wine to make up 5% alcohol content + 2% NaCl, 0.01% anise and 1% red pepper
8. Kao-Liang wine to make up 10% alcohol content + 0.01% benzaldehyde + 2% NaCl
9. Kao-Liang wine to make up 10% alcohol content + 1% acetic acid + 4% NaCl
10. Kao-Liang wine to make up 10% alcohol content + 5% NaCl
11. 10% Ethyl alcohol + 5% NaCl

The fermented pehtze were dipped in the above solutions and preserved in closed

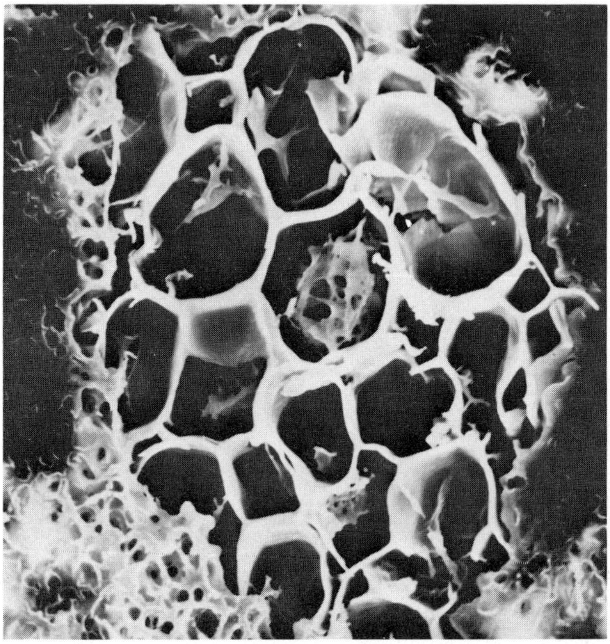

FIGURE 7. The interior fine structure of fresh soybean curd. (Magnification × 2000.)

glass containers. Solutions (8), (9), (10), and (11) were appropriate for preservation of pehtze both at room temperature (20 to 35°C) and at 10°C for more than 6 months without changes in organoleptic qualities. Addition of sesame oil to the alcoholic saline solutions is reported to improve the preservation of pehtze.

Lin[18] studied the effect of salting on the internal structural change of soybean curd and sufu. Scanning electron microscopic observation showed that the internal microstructure of the soybean curd exhibits an aggregate of protuberant globules, and a gradient of hydrolysis (Figure 7). Gradual structural changes occur internally in soybean curd after salting and subsequent aging. The protuberant globules dissolve and the characteristic small globular net structure disappears when the soybean curd is submerged in saline solution (Figure 8)[18].

4. Processing and Aging

Generally, fermented pehtze is processed and aged in earthenware jars containing a dressing mixture. The dressing mixture varies with the type of sufu. The most common dressing mixture used consists of 2 kg salt (NaCl), 1 kg soy mash, 0.6 kg red koji, 0.6 kg jaggery, and 6 kg water. Additional essence can be added to the dressing mixture to give a special flavor. For example, rose essence is mixed with the dressing mixture to prepare rose sufu. For preparing tao sufu, residue from rice wine, cloves, and orange peels are added to the dressing mixture.

For aging, alternate layers of pehtze and dressing mixture are packed into the jar to 80% of its volume, then the jar is filled with brine containing 12 to 20% salt (NaCl). The mouth of the jar containing the pehtze, dressing mixture, and brine solution is wrapped with sheath leaves of bamboo shoots and sealed with clay. The sealed jars are aged for 1 to 3 months or longer for further fermentation[10,13] before packing into small containers for marketing. Shaoshing wine is usually added to sufu, mainly for the purpose of preservation.

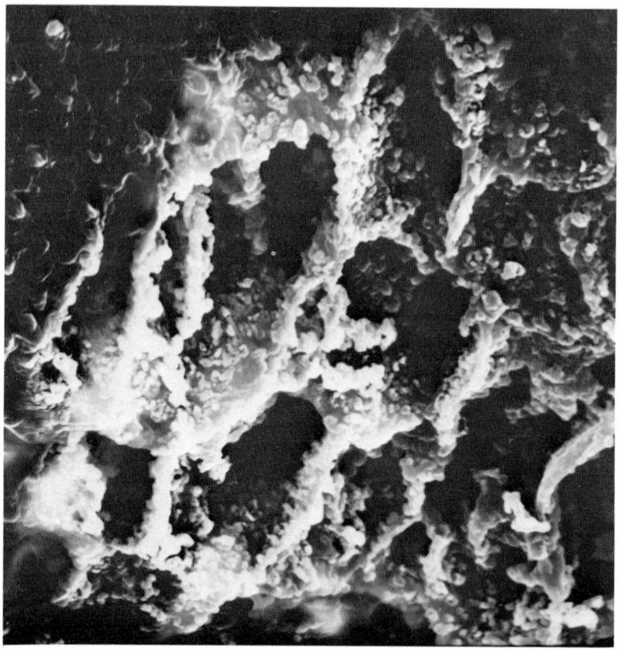

FIGURE 8. Inside of the soybean curd submerged in saline solution for 20 days. (Magnification × 2000.)

Table 1
COMPOSITION OF TOFU, PEHTZE, AND SUFU[20]

	Tofu		Pehtze		Sufu	
Component (%)	Fresh	Dried	Fresh	Dried	Fresh	Dried
Moisture	75.8	—	70.0	—	59.7	—
Protein	16.0	66.0	17.9	59.7	15.9	39.4
Fat	7.2	29.7	9.8	32.8	20.3	50.4
Carbohydrate	0.1	0.4	0.5	1.7	0.0	0.0
Fiber	0.0	0.0	0.4	1.3	1.1	3.7
Ash	0.9	3.9	1.4	4.5	3.0	7.4

III. BIOCHEMICAL CHANGES AND COMPOSITION

The chemical composition of fresh and dried tofu, pehtze, and sufu is presented in Table 1. Fresh tofu contains more moisture than pehtze and sufu. Sufu has a high fat content compared to tofu and pehtze (Tabel 1). The changes in nitrogenous compounds of tofu, pehtze, and sufu are presented in Table 2. Enzymes such as invertase, emulsin, trypsin-like protease, pepsin-like protease, oxidase, and catalase are found in pehtze.[20] These enzymes are produced by the mold during fermentation and act on their respective substrates.

Soybean proteins are hydrolyzed into peptides and amino acids by the mold proteases. Free amino acid, such as aspartic acid, glutamic acid, serine, alanine, and leucine/isoleucine, are found in the water-soluble fraction.[15] Liu[21] also found that the amount of soluble protein and nonprotein nitrogenous compounds increased in pehtze as a result of the proteolytic action of molds. Many of the hydrolyzed products were

Table 2
CHANGES IN NITROGENOUS COMPOUNDS OF TOFU, PEHTZE, AND SUFU[20]

Nitrogen compound (%)	Tofu	Pehtze	Sufu	Solution
Protein nitrogen	99.1[a]	64.0	83.5	5.7
Formol nitrogen	1.4	18.8	17.8	54.5
Ammonia nitrogen	0.04	7.4	0.8	10.4

[a] Percent of total nitrogen.

Table 3
AMINO ACID CONTENT OF SUFU[22]

Amino acid	Content (g/100 g protein)
Alanine	10.0
Arginine	2.1
Aspartic acid	5.1
Cystine	0.4
Glutamic acid	0.6
Glycine	4.4
Histidine	1.4
Isoleucine	4.8
Leucine	8.8
Lysine	7.0
Methionine	0.7
Phenylalanine	4.6
Proline	2.4
Serine	2.3
Threonine	2.0
Tryptophan	0.6
Tyrosine	2.2
Valine	5.3

leached into the mash or brine solution during the aging process. As a result of leaching, one can expect a reduced protein nitrogen content in pehtze and sufu (Table 2). The mash or brine solution contains more formol and ammonia nitrogen compared to tofu, pehtze, and sufu.

The amino acid composition of commercial sufu is presented in Table 3. The cystine content of sufu may be low because it was determined by a bioassay method. However, cystine content of sufu may in fact be low because of its degradation or conversion to other sulfur compounds during fermentation and subsequent aging.

The release of proteinase enzyme from the mycelium of *Mucor hiemalis* grown in soybean medium was reported by Wang.[23] Only a small fraction of the proteinase produced by the mold appears in the culture filtrate, whereas most of the enzyme is bound to the mycelial surface. Addition of sodium chloride or other ionizable salts to the growth medium resulted in the release of the loosely bound enzyme from the mycelium. Optimum pH of the enzyme proteinase ranges from 3.0 to 3.5. Wang[23] reported that maximum release of proteinase from the mycelium can be achieved at a sodium chloride concentration of up to 0.3 *M*. He further sugested that enzyme proteinase is loosely attached (as ionic linkage) to the mycelium and can be released from the mycelium by rupturing. When the mold is grown in a medium with no salt, a small fraction of proteinase appears in the culture filtrate and most of it (proteinase) remains

Table 4
COMPOSITION OF MAJOR TYPES OF SUFU IN TAIWAN[10]

Component	Tsao sufu	Red sufu	Kwantung sufu	Rose sufu	Yunnan sufu
Moisture	69.03	61.25	74.46	59.99	64.77
Crude protein	12.87	14.89	12.42	16.72	12.16
Fat	12.89	14.31	6.39	13.74	14.23
Crude fiber	0.13	0.42	0.11	0.14	0.27
Soluble nonnitrogen	trace	trace	trace	trace	trace
Ash	5.08	9.13	6.61	9.41	8.56
Total nitrogen	2.06	2.38	1.99	2.68	1.95
Protein nitrogen	1.30	1.56	1.27	1.81	1.30
Nonprotein nitrogen	0.76	0.82	0.72	0.87	0.64
Ammonia nitrogen	0.20	0.16	0.18	0.18	0.18
Amino acid nitrogen	0.27	0.27	0.24	0.31	0.19
Other nitrogen	0.30	0.39	0.30	0.38	0.36

Note: Expressed in percentages.

attached to the mycelium. The mycelium-bound proteinase can be easily eluted by sodium chloride or other ionizable salts from the intact resting mycelium. Nonionizable substances such as urea, glucose, and sucrose were reported to be unsatisfactory in eluting the proteinase from the mycelium and growing culture.

The flavor and aroma of sufu develop during the brining and aging process. Traditionally, it is believed that the flavor and aroma of sufu improves as the aging process progresses. During the first 10 days of aging, the mold enzymes act on substrates such as proteins and lipids to produce various hydrolyzed products. These hydrolyzed products provide the principal constituents for development of mild and characteristic flavors in sufu.[9] The lipids in sufu are partially hydrolyzed to fatty acids during aging. The alcohol in the dressing mixture reacts with the free fatty acids of sufu to form esters, and these esters subsequently give pleasant flavor to the product sufu. The alcohol in the dressing mixture also prevents the growth of undesirable spoilage microorganisms and pathogens.

The salt in the dressing mixture imparts a salty taste to the sufu and retards the growth of contaminating microorganisms and mold.[23] Salt in the dressing mixture causes a release of the mycelia-bound protease. The released protease enzyme penetrates into the molded cubes during aging and hydrolyzes proteins.[23]

Shelf-life is a major problem with sufu. Alcohol in the dressing mixture helps to preserve sufu; however it gives a harsh flavor to the sufu. Therefore it is important to find a method for preservation of fermented pehtze and sufu without the use of alcohol in the dressing mixture. Another drawback is the use of soy mash which imparts an excessively salty flavor. Wai[10] used paraffin for coating pehtze cubes in an attempt to overcome the sufu shelf-life problems. In this study, pehtze was salted with 7% sodium chloride and then coated with a layer of melted paraffin. The paraffin coated pehtze was stored in a glass container at room temperature. The pehtze was found to be satisfactory even after storing at room temperature for one month.

There are several types of sufu available in Taiwan. The composition of these sufu types is shown in Table 4. Except for differences in taste and flavor, the types of sufu are generally similar in composition. Nyufu is a type of sufu that is produced by partial dehydration of soybean curd followed by inoculation with *Actinomucor repens*. The mold grown curd (nyufu) is soaked in soy sauce moromi mash for aging. During aging nyufu moisture content decreases and the salt, reducing sugar, and amino-nitrogen content increase.[24] Formic acid and acetic acid were detectable at every step, whereas propionic acid and isobutyric acid were detected only in moromi mash and nyufu.[24] Finally, butyric acid is found in nyufu only after mold inoculation.

Table 5
NUTRITIONAL COMPOSITION OF SUFU[25]

Component	Content[a]
Moisture (g)	53.3
Protein (g)	15.6
Fat (g)	10.1
Carbohydrate (g)	7.1
Crude fiber (g)	0.1
Ash (g)	11.2
Food energy (cal)	168
Calcium (mg)	231.0
Phosphorus (mg)	301.0
Iron (mg)	7.5
Thiamin (mg)	0.04
Riboflavin (mg)	0.13
Niacin (mg)	0.50

[a] per 100 g sufu.

IV. NUTRITIONAL QUALITY

From a nutritional viewpoint, sufu contains a higher content of protein-nitrogen than other oriental soybean products, such as miso and natto.[16] Nutritionally, soybean curd and soybean milk have the same importance to the people of Asia as cheese and milk from cows do to the people of the Western Hemisphere. Asians prefer sea salt-precipitated curd because it gives the desired texture to the sufu and serves as an important source of calcium.[1,9,11]

Analyses of sufu products have been made to ascertain their nutritive value and to provide the necessary data for the proper evaluation of the diet of various groups of Chinese people. The analyses were made on samples of sufu products purchased in Taipei markets. The average nutrient composition of sufu is presented in Table 5. Sufu has appreciable amounts of vitamins (riboflavin and niacin) and minerals (calcium and phosphorus).[25] It is also a good source of hydrolyzed proteins and free amino acids and free fatty acids.

V. TOXICOLOGY

Tang[26] conducted an extensive survey on the toxicity and safety of fermented foods. She collected 80 samples of sufu from the markets throughout Taiwan. Only a small percentage (1.3%) of samples were found to be contaminated with other microorganisms. No aflatoxin B_1 was detected in these products.

When Sprague-Dawley mice were fed 30 g sufu per kg body weight, the mice did not die, and there were no toxic or growth effects.[11] If properly fermented, sufu and its related products are not hazardous to health. The microorganisms involved in the sufu fermentation are not known to produce harmful toxins.

VI. CONCLUSIONS

Sufu is widely consumed as a relish by Chinese people. It is a creamy, cheese-like product and has a mild flavor. Further, sufu can be used in the same way as cheese. However, sufu is still confined to the conventional manufacturing process and aged naturally by soaking the fermented tofu in a variety of ways. Sufu is not popular in the Western Hemisphere. Nevertheless, if further improvements in taste, flavor, and texture are made, it may become more widely popular.

Although a pure culture method for preparing sufu has been developed, future studies are needed to develop uniform high quality products and well-defined economical processes to manufacture sufu. Sufu contains more protein-nitrogen than other soybean fermented foods such as miso and natto. Future studies should be directed towards the use of other legumes, such as winged beans, mung beans, etc. in preparation of sufu. Nutritional-quality evaluation of sufu needs further investigation. Su and his co-workers[27,28] isolated and identified two fungi from sufu namely, *Mucor mucedo* and *Mucor pusillus*. These two fungi were found to produce a potent milk-clotting enzyme. The mold-produced milk-clotting enzyme can be used satisfactorily in cheese making.[28] However, production of milk-clotting enzyme from the sufu mold and its subsequent use in preparation of various types of cheese need further studies.

REFERENCES

1. Steinkraus, K. H., Ed., *Handbook of Indigenous Fermented Foods,* Marcel Dekker, New York, 1983.
2. Su, Y. C., Sufu (tao-hu-yi) production in Taiwan, presented at the Symp. on Indigenous Ferment. Foods, Bangkok, Thailand, 1977.
3. Su, Y. C., Fermented foods in Taiwan, *Ferment. Ind.,* (Jpn.), 37, 102, 1979.
4. Su, Y. C., Traditional fermented foods in Taiwan, in Proc. Oriental Ferment. Foods, Food Industry Research and Development Institute, Hsinchu, Taiwan, 15, 1980.
5. Ohta, T., Nyu-fu I, *J. Brew. Soc.,* (Jpn.), 60, 588, 1965.
6. Ohta, T., Nyu-fu II, *J. Brew. Soc.,* (Jpn.), 60, 695, 1965.
7. Yamazaki, M., Chinese fermented foods, in *Sino-Research,* No. 1, 218, 1920.
8. Hesseltine, C. W., Fermented products — miso, sufu, and tempe, in Proc. Int. Conf. Soybean Protein Foods, Peoria, Illinois, 170, 1967.
9. Wang, H. L. and Hesseltine, C. W., Sufu and lao-chao, *J. Agric. Food Chem.,* 18, 572, 1970.
10. Wai, N. S., Soybean cheese, *Bull. Inst. Chem. Acad. Sinica.,* 9, 74, 1964.
11. Su, Y. C., Production and consumption of fermented foods in Taiwan, presented at Annu. Meet. Chinese Inst. Food Sci. Tech., Taipei, Taiwan, 1983.
12. Wang, H. L. and Hesseltine, C. W., Mold-modified foods, in *Microbial Technology, Vol. 2,* 2nd ed., Peppler, H. J. and Perlman, D., Eds., Academic Press, New York, 1979, 95.
13. Wai, N. S., A new species of mono-mucor, *Mucor sufu.* in Chinese soybean cheese, *Science,* 70, 307, 1929.
14. Wai, N. S., A new species of mono-mucor isolated from Chinese soybean cheese, *Gakugei, (Jpn.),* 9, 113, 1930.
15. Wai, N. S., Investigation of the various processes used in preparing Chinese cheese by the fermentation of soybean curd with *Mucor* and other fungi, Final Tech. Rep., UR-A6-(40)-1, U.S. Department of Agriculture, Washington, D.C., 1968.
16. Hesseltine, C. W., A millenium of fungi, food and fermentation, *Mycologia,* 57, 149, 1965.
17. Lin, L. P., Biochemical changes and fungal growth during fermentation of sufu, 13th Int. Cong. Microbiol., Boston, 1982.
18. Lin, L. P., Chen, S. H., and Chan, H. C., Relationships of mold growth, chemical components and microstructure during manufacture of sufu, Mem. Coll. Agri., Natl. Taiwan Univ., 22, 76, 1982.
19. Wang, H. L., Environmental control in Chinese traditional food processing, presented at the Symp. Indigenous Ferment. Foods, Bangkok, Thailand, 1977.
20. Liu, P. W., On some enzymes of pehtze, *J. Agric. Chem. Soc., (Jpn.),* 8, 273, 1932.
21. Liu, P. W., Chemical changes during the preparation of sufu, *J. Agric. Chem. Soc., (Jpn.),* 8, 162, 1932.
22. Watanabe, T., Ebine, H., and Ohta, T., *Soybean Foods,* Kolin Press, Tokyo, 1959.
23. Wang, H. L., Release of proteinase from mycelium of *Mucor hiemalis, J. Bacteriol.,* 93, 1794, 1967.
24. Okada, C., Katoaka, E., and Togano, T., Analysis of some component amino acids and volatile acids of nyufu (fermented soybean curd), *Eiyo to Shokuryo, (Jpn.),* 27, 309, 1974.

25. Su, Y. C., Tofuzu, in *Agricultural Products Processing,* Kuoli Pienikuan, Taipei, 1975.
26. Tang, S. Y., Dietary aflatoxins and liver cancer, Proc. of Workshop on Food Safety, Food Industry Research and Development Institute, Hsinchu, Taiwan, Oct. 17 to 19, 1977, 193.
27. Su, Y. C. and Chen, W. P., Studies on milk-clotting enzymes from microorganisms. I. Screening tests and the production of the enzymes, *J. Chin. Agric. Chem. Soc.,* 8, 73, 1970.
28. Su, Y. C., Chen, W. P., Yeh, S. C., Yang, F. M., and Lin, C. W., Studies on milk-clotting enzymes from microorganisms. II. The purification and the properties of the enzymes and their application in cheese-making, *J. Chin. Agric. Chem. Soc.,* 9, 80, 1971.

Chapter 5

NATTO

T. Ohta

TABLE OF CONTENTS

I. INTRODUCTION

A. Types of Natto

Natto is a popular soybean fermented food in the Japanese diet. Similar products are called "tu-su" by the Chinese and "tao-si" by the Philippinos. It is eaten with boiled rice and often used as a flavoring agent in cooked meats, vegetables, and seafoods. There are three major types of natto produced in Japan and each has its own method of preparation.[2] They are Itohiki-natto, Yukiwari-natto, and Hama-natto. Itohiki-natto (sticky natto) is produced by fermenting whole cooked soybeans with *Bacillus natto* and is produced in large quantities, i.e., accounts for more than the total production of the other two. Yukiwari-natto is made by mixing Itohiki-natto with rice koji and salt, and then aging. Hama-natto is prepared using the koji mold *Aspergillus oryzae.* Production of Hama-natto is limited to the Hamanatsu and Kyoto areas of Japan where it is sold as a local souvenir item.

The literature on Itohiki-natto is rather scanty. The earliest record for the term "natto" appeared in 1068 in Sin-Sarugaku-Shiyu,[1] but there was no description of the preparation method. Itohiki-natto has been the stock food of village farmers and for Buddhist temples in the wintertime for many years. Natto, during early periods of use, was prepared simply by wrapping cooked soybeans with rice-straw bundles, and fermenting the wrapped soybeans at ambient temperature. Modern techniques involving the use of starter cultures such as *Bacillus natto* developed after the 1920s.

B. Production and Consumption

The production of Itohiki-natto was approximately 170,000 tons (equivalent to 85,000 tons of soybeans) in 1982. This was nearly a 10% increase from the production in 1980. This increase is extraordinary among the traditional fermented foods of Japan and may be due to: (1) growing concerns of the public on health and diet relations, i.e., excess intake of animal fats and salt, (2) the nonsalty nature of the product, and (3) the excellent quality maintenance and shelf-life of natto products, i.e., extended use of low-temperature distribution systems from natto factories to households. If natto is not stored at low-temperatures, there may be quality deterioration due to high viable bacteria counts. The majority of natto-producing factories in Japan are small family businesses that produce about 300 kg of natto per day. Freshly produced natto is distributed locally. However, there is a growing number of large natto factories which produce more than 3000 kg/day. Natto products from large factories are sold through large supermarkets since they have the low-temperature storage facilities.

The annual per capita consumption of natto is 760 g in Japan. Until the 1950s, the preference of consumers for natto was limited to eastern regions of Japan. This localization of production and consumption recently changed due to the improved acceptance of natto in the western regions of Japan. In Japanese homes, natto is eaten with thinly sliced Chinese onion, nori seaweed, and mustard mixed together with a small amount of soy sauce, along with a bowl of cooked glutinous rice. Natto is also used as an ingredient of nori-wrapped sushi and in noodle soup. Natto is often eaten for breakfast and dinner.[2]

II. METHOD OF PREPARATION

A. Raw Materials

Soybeans of small to medium size, with a white to pale yellow hilum and smooth seed coat are preferred. Further, soybeans for natto should be soft, smooth and sweet when cooked. Often, they have a high water-absorbing capacity, high carbohydrate content, and low oil content. Japanese domestic soybeans of uniform size are preferred for natto preparation. Soybeans of uniform size are often imported from mainland China for natto production because Japanese soybeans are expensive and not available in large quantities. Soybeans imported from the U.S. and Canada are not preferred, because they don't have the desired flavor and they require more fermentation time. In general, small, uniform sized (6.0 to 6.6 mm diameter) soybeans are preferred since they ferment more uniformly than mixed sizes, and they give a better appearance to the finished product. Natto made from small soybeans is more acceptable for eating with cooked rice. Small size soybeans are cultivated in Ibaraki Prefecture of Japan. They are generally sold for 1.5 times the price of medium-sized (6.6 to 7.2 mm diameter) soybeans.

B. Preparation of Itohiki-Natto

The method for commercial preparation of Itohiki-natto will be described with some reference to the less common Hikiwari-natto. Whole soybeans are used for preparing Itohiki-natto. Hikiwari-natto is prepared in the same manner as Itohiki-natto except that dehulled soybeans cracked into 2 to 4 pieces are used. Natto made from whole soybeans is reported to be superior in palatability and higher in ammonia content than natto made from defatted soybeans.[3]

1. Soaking and Cooking of Soybeans

The small, uniform-sized soybeans are cleaned and soaked in water overnight at ambient temperature (10 to 15°C). The soaking time can be shortened by increasing the temperature of soak water. During the overnight soaking period, the soybeans absorb water to twice their original weight and become soft and smooth on the surface. Soaked soybeans are cooked for 20 to 30 min at a steam pressure of 1.0 to 1.5 kg/cm^2. This inactivates much of the natural microbial load. Soybeans for Hikiwari-natto are cooked for 8 to 10 min at a steam pressure of 1.0 kg/cm^2. Continuous cookers with rotary bulbs and belt conveyors are usually used in large-scale manufacturing plants. The desired hardness of cooked soybeans for natto preparation varies from 80 to 100 g as the weight required to break the cooked soybeans.

2. Inoculation and Packaging

The cooked soybeans are removed from the cooker, cooled to 45°C, and inoculated with a spore suspension of the natto starter culture. The starter culture, *Bacillus natto,* is inoculated at 10^7 to 10^8 spores per 10 kg cooked soybeans. The inoculated beans are mixed thoroughly to give a uniform distribution of spores. Natto starter cultures are commercially available in Japan.

The inoculated beans are weighed, filled into containers, and packaged before being placed in fermentation rooms. The most commonly used container holds 50 to 100 g of natto. This is the usual consumption amount for one household. The packaging containers are usually polystyrene foam, whereas the traditional packaging material includes rice-straw.

3. Fermentation

The packaged beans are fermented for 18 to 20 hr in a fermentation room controlled at a temperature of 40 to 42°C and a relative humidity of 80 to 90%. In the first two hours of incubation, the temperature of beans increases to 48°C as a result of the heat produced by microbial growth. Some aeration is provided during fermentation to prevent overheating. Within 6 to 8 hr of incubation, the surface of the soybeans changes to a whitish color due to bacterial growth and there is a fermentation odor. Water content of cooked beans, incubation temperature, and relative humidity influence the quality of the natto. Takahashi and Shimikawa[4] reported that the best quality natto can be produced by incubating the inoculated beans at 40°C and 85° relative humidity for 12 to 16 hr. During fermentation, *Bacillus natto* produces polymers of glutamic acid which cause the surface of the final product to have a viscous appearance and texture. High quality natto characteristically is covered with a large amount of very viscous polymers.

4. Maturation and Stabilization

The packages containing fermented natto are moved from the fermentation room to a refrigerated storage room for cooling and maturation. The packages are held for 1 to 2 nights at 0 to 5°C. Keeping the natto packages in refrigerated storage below 10°C stabilizes the quality of the fermented product by preventing further biological activity of the natto microorganisms. The commercial production of natto has been completely automated with automatic weighing of soybeans and filling of containers, as well as computerized temperature control.

C. Preparation of Yukiwari-Natto and Hama-Natto

Yukiwari-natto[5] is prepared by mixing Itohiki-natto with salt and rice koji, and then aging the mixture at 25 to 30°C for about two weeks. *Aspergillus oryzae* is used to prepare the rice koji.

To prepare hama-natto,[5] soybeans are soaked in water for 4 hr and then steamed without pressure for 1 hr. The cooled beans are inoculated with a koji prepared from roasted wheat or barley and fermented for about 20 hr, or until covered with the green mycelium of *A. oryzae*. The fermented beans are dried in the sun or by forced warm air to reduce the moisture content to about 12%. The dried beans are submerged in a salt brine along with strips of ginger and shiso leaves (*Perilla ocymoides*) and allowed to age for 6 to 12 months. During aging, extensive breakdown of proteins, carbohydrates, and lipids occurs. This contributes to desirable sensory qualities of Hama-natto. A more detailed description of Hama-natto is given in the previous chapter.

Ama-natto is another type natto, but it is not a fermented food. This product is prepared from cooked adzuki-beans, kidney beans, cowpea, pea, and broad beans with sugar.

D. Fermentation Microorganism

The earliest scientific studies on the microorganisms involved in natto fermentation were reported in 1894.[6] Later in 1905, Sawamura[7] demonstrated that natto fermentation occurs as a result of action of *B. natto*, a strain of *Bacillus Subtilis*. *B. natto* has an optimum growth temperature of around 40°C when grown on peptone-dextrose media. The majority of spores of this microorganism germinate at a temperature between 40 and 45°C in 2 hr. Sakurai and Nakano[3] conducted studies on the microbial counts of natto dried at various temperatures (60, 50, 105°C) for 3 to 7 hr. They reported that *Bacillus natto* survives high drying temperatures (150°C for 7 hr).

One of the unique features of natto bacterium (*Bacillus natto*) is the formation of sticky viscous material. This viscous material gives natto its unique characteristics as food, including its specific odor.

III. COMPOSITION AND PHYSICAL PROPERTIES

A. Chemical Composition

The chemical composition of natto is presented in Table 1. Natto is a highly nutritious food. It contains about 16.5% protein. A change in the nitrogenous compounds of natto occurs during fermentation (Table 2). There is general agreement that the riboflavin content of natto is higher than that of cooked soybeans; however, there are

Table 1
COMPARATIVE CHEMICAL COMPOSITION OF NATTO AND COOKED SOYBEANS

Component	Cooked soybeans	Natto
Moisture (%)	63.5	59.5
Protein (%)	16.0	16.5
Fat (%)	9.0	10.0
Carbohydrate (%)	7.6	9.8
Crude fiber (%)	2.1	2.3
Ash (%)	1.9	1.8
Calcium (mg/100g)	70.0	90.0
Phosphorus (mg/100g)	190.0	190.0
Iron (mg/100g)	2.0	3.3
Sodium (mg/100g)	1.0	2.0
Potassium (mg/100g)	570.0	660.0
Thiamin (mg/100g)	0.22	0.07
Riboflavin (mg/100g)	0.09	0.56
Energy (kcal/100g)	180.0	200.0

Table 2
CHANGE IN NITROGEN COMPOUNDS DURING FERMENTATION OF NATTO

Type of nitrogen compound	Cooked soybeans	Natto (Itohiki)
Water-soluble nitrogen (%)	17.0	55.0
Amino nitrogen (%)	1.0	9.0
Ammonium nitrogen (%)	0.3	3.0

conflicting reports relating to the effect of fermentation on thiamin. Sano[8] reported that the thiamin and riboflavin content of natto was about threefold higher than that of the original soybeans and the vitamin B-12 content increased nearly fivefold. Hayashi[9] however, reported that boiling markedly decreased the thiamin content of soybeans, but fermentation by *B. natto* increased the thiamin content to about the same level as that in raw soybeans. In another study, Arimoto[10] found an increased water-soluble protein and vitamin B-2 content in natto after fermentation with *B. natto.* The total amino acid content of raw soybeans, steamed soybeans, and fermented natto appears to be quite similar, except for a possible lower content of arginine and proline in natto (Table 3).

Hamma-natto contains about 38% moisture, 25% protein, 25% carbohydrates, and 11 to 13% salt. Kiuchi et al.[2] studied the lipid content and composition of Itohiki-natto, Yukiwari-natto, and Hama-natto. The total lipid content of the finished products is 2.8, 10.9, and 6.4%, respectively. The pattern of the lipids of Itohiki-natto is similar to that of soybeans. The koji mold produces lipase during fermentation which results in production of free fatty acids. The free fatty acid content of Hama-natto is about 12% of the net weight, and 78% of the total lipid weight.[2] These free fatty acids play an important role in the flavor development of Yukiwari- and Hama-natto.[2,12] Yukiwari-natto contains 5 to 18% free fatty acids. The fatty acid composition of Itohiki- and Hama-natto are similar to that of raw soybeans but Yukiwari-natto contains lauric and myristic acids in addition to the other acids. Goto[11] reported that the pre-

Table 3
AMINO ACID PROFILE OF RAW SOYBEANS, STEAMED SOYBEANS, AND FERMENTED NATTO
(g amino acid/100 g protein)[3]

Amino acid	Raw soybeans	Steamed soybeans	Fermented natto (16 hr)
Glycine	3.6	3.3	3.5
Alanine	4.7	4.4	4.5
Valine	5.8	5.6	5.7
Isoleucine	5.1	5.2	5.2
Leucine	8.6	8.6	8.6
Aspartic acid	11.0	11.0	11.0
Glutamic acid	18.0	18.0	18.0
Lysine	6.9	6.8	6.8
Arginine	6.8	5.6	5.0
Histidine	3.3	3.3	3.1
Phenylalanine	4.9	5.1	5.2
Tyrosine	2.8	2.8	2.9
Proline	8.5	8.3	8.0
Tryptophan	1.0	1.0	1.1
Methionine	1.0	1.1	1.2
Cystine	0.7	0.7	0.7
Serine	6.3	6.0	6.5
Threonine	4.7	4.5	4.6

dominant fatty acid in natto is linoleic acid followed by oleic, linolenic, and stearic acid. Further, he reported that the total lipids of natto consist of at least eight components. The main components are triglycerides, free fatty acids, diglycerides, monoglycerides, and traces of polar lipids. In general, amylolytic, proteolytic, and lipolytic activities of microorganisms result in extensive breakdown of soybean components during fermentation.

B. Physical Properties

1. Mucous Materials

Mucous material is produced by *B. natto* during fermentation of the cooked soybeans. This gives natto a unique flavor and odor. This material is about 0.1 to 0.8% (dry weight basis) of natto. The mucous material is a polypeptide of glutamic acid (D and L) and a fructan (polymerized fructose). The quality of the natto is greatly influenced by the amount and nature of this material. Generally, good quality natto is very sticky. The formation of mucous materials during the preparation of natto is stimulated by the glutamic acid or L-arginine, DL-alanine, L-aspartic acid and D-proline. Mucous formation is also stimulated by the sucrose in the soybeans and by an undisturbed fermentation for a few hours at the final stages. Mucous material quality and quantity in natto can be increased by the addition of 0.1 to 0.5% monosodium glutamate to cooked soybeans.

Ishikawa et al.[13] studied the spinnability of a natto mucin. Purified mucin solution isolated from natto beans contains 22.1% fructan and 77.6% poly-DL-glutamic acid and has a high viscosity. They reported that mucin was spinnable due to formation of network structures of randomly coiled poly-DL-glutamic acid through intermolecular H-bondings in the presence of fructan. Iso et al.[14] investigated the viscometric behavior of natto mucin in solution. A 2% solution of natto mucin was thixotropic at pH 5.7 and apparently Newtonian at pH 4.3. The viscometric behavior of a 6% solution of mucin at pH 4.3 was almost the same as that at pH 5.7. The results were explained in terms of the conformational change of the molecule according to the pH change, i.e., the mucin molecule was randomly coiled at pH 5.7 and was a rodlike molecule at pH 4.3.

2. Spots on Natto

Spots that are white to pale yellow in color sometimes appear on the surface of natto if the product is kept for long periods of time. Sometimes these spots may look like mold growth and are mistaken for contaminants of microbial or chemical origin. These spots on the surface of the natto are crystals of L-tyrosine.[15] The crystals are formed during storage of natto at refrigerated temperatures because of the low water solubility of tyrosine. Tyrosine is a hydrolysis product of soybean proteins and is not toxicogenic. The unpleasant appearance of the spots reduces the commercial value of natto.

3. Organoleptic Properties

Natto has a mild flavor and taste. Some of the flavor originates from the hydrolysis of soybean protein to peptides and amino acids. The unique flavor of natto has not been completely identified chemically; however, it is thought to be related to the diacetyl content.[16] Kosuge[17] characterized tetra-methyl pyrazine as the flavor component in natto. Some of the free fatty acids in natto are reported to produce undesirable odors. Butyric acid and isovaleric acid can contribute to the undesirable odors in natto.[17] The large amount of free fatty acids in Hama-natto (78% of total lipids) gives this product a strong harsh taste.[2] However, sweet flavor factors produced from amino acids and sugars during fermentation modify the harsh taste which results in a milder flavor.

An ammonia-like odor is directly involved in the quality of natto flavor.[19] The degree of ammonia evolution increases as the storage temperature of natto increases. At 5°C, very little ammonia is produced in natto. The time required for production of 200 mg/100 g ammonia is 14 days at 5°C compared to 2 days at 20°C and 1 day at 20°C. Ammonia production also depends on the composition of the soybeans. Soybeans with a high carbohydrate content produce less ammonia than the soybeans with a lower carbohydrate content, i.e., soybeans with higher carbohydrate content from Hokkaido produce less ammonia than the soybeans from the U.S.

IV. NUTRITIONAL QUALITY

Natto has a high nutritional value with improved digestibility and biological value of between 38.4 and 54.7.[20,21,25] In animal feeding studies, Ebine[22] reported that the absorption rate and biological value of natto were 83% and 63%, respectively. He further found a significant improvement in the biological value of natto compared to raw soybeans.

Hayashi and Nagao[23] prepared natto in the usual manner as well as other ways (prepared by immersing raw soybeans either in $CaCl_2$ solution or $CaCO_3$ emulsion). Each type of natto was stored at 5°C for 5 days and the nutritional value compared in feeding experiments with rats. There was an increase (23%) in the body weight of rats receiving natto prepared from soybeans immerced in $CaCl_2$ over the rats receiving normal natto. However, the increase in body weight of rats fed natto prepared from soybeans immersed in $CaCO_3$ was not significantly different than that of rats fed the typical natto. Hayashi et al.[24] found that blood pressure was lowered in subjects fed natto. The substance reponsible for lowering blood pressure is mainly present in the natto bacilli.

V. TOXICOLOGY

Like many other traditional fermented foods of the Orient, no comprehensive toxicological studies have been conducted with natto and *Bacillus natto*. However, there are no reports to indicate any possible toxic effects due to feeding or consumption of natto.

Some of the natto preparations have been tested and marketed as animal feed additives. These studies have been conducted with hogs and other domestic animals. The use of natto as an animal feed additive stimulated the intestinal *Bifidobacterium* growth and animal growth.

VI. CONCLUSIONS

The early Japanese who were not accustomed to consuming meat, developed the technology to make foods such as natto from soybeans. Foods made from soybeans serve as a major source of protein and fat in the Japanese diet.

Natto is prepared by fermenting cooked soybeans with *Bacillus natto*. Natto has a high nutritional value, improved digestibility, and appreciable amounts of certain vitamins as a result of fermentation. Natto is also a good source of fiber and free fatty acids. Experiments with farm animals (hogs and poultry) have demonstrated that natto stimulates the intestinal microflora growth in animals and animal growth. Further research is needed on the nutritional value of natto. There are occasional incidences of food poisoning from natto-containing foods; however, the cause of such poisoning is not due to the natto itself, but due to other associated ingredients, such as eggs that may be contaminated with *Salmonella*. Research should be conducted on possible use of other dry beans in preparation of natto. This may help to better utilize other inexpensive dry beans.

REFERENCES

1. Fugiwara, A., Sin-Sarugaku-Shiyu, 1068.
2. Kiuchi, K., Ohta, T., Otoh, H., Takabayashi, T., and Ebine, H., Studies on lipids of natto, *J. Agric. Food Chem.*, 24, 404, 1976.
3. Sakurai, Y. and Nakano, M., Production of high protein foods from fermented soybean products, in Progress in Meeting Protein Needs of Infants and Preschool Children, Publ. No. 843, National Academy of Sciences, National Research Council, Washington, D.C., 1961, 251.
4. Takahashi, S. and Shimakawa, T., Studies on the cause of bitter taste and ammonia in manufactured soybeans, *Ehime-Ken Kogya Shikento Kenkjo Hokou*, 14, 13, 1976.
5. Beuchat, L. R., Indigenous fermented foods, in *Biotechnology, Vol. 5*, G. Reed, Eds., Verlag and Chemie, Weinheim, West Germany, 479, 1983.
6. Yabe, K., On the vegetable cheese, natto, *Bull. College Agric. Tokyo Univ.*, 2, 68, 1894.
7. Sawamura, S., On the microorganisms of natto, *Bull. Coll. Agric. Tokyo Univ.*, 3, 107, 1905.
8. Sano, T., Feeding studies with fermented soy products (natto and miso), in Progress in Meeting Protein Needs of Infants and Preschool Children, Publ. No. 843, National Academy of Sciences, National Research Council, Washington, D.C., 1961, 274.
9. Hayashi, U., Japanese natto, *Jpn. J. Nation's Health*, 28, 568, 1959.
10. Arimoto, K., Nutritional research on fermented soybean products, in Progress in Meeting Protein Needs of Infants and Preschool Children, Publ. No. 843, National Academy of Sciences, National Research Council, Washington, D.C., 1961, 269.
11. Goto, K., Lipid-origin flavor constituents of natto (fermented soybeans). I. Fatty acid composition and fractionation of total lipids, *Obihiro Chikusan Daigaku Gakyiutsu Kenkyu Hokoku, Dia-1-by*, 8, 3, 1974.

12. Itoh, H., Mikkuni, S., and Ebine, H., Japanese Natto, unpublished data, 1976.
13. Ishikawa, H., Okubo, K., and Oki, T., Characteristic spinnability of a natto mucin solution, *Nippin Kolgku Kaishi,* 11, 2171, 1972.
14. Iso, N., Mizuno, H., Saito, T., Takeuchi, H., Suyama, Y., Kawamura, S., and Ogawa, H., The viscometric behavior of natto mucin in solution, *Agric. Biol. Chem.,* 40, 1871, 1976.
15. Ohta, T., General food industry, *Koseikaku,* 563, 1970.
16. Obata, Y., Flavor components of natto, *J. Ferment. Technol. (Jpn.),* 37, 159, 1959.
17. Kosuge, T., A component of natto flavor, *Nippon Yakugaku Kaishi,* 82, 190, 1962.
18. Kihara, Y., Component of natto, *Nippon Nogei Kagaku Kaishi,* 38, 257, 1961.
19. Ohta, T., Natto and cold chain, *Cold Chain Res.,* 6, 110, 1980.
20. Steinkraus, K. H., Ed., *Handbook of Indigenous Fermented Foods,* Marcel Dekker, New York, 1983.
21. Reddy, N. R., Pierson, M. D., Sathe, S. K., and Salunkhe, D. K., Legume-based fermented foods: their preparation and nutritional quality, *CRC Crit. Rev. Food Sci. Nutr.,* 17, 335, 1982.
22. Ebine, H., Fermented soybean foods in Japan, in Trop. Agric. Res. Ser. No. 6, Symp. Food Legumes, Japan, 1972.
23. Hayashi, U. and Nagao, K., Experimental studies on the nutritive value of natto during sealed preservation, *Teikoku-Gakuen Kiyo,* 1, 13, 1975.
24. Hayashi, U., Nagao, K., Tsa, Y., and Yoshioka, Y., Experimental studies on the nutritive value of "Natto": relationship between food containing "natto" and blood pressure of SJR, *Bull. Teikoko-Gakuen Koyo,* 2, 9, 1976.
25. Standal, B. R., Nutritional value of proteins of oriental soybean foods, *J. Nutr.,* 81, 279, 1963.

Chapter 6

TEMPE

F. G. Winarno and N. R. Reddy

TABLE OF CONTENTS

I. INTRODUCTION

Tempe is one of the most popular fermented foods of Indonesia. It is consumed by all socio-economic groups and serves as a major source of protein, calories, and vitamins in the diet. Tempe has been produced and consumed in Indonesia for centuries, but there are no written records of its origin. The central and east Java areas of Indonesia are the major tempe producers; therefore, tempe probably originated in these areas. Besides these areas, it is produced and consumed in other parts of Indonesia. Tempe is relatively unknown in surrounding countries such as Thailand, China, and Japan, where soybeans form an important part of the diet. Tempe is produced in small quantities and consumed by immigrants from Indonesia in Malaysia, Surinam, Canada, and the Netherlands. It is becoming a popular food for many vegetarians in the U.S. Tempe experienced a sudden rise in popularity in the U.S. in 1975, although it was studied extensively and known in the U.S. as early as 1950.[1] In 1971, soybeans and the fermented food tempe were introduced in the Republic of Zambia for improving the nutritional status in that country.[2] Tempe has been introduced recently in Sri Lanka through trainees, who received special training on production and processing technologies for tempe in Indonesia.

Geerligs[3] described in 1895 some Chinese foods made from soybeans and reported a method used for preparation of tempe in Java, Indonesia. This is the first known reference for tempe. Boorsma[4] described extensively production methods and chemical analyses of native foods of Netherlands including tempe. Later several researchers[5-7] reported extensive and detailed information about tempe. In the U.S., one group led by Professor K. H. Steinkraus at the New York State Agricultural Experiment Station, Geneva, New York, and another group led by Dr. C. W. Hesseltine at the Northern Regional Research Laboratory, Peoria, Illinois, extensively studied the preparation and chemical composition of tempe, molds involved in the fermentation of tempe, and isolated and identified the molds involved in tempe fermentation.[8]

Tempe was formerly considered an inferior food in part because of its cost, compared to other protein foods such as meats, fish, and eggs. Over the last 15 years the attitude towards tempe has changed. Today, more attention has been given to tempe because it is an inexpensive source of proteins, vitamins, and calories. The total annual production of tempe is about 500,000 tons.[9] In Indonesia, tempe production is still a household art. Most of the 41,000 small cottage industries that make fresh tempe daily are family run and employ about 128,000 workers.[1,10] Each small cottage industry employs about 3 workers and uses approximately 11 lb (5 kg) of dry soybeans per day to produce 21 lb (10 kg) of fresh tempe. The larger cottage industries employ 10 to 20 workers and use 600 to 1,100 lb (500 kg) of dry soybeans per day to produce tempe.[1] The average retail price of tempe is about US $0.25 per kg.

A well-made tempe by definition is a compact cake completely covered and penetrated by the white mold mycellium of *Rhizopus* sp. In describing tempe, an additional word is added to tempe to indicate the raw materials from which it was prepared. For example, *tempe kedelai* is made from kedelai (soybeans) and *tempe lamtoro* is made from lamtoro seeds.[10] However, in this chapter the word tempe (without suffix) will be used instead of *tempe kedelai.*

II. METHODS OF PREPARATION

A. Raw Materials

About 64% of the total soybean production in Indonesia is used for tempe preparation. Soybeans are one of the major substrates used for tempe making. However, tempe can also be prepared by substituting for soybeans various other substrates such

Table 1
FUNGI ISOLATED FROM INDONESIAN TEMPE[23]

Sample	Place of origin	Region	Species
Ragi-tempe	Malang	Eastern Java	*Rhizopus oryzae* *Rhizopus arrhizus* *Rhizopus oligosporus* *Mucor rouxii*
Ragi-tempe	Surakarta	Central Java	*Rhizopus oryzae* *Rhizopus stolonifer*
Tempe	Djakarta	Western Java	*Mucor javanicus* *Trichosporon pullulans* *Aspergillus niger* *Fusarium* spp.

as peanuts, benguk (*Mucuna pruriens* DC, a type of bean), lamtoro (*Leucaena leucocephala* Benth), solid residue of the tahoo (soybean tofu), wheat, and several other soybean and cereal mixtures.[8,11,12]

B. Microorganisms Involved

Different mold species have been reported in the literature as the microorganisms responsible for the fermentation of soybeans to tempe.[3,4,13] However, most earlier researchers[14,18] indicated *Rhizopus oryzae* Went et Geerligs as the mold responsible for fermentation of tempe. Steinkraus et al.[18] studied tempe fermentation using several pure cultures of *R. oryzae* including NRRL 395, 1526, 1528, and 2549 and a pure culture of *Rhizopus oligosporus* NRRL 2710. They reported that the inoculation of cooked soybeans with individual strains of *R. oryzae* produced a satisfactory tempe in 48 hr compared to 24 hr with *R. oligosporus* NRRL 2710. Later on, *R. oligosporus* NRRL 2710 was tentatively identified as the principal mold (*R. oligosporus* saito NRRL 2710) used in Indonesia for preparation of tempe.[12,19,20] Ko and Hesseltine[11] indicated that at least four species of *Rhizopus* (*R. oligosporus* saito, *R. oryzae, R. stolonifer* (Ehrenbex Fries) Lind, and *R. arrhizus* Fischer) could be used in the preparation of tempe. At least two species of *Rhizopus* (*R. oryzae* and *R. arrhizus*) were successfully isolated from tempe of Holland.[11]

In addition to the mold *R. oligosporus,* numerous bacteria (both spore-and non-spore-forming types) and yeasts are reported to exist in tempe.[16,21,22] The presence of bacteria in the tempe fermentation contributes to off-odors and flavors. Dwidjoseputro and Wolf[23] and Yeoh and Merican[24] isolated a variety of molds, yeasts, and bacteria from tropical tempe and tempe inocula of Malaysia and Indonesia (Tables 1 and 2). However, the significance of these mixed cultures in the tempe fermentation is not clearly known.

C. Starter Culture

The availability of an appropriate starter culture is essential for producing a good quality tempe.[25] Traditionally, Indonesian people prepare the starter culture by collecting small pieces of previously fermented tempe. The collected pieces of tempe are air dried or sundried, ground to a smooth powder, and used as the inoculum. In some cases, the surface of a previously fermented tempe cake, where most of the mycellium is found, may be sliced, sundried, ground, and used as the inoculum.[11]

Usar (also called *waru* or *laru*) is the most popular starter culture for tempe preparation in Indonesia. *Usar* is prepared of leaves of *Erythrina* sp., *Hibiscus dimilis* B_1 or *Hibiscus tiliaceus* Linn whose local name is *waru putih* in Java. The lowersides of these leaves are covered with downy hairs (*trichomos*) to which the mold mycellium and

Table 2
MICROORGANISMS ISOLATED FROM SEVERAL MALAYSIAN TEMPE SAMPLES[24]

Sample	Town	State	Species
Tempe	Kajang	Selangor	*Rhizopus* spp.
			Aspergillus spp.
			Penicillium spp.
			Micrococcus spp.
Tempe	Kuala Lumpur	Selangor	*Rhizopus* spp.
			Mucor spp.
Tempe	Bangi	Selangor	*Rhizopus* spp.
			Aspergillus spp.
			Bacillus spp.
			Micrococcus spp.
			Trichosporon spp.

FIGURE 1. *Usar* made traditionally using *hibiscus* leaves.

spore can adhere. *Usar* is prepared as follows. A hibiscus leaf is placed on a tray with the lower hairy side facing up and the leaf surface is sprinkled with 30 to 40 inoculated soybeans. Next, a second leaf of about the same size is placed with the lower side down on top of the first leaf to form a sandwich with inoculated soybeans in the middle. Likewise, several sandwiches are prepared and wrapped in a perforated plastic wrap and placed in a wooden tray. The wooden tray containing these sandwiches is covered with a gunnysack (cloth) and fermentation is allowed to proceed for 5 to 6 hours. During fermentation, the mold will grow on the soybeans and hairy lower sides of leaves. The leaves are removed, sundried, and stored until they are used for inoculation (Figure 1). A small leaf of *Usar* inoculates about 3 kg of dry soybeans that have been previously soaked, dehulled, and cooked.[1] This method is more popular because it produces the pure traditional starter culture. Further, growing the mold between leaves eliminates contamination of microorganisms. Research on the unique nature of usar has not been published. In some parts of Indonesia, other leaves with hairs, especially teak (*Tectona grandis*), are used for producing mold starter culture.[1,11,13]

Steikraus et al.[15,26] introduced a method for large-scale production of tempe starter. This method involved growing pure mold cultures isolated from scraps of previously fermented tempe onto cooked, dehulled soybeans. The inoculated soybeans are incubated at 37°C for 4 days to allow mold growth and sporulation. The full grown cultures are freeze-dried, pulverized in a sterile laboratory mill, and used as a starter for a new batch of tempe fermentation. Other researchers[27-30] have also reported on mass production of spores; however, expense and time involved are increased.

The amount of inoculum required to make a satisfactory tempe was suggested to be 1×10^6 mold spores per 100 g of cooked soybeans.[25] Fermentation time becomes too critical if the amount of inoculum is larger. However, too small an amount of inoculum may provide a chance for growth of contaminating bacteria.[8,25]

D. Preparation

A broad range of tempe preparation methods have been reported for different places and countries.[4,8] However, the major steps for preparation of tempe are the same regardless of the places and countries of production. These methods yield an organoleptically satisfactory tempe. From a nutritional viewpoint, the more prolonged boiling and soaking processes are unnecessary since they may lead to significant nutrient losses, more energy consumption, and increased time. Two methods (traditional and pilot plant production) for production of tempe are presented here. For an extensive coverage of tempe preparation methods, the reader is referred to the publications of Steinkraus[8] and Shurtleff and Aoyagi.[1]

1. Traditional

The essential steps for traditional preparation of tempe are presented in Figure 2. First, the soybeans (Figure 3A) are soaked and cooked for 30 min in boiling water to loosen the soybean seed coats. The seed coats of cooked soybeans are hand removed or rubbed with feet to loosen the seed coats and washed with water to separate dehulled beans (Figure 3B) from seed coats. The dehulled beans are again soaked overnight to hydrate and allow bacterial acid fermentation. The soaked dehulled beans are cooked again, drained, cooled, and inoculated with a starter culture or an inoculum from a previous batch, wrapped in banana leaves or perforated plastic bags, and incubated for up to 48 hr at room temperature. A variety of other suitable leaves such as delina (*Delina indica*), golden berry (*Baccaurea motiyana*), and almond (*Terminalia catappa*) leaves are used as wrappings for small packets of fermenting tempe. Wrapping of soybeans in leaves keeps the soybean cotyledons moist during the fermentation and allows gases to exchange. A photograph of a traditional tempe from Indonesia is shown in Figure 4.

The soybeans undergo a natural bacterial acid fermentation during the hydration step prior to cooking. This protects the tempe from the growth of undesirable microorganisms.[18,31]

When large amounts of tempe need to be prepared, the soaked whole soybeans are placed in bamboo baskets at the edge of a stream and rubbed by foot to float seed coats away from the dehulled beans in the running water. The dehulled beans are then boiled without pressure in excess water for 30 min.[8] The cooked dehulled soybeans are drained, cooled, and inoculated with a starter culture or an inoculum from a previous batch. The mass of inoculated soybeans are wrapped in wilted banana leaves or perforated plastic bags and tied with a string (made of bamboo) and incubated for up to 48 hr. Freshly prepared tempe is sold daily in Indonesia.[12]

Whole, clean soybeans

↓

Boil 30 minutes to partially hydrate the beans to facilitate hull removal

↓

Dehull by hand and wash

↓

Soak overnight to hydrate and allow bacterial acid fermentation

↓

Boil 60 minutes in soak water (partial cook)

↓

Drain and cool

↓

Inoculate with a tempe cake from previous batch

↓

Wrapped in clean, wilted banana leaves or perforated plastic wrap and fermented for up to 48 h

↓

Tempe

FIGURE 2. Flow sheet: Indonesion household tempe process.

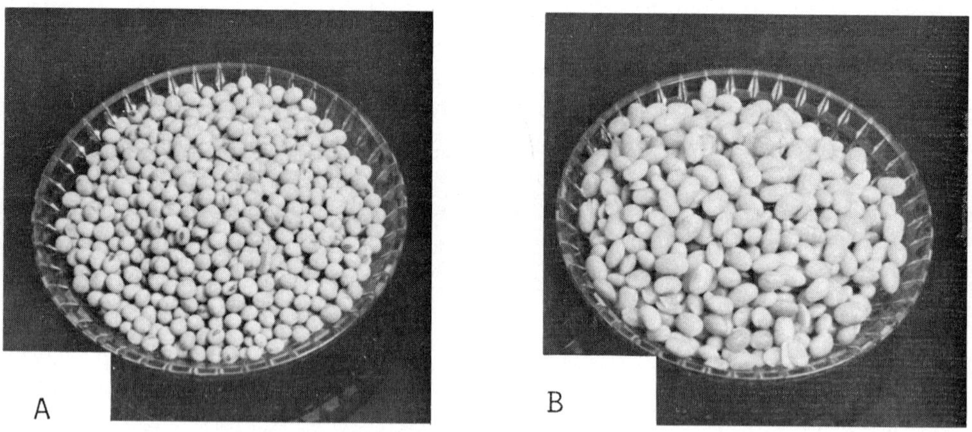

FIGURE 3. Soybean as raw material for making tempe: (A) uncooked whole soybeans, and (B) cooked dehulled soybeans.

Several researchers[26,32] have studied and reported on the use of soybean grits for preparation of tempe. This involves only minor modifications in the process.

2. Pilot Plant Production

Steinkraus et al.[26] developed a pilot plant method for production of tempe (Figure 5). This pilot plant method involves size grading, dry dehulling, hydration/acid fermentation, cooking, draining, cooling, and inoculation with mold and fermentation for 18 hr at 35 to 38°C and 75 to 85% relative humidity.

FIGURE 4. Traditional Indonesian tempe: (A) wrapped in banana leaves, and (B) wrapped in perforated plastic bags.

a. Dehulling

The graded soybeans can be dehulled by two processes (1) dry dehulling and (2) wet dehulling. In the dry dehulling process, the soybeans are preheated for 10 min at 93°C to shrivel the cotyledons and loosen the seed coats. Preheat treatment can be accomplished by using a hot air dryer or by sundrying the soybeans for 1 to 2 hr. The heated soybeans are cooled and passed through a properly spaced Burr mill to crack the shrivelled seeds. The seed-coats are later separated from cotyledons by passing the mixture

	Yield	Solids Loss (%)
Size-grade raw dry soybeans ↓	1000 g dry beans	
Heat 10 min at 93°C ↓		
Dehull in burr mill ↓		
Gravity-separate hulls ↓	80 g hulls	8% hulls
Soak cotyledons 2 h at 25°C or 30 min at 100°C in dilute lactic acid ↓	920 g cotyledons	
Boil 90 min in soak water ↓	798 g dry solids	12.2% soak/cook
Drain and cool to 38°C ↓		
Mix with mold inoculum ↓		
Spread on dryer trays ↓		
Cover with a sheet of waxed paper ↓		
Incubated at 35–38°C (75–85% Relative Humidity) for 18 h ↓		1.7%
Tempe ↓		
Dehydrate ↓	781 g dry tempe (78.1% recovery of starting solids)	
Package in plastic bags		
		21.9% total solids loss

FIGURE 5. Pilot plant process for production of tempe. (From Steinkraus, K. H., Van Buren, J. P., Hackler, L. R., and Hand, D. B., *Food Technol.*, © Institute of Food Technologists, 19, 66, 1965. With permission)

over a gravity separator or by use of an aspirator. Dry dehulling is the most efficient and continuous method.

Wet dehulling is generally done after a precooking which facilitates hydration of beans. The hydrated beans are dehulled by hand or feet as described earlier. Wet dehulling requires no machines. This method may not be feasible for large scale tempe production. Wet dehulling is only for production of tempe at the village level as in Indonesia.

b. Hydration and Acid Fermentation

Hydration is accomplished by soaking the dehulled beans in acidified water for 2 hr at 25°C or 30 min at 100°C (Figure 5). Traditionally, in Indonesia, the dehulled beans are soaked in water overnight at room temperature to allow bacterial fermentation. During soaking, the beans absorb an amount approximately equal to the weight of the water. Traditional soaking permits bacterial growth which results in acid fermentation and a drop of pH to 4.5 to 5.3. To date, the bacteria responsible for acidification have not been identified and characterized.[8] Further, a drop in pH prevents the growth of undesirable microorganisms that might contaminate tempe. Soaking and discarding the soak water helps in partial elimination or reduction of antinutritional and/or toxic factors (chymotrypsin and trypsin inhibitors, phytic acid, tannins, flatulence-producing factors, and others). This subsequently improves the nutritional value of tempe.[11,33] Sometimes the presence of water-soluble antinutritional factors in the soybeans inhibit the mold growth.[11,34,35]

c. Cooking, Draining, and Cooling

The soaked beans are cooked for up to 2 hr in boiling water, or the beans are boiled for up to 90 min in the soak water. Cooking partially destroys the antinutritional factors such as trypsin and chymotrypsin inhibitors and releases some of the nutrients required for mold growth. Traditional cooking times vary up to 3 hr in Indonesia.

The cooked beans are drained thoroughly using a wire mesh or woven basket and cooled to 38°C before inoculation. Excess water would favor bacterial growth and spoilage of tempe during mold fermentation. Sometimes in Malaysia the drained beans are rolled in a cloth prior to inoculation.

Indonesians frequently spread the cooked beans on flat woven bamboo trays. The excess water trickles through the bottom of the tray and the surface of the cooked beans dry as they cool.

d. Inoculation and Incubation

About 3 g of dried, pulverized tempe mold (*R. oligosporus*) culture containing both mycellium and spores are used to inoculate 1 kg of cooked, drained beans. The inoculum should be thoroughly mixed with the beans. In Indonesia, a dried powdered tempe inoculum or *usar* is sprinkled over the surface of the cooked, drained beans and thoroughly mixed to distribute the mold spores over the surface of all cooked, drained beans.

The inoculated beans are spread on stainless steel dryer trays (38 × 81 × 1.3 cm), covered with a sheet of waxed paper to maintain moisture in the beans and prevent excessive aeration, and incubated at 35 to 38°C, 75 to 78% relative humidity for up to 18 hr in a fermentation room. Covering of inouclated beans with waxed paper helps in the quick growth of mold and development of a typical tempe cake. During fermentation, considerable heat is produced as a result of mold growth which results in a higher temperature,[20] 10°C above incubation temperature. A rise in temperature during fermentation may cause damage to mold growth. A rise in temperature during fermentation should be prevented.

e. Harvesting and Storage

The tempe should be harvested as soon as the bean cotyledons have been completely overgrown and knitted into a compact cake (Figure 4). Freshly made tempe can be stored for a day or two at room temperature without changing many of its qualities and flavor characteristics. Traditionally in Indonesia, tempe is consumed on the day it is made. If the fresh tempe is stored for longer periods of time (more than two days) at room temperature it becomes unfit for consumption because of off-flavors and odors produced during storage by contaminating bacteria and other microorganisms.[36]

American vegetarians like "tempe Burgers".[37] These are prepared by cutting a fresh

Table 3
COMPARATIVE EFFECT OF FREEZE-DRYING VERSUS HOT AIR DEHYDRATION (69°C) ON TEMPE[18]

Tempe	pH	Reducing substances (%)	Soluble solids (%)	Soluble nitrogen (%)
Fresh	6.3	0.71	17.6	2.31
Lyophilized	6.2	0.41	19.5	1.19
Hot air dried	5.3	0.28	13.8	0.61

tempe cake into either square or circle shaped pieces of about 1.5 cm thick. The cut pieces are placed in a covered skillet and steamed for 15 min with sufficient water to cover the bottom of the skillet. The steamed tempe pieces are then fried in a cooking oil until both sides are golden brown. The fried tempe is served in a bun with tomato, onion slices, and lettuce.[37]

Storage stability of tempe can be extended by drying, drying and frying, dehydration, or freezing. In Indonesia, the tempe cake is cut into thin strips (Figure 4) and sundried.[15] The sundried strips are dipped in a spiced thin rice dough, deep fried in a cooking oil, and packed in sealed plastic bags for later consumption. These deep fried strips remain tasty and acceptable for several weeks or months when stored at room temperature. Tempe is also used in the preparation of the Indonesian dish *Sambel goreng kering tempe,* where the half-sundried thin tempe strips are mixed with hot pepper and brown sugar, and deep fried in a cooking oil. This product also remains tasty and acceptable for several days to weeks.

Steinkraus et al.[26] developed a method for production of dehydrated tempe. They cut freshly prepared tempe into small squares (2.5 cm) and dehydrated these in a circulating hot air dryer at 69°C for 90 to 120 min. During drying, the moisture content of the tempe squares is reduced to 2 to 4%. Dehydrated tempe can be stored in a plastic storage bag for months at room temperature without changes in product color or flavor. However, Steinkraus et al.[26] found changes in reducing substances, soluble solids, and soluble nitrogen content of tempe due to hot air dehydration (Table 3). It is not known whether the reduction in these solubles represent a loss in nutrient value of tempe.

Hesseltine et al.[19] suggested that tempe be stored by a freezing method. This method involved blanching and freezing. The freshly prepared tempe is cut into thin slices and blanched for 5 min in a boiling water to inactivate the mold and proteolytic and lipolytic enzymes. The blanched slices are placed in cellophane packages and frozen. They further reported that the blanched tempe slices can be stored for up to 100 days by the freezing method without significant changes in organoleptic characteristics (appearance, color, and taste).

Iljas et al.[38] studied the storage stability of tempe in sealed cans. The can containing blanched tempe slices can be stored for up to 10 weeks without significant changes in acceptability.

Another method to extend the shelf life of tempe is to defer the fermentation.[28,39] In this method, plastic tubing (5 × 15) is filled with pre-inoculated tempe and kept in a freezer to defer the fermentation until tempe is needed. Whenever tempe is needed, a small part of the plastic tubing containing preinoculated tempe is cut, perforated for aeration, and allowed to ferment in a warm place (30°C) for 20 to 22 hr. Similarly, preinoculated beans in small packages can be stored in a freezer in retail stores and sold as a ready-to-ferment tempe.

Table 4
SOLIDS LOSSES FROM TEMPE PROCESSING OF VARIETIES OF WHOLE SOYBEAN AND SOYBEAN GRITS[18,27,40]

Material and procedure	Soybean varieties Harosoy (%)	Hawkeye (%)	Seneca (%)
Whole Soybeans			
Soaking	4.7	1.6	4.9
Dehulling	7.9	7.9	17.1
Cooking	11.0	14.0	1.6
Fermentation	3.4	1.0	3.9
Total loss	27.0	24.5	27.5
Soybean Grits			
Mechanical dehulling	9.5	9.5	
Soaking	7.2	5.1	
Cooking	27.1	13.9	
Fermentation	4.5	4.0	
Total loss	48.3	32.5	

f. Losses and Yield

The steps (dehulling, soaking, cooking, draining, and fermentation) used in the preparation of tempe may contribute to losses of soybean solids, protein, and nutrients.[18,27,40] Solids loss that occurs during preparation of tempe is outlined in Table 4. The total loss of solids ranges from about 24 to 48% depending on soybean variety and whether whole beans or grits are used (Table 4). The use of soybean grits in the preparation of tempe results in greater losses of solids compared to whole soybeans. The solids losses of soybeans may be partially reduced by adding an equal amount of water to soak the beans (beans-to-water ratio of 1:1 w/v), and not discarding the cooking water. However, this procedure may result in poor quality of tempe cake due to possible bacterial growth during fermentation. The solids losses and yield that occur during tempe production are shown in Figure 5. On a dry weight basis, the solids recovery is 78%. The total solids loss during tempe production is about 22% and is attributed to various production steps: 8% due to hulls removal, 12.2% during soaking and cooking, and 1.7% during fermentation.[26]

E. Organoleptic Characteristics of Tempe

Freshly made tempe is a white compact cake, has a yeasty odor and does not have a beany flavor. Tempe is used as a main dish in Indonesia. Tempe is often served in various ways, such as fried in oil, baked, or as a soup. Usually, the cake is sliced thin, dipped in a salt solution, and deep-fat fried in coconut oil until its surface is crisp and has a golden brown color.[33] Following deep-fat frying, the flavor becomes nut-like and peppery due in part to the presence of free fatty acids. Deep-fat fried tempe is highly acceptable even to people of the western world. Overripened tempe is often used in small amounts as a seasoning in soups and dishes such as *Sambal tumpang* and *sayur loden.*[1] Ching[41] reported on the preparation of a shrimp flavored tempe using a thermo-tolerance *Rhizopus* sp. T-3 isolated from a native tempe collected from Bali Island.

Table 5
COMPOSITION OF NUTRIENTS PER 100 G OF TEMPE[11,32,38]

Tempe type	Food energy (cal.)	Moisture (%)	Protein (%)	Fat (%)	Carbo-hydrate (%)	Fiber (%)	Ash (%)
Soy tempe fresh	157	60.4	19.5	7.5	9.9	1.4	1.3
Soy tempe, sundried		8.9	43.1	18.0	26.2	3.8	3.8
Soy tempe, freeze-dried		1.9	46.2	23.4	25.8	2.7	2.7
Soy tempe, dry basis		0.0	54.6	14.1	27.9	3.1	3.5
Soy tempe, deep fried		50.0	23.0	18.0	8.0	2.0	1.0

Table 6
AMINO ACID COMPOSITION OF TEMPE COMPARED WITH THE FAO/WHO REFERENCE PATTERN (EXPRESSED IN MG/G OF NITROGEN)[32,62]

Amino acids	FAO/WHO pattern	Soybean tempe	Tempe as percent of FAO/WHO pattern	Soybeans	Eggs
Methionine-cystine	220	171	78	165	342
Threonine	250	267	107	247	302
Valine	310	349	113	291	437
Lysine	340	404	119	391	417
Leucine	440	538	122	494	547
Phenylalanine-tyrosine	380	475	125	506	588
Isoleucine	250	340	136	290	378
Tryptophan	60	84	140	76	106

III. NUTRIENT COMPOSITION

Tempe is an excellent source of proteins, vitamins, and minerals. It is often used in the diet of diabetics because of the low utilizable carbohydrate content.[1] During a 72-hour fermentation of tempe, total soluble solids and soluble nitrogen respectively increase from 13 to 28% and 0.5 to 2.5%, while the total nitrogen remains fairly constant.[18] The nutrient composition of tempe is presented in Tables 5 through 9.

A. Proteins and Amino Acids

The protein content of fresh tempe is about 19.5% and more than 50% on a dry weight basis. Due to its high protein content, tempe makes a good substitute for meat in the diets of people from Indonesia. Of the 19.5% protein, 56.0% may be available for utilization in humans.[42] About 100 g of fresh tempe supplies 10.9 g of available protein or more than 25% of the daily adult male requirement for protein.

The amino acid composition of tempe along with the FAO/WHO reference pattern is presented in Table 6. Methionine is the first limiting amino acid in tempe. The high lysine content in tempe compliments cereal foods that are low in lysine. Murata et al.[43] studied the nutrient content of tempe and found no difference in protein and ash during fermentation. However they reported an increase of up to 85% in the free amino acids content of tempe over unfermented soybeans. It is not known whether such an increase in free amino acids contributes to improved nutritional quality. Several

Table 7
AMINO ACIDS (g/16 g NITROGEN) AND NUTRITIVE INDEX OF RAW AND FERMENTED SOYBEANS AND THE MOLD *RHIZOPUS OLIGOSPORUS*

Component	Raw soybeans	Length of soybean fermentation (hr)				Mycellium of *R. oligosporus*
		0	24	36	72	
Lysine	6.08	5.92	5.79	5.51	5.54	4.07
Histidine	2.50	2.53	2.47	2.27	2.46	1.52
Arginine	7.13	7.07	6.57	6.14	6.17	2.47
Aspartic acid	11.30	11.10	10.60	10.60	10.50	4.82
Threonine	3.76	3.82	3.73	3.71	3.62	2.56
Serine	5.67	5.92	5.55	5.81	5.55	2.95
Glutamic acid	16.90	17.10	15.70	16.10	15.80	4.72
Proline	4.86	5.10	4.80	4.54	4.52	1.96
Glycine	4.01	4.07	3.91	3.92	3.86	2.68
Alanine	4.23	4.30	4.27	4.34	4.29	2.77
Valine	4.59	4.81	4.51	4.48	4.43	3.41
Cystine	1.70	1.70	1.56	1.61	1.55	1.03
Methionine	1.22	1.30	1.34	1.28	1.25	0.89
Isoleucine	4.62	4.83	4.65	4.65	4.59	3.22
Leucine	7.72	8.07	7.69	7.78	7.65	4.61
Tyrosine	3.39	3.50	3.63	3.58	3.48	1.85
Phenylalanine	4.84	5.02	4.96	4.92	5.06	2.89
Tryptophan	1.24	1.03	1.32	1.29	1.23	0.50
Ammonia	1.85	1.87	2.22	2.04	2.68	4.37
Glucosamine	0.187	0.199	0.646	0.881	1.23	27.30
Essential amino acid index	74.4	74.4	74.6	73.5	72.8	46.2

From Stilling, B. R. and Hackler, L. R., *J. Food Sci.*, © Institute of Food Technologists, 30, 1043, 1965. With permission.

Table 8
VITAMINS AND MINERALS IN FRESH SOY TEMPE[35,43,45,47]

Nutrient	Amount/100 g fresh tempe	U.S. Recommended Daily Allowance (RDA)	100 g tempe as a % of RDA
Vitamin A	42 I.U.	5000 I.U.	1
Thiamin	0.28 mg	1.50 mg	19
Riboflavin	0.65 mg	1.70 mg	28
Niacin	2.52 mg	20.00 mg	13
Pentothenic acid	0.52 mg	10.00 mg	7
Pyridoxine	830.00 μg	2000.00 μg	42
Folacin	100.00 μg	400.00 μg	25
Vitamin B-12	3.90 μg	3.00 μg	130
Biotin	53.00 μg	300.00 μg	18
Calcium	142.00 mg	1000.00 mg	14
Phosphorus	240.00 mg	1000.00 mg	24
Iron	5.00 mg	18.00 mg	28

Table 9
DISTRIBUTION OF FREE FATTY ACIDS DURING SOYBEAN TEMPE FERMENTATION

Sample	Free fatty acids (mg/100 g of tempe) Palmitic acid	Stearic acid	Oleic acid	Linoleic acid	Linolenic acid	Total g/100g of tempe	% of total ether extract
Cooked soybean, free fatty acids	41	31	127	0	0	0.26	1.09
24 hr tempe	420	175	713	2510	293	3.59	13.87
30 hr tempe	771	202	802	2543	204	4.77	18.93
48 hr tempe	665	202	1359	4138	304	6.93	30.00
69 hr tempe	863	367	1671	5032	302	8.19	35.11

From Wagenknecht, A. C., Mattick, L. R., Lewin, L. M., Hand, D. B., and Steinkraus, K. H., *J. Food Sci.*, © Institute of Food Technologists, 26, 373, 1961. With permission.

others[32,40] reported similar results for the free amino acid content in tempe. Stillings and Hackler[44] observed that, in general, the amino acid content of tempe remained unchanged or declined slightly during 72 hr-fermentation (Table 7). However, these researchers reported that the tryptophan values were increased significantly during the 24 hr-fermentation period and decreased thereafter (Table 7). There is no change in the essential amino acid index through a 24 hr fermentation. Steinkraus et al.[45] observed that after 36 hr fermentation of soybeans with *R. oligosporus,* there was about a 10% decrease in lysine, and after 60 hr, more than a 25% decrease.

Stillings and Hackler[44] reported that steaming tempe for 120 min at 100°C and deep-fat frying had little effect on the amino acid composition of tempe. However, they found a decrease in lysine and cystine content after deep-fat frying of tempe for 7 min.

B. Vitamins and Minerals

The vitamin and mineral content of tempe is presented in Table 8. The ash content of tempe decreases during fermentation.[31,46] However, tempe can be considered a good source of minerals (calcium, phosphorus, and iron) and vitamins (thiamin, riboflavin, pyridoxine, folic acid, and vitamin B-12).[71] One hundred grams of fresh tempe supplies about 18 to 30% of recommended daily allowance (RDA) of important vitamins (Table 8).

An increase in the content of riboflavin, niacin, and vitamin B-12 during fermentation of tempe has been reported.[47] Steinkraus and Van Veen[48] reported a slight decrease in pantothenic acid during tempe fermentation with *R. oligosporus* compared to a two to six times increase reported by Murata et al.[32] Data on nutritional components, especially vitamins, are conflicting. Tempe made with a pure culture of *R. oligosporus* contains very little vitamin B-12 compared to tempe made by the traditional methods.[35] About 100 g of tempe made with pure mold contains 0.047 μg of vitamin B-12 and supplies less than 2% of RDA of vitamin B-12. Later it was found that a specific bacterium is present in the commercial samples of tempe (made in Toronto, Canada) prepared by traditional methods, which synthesize vitamin B-12 during fermentation.[35] Curtis et al.[49] identified this bacterium as *Klebsiella pneumoneae* from traditional Indonesian tempe. It is commonly present in plant materials. The vitamin B-12 content of tempe can be increased to as high as 14.8 μg/100 g by adding *K. pneumoneae* to the starter culture. The presence of this microorganism in commercial tempe has not been reported to have pathogenic effects, probably because this bacterium can be killed by ordinary cooking and/or deep-fat frying. The significance and confirmed presence of *K. pneumoneae* in tempe need to be examined further. The vitamin B-12 content of tempe presently prepared in North America ranges from 1.5 to 6.3 μg/100 g and is sufficient to meet the RDA requirement for vitamin B-12. Tempe prepared in Indonesia contains 30 μg/g of vitamin B-12 and provides 60% of the daily vitamin B-12 requirement if one consumes 60 g tempe daily. The discovery of *Klebsiella* in tempe offers a possibility of producing a commercial tempe starter containing a similar bacterium, which can be used at home or in factories in producing tempe rich in protein and vitamin B-12.

C. Lipids

Tempe mold posses strong lipase activity[30] and that in turn liberates free fatty acids from soybean lipids during fermentation.[54] Changes in soybean lipids during tempe fermentation was observed by Wagenknecht et al.[50] They found that more than one-third of the neutral fat of soybeans was hydrolyzed during a 3-day fermentation with *R. oryzae* (Table 9). They identified the free fatty acids as palmitic, stearic, oleic, linoleic, linolenic, with linoleic acid predominating. No free linoleic or linolenic acid was found in the cooked unfermented soybeans (Table 9). However, Murata et al.[32] reported an increase in the acid value and oleic acid content, and a decrease in linoleic acid content, after fermentation of soybeans with *R. oligosporus.*

The lipids in soybean tempe are reported to be more resistant to autooxidation than those in soybeans.[51,52] Ikehata et al.[53] found that the peroxide value of lyophilized tempe stored at 37°C for 5 months increased from 6 to 12 compared to an increase from 6 to 426 in unfermented soybeans. Antioxidants produced by the tempe mold have been extensively studied.[43,51-57] Stored tempe does not develop rancidity because it contains an antioxidant, 6,7,4-trihydroxyisoflavone (Factor-2) that is produced by the tempe mold.[56-57] However, this antioxidant is not effective in preventing autooxidation of soybean oil and/or flour when mixed.[56] Gyorgy et al.[56,57] speculated that the insolubility of the isoflavone in the oil, and the difficulty of isoflavone dispersion in soybean flour may in part account for its failure to prevent autooxidation.

Wang et al.[58] isolated and identified active antitryptic compounds from tempe. These were the free fatty acids: oleic, linoleic, and linolenic acids. Free fatty acids have been previously reported to inhibit various enzymes such as glycolytic, gluconeogenic, lipogenic, and proteolytic types. They further found that the inhibition of trypsin by free fatty acids was nonspecific.

D. Carbohydrates

Raw soybeans and cooked soybeans, respectively, contain about 8.0% and 3.8% of sugars (sucrose, raffinose, and stachyose).[59] Raffinose and stachyose are reported to cause flatulence in humans. The tempe mold, *R. oligosporus* hydrolizes most of the soybean sugars during fermentation. The hydrolysis of flatulence-producing sugars occurs at a slower rate during tempe fermentation by *R. oligosporus.*[59]

E. Antibacterial Compound Production

An antibacterial compound produced by *R. oligosporus* during tempe fermentation has been reported by Wang et al.[60] This compound does not exhibit a broad spectrum of activity and is very active against Gram-positive bacteria including *Clostridium botulinum, Colstridium sporogenes, Bacillus subtilis,* and *Staphylococcus aureus.* Antibacterial tests indicated that this compound is glycopeptide in nature.

Table 10
PER[a] AND DIGESTIBILITY COEFFICIENT OF TEMPE BEFORE AND AFTER FERMENTATION[36,40,42,46]

Tempe type	PER		Digestibility coefficient (%)	
	Before	After	Before	After
Soybean tempe	2.63	2.56	86.9	86.2
Wheat tempe	1.28	1.71	—	—
Wheat: Soybean tempe (1:1)	2.49	2.79	—	—

[a] Protein Efficiency Ratio.

IV. NUTRITIONAL QUALITY

Tempe is usually consumed in various forms such as deep-fat fried, dried, steamed and fried, and as a supplement to soups. The nutritional value of tempe depends on the types of processes given to tempe before consumption. Several researchers[7,40,42,43,45,46,61] reported an improvement, or no improvement, in the nutritional value and PER (Protein Efficiency Ratio) of tempe over unfermented beans (see Table 10 for PER values). The improvement of PER of tempe during fermentation reported by some researchers may be attributed to (1) better availability of amino acids from the tempe and greater digestibility of tempe protein, and (2) differences in the substrates and substrate mixtures, and conditions employed.[38,40]

Hackler et al.[42] studied the nutritional value of tempe fermented for various periods of time (0, 12, 24, 36, 48, 60, and 72 hr). They found that the nutritional value of tempe is reduced with an increase in fermentation time. Further, they reported that PER value remained fairly constant throughout the course of fermentation. Recently Zamora and Veum[61] reported that fermentation of cooked soybeans with *R. oligosporus* greatly improved the apparent digestibility and net protein utilization in rats fed diets containing tempe. The digestibility coefficient of tempe is about 86.2%, which is lower than that of tofu (95%). Most of the foods prepared from dry beans are difficult to digest because they contain a variety of antinutritional and/or toxic factors that interfere in the availability of minerals and amino acids. Tempe has a biological value of 58.7.[63]

Deep-fat frying may affect the digestibility of proteins and the nutritional value of tempe. Stilling and Hackler[44] reported that deep-fat frying of tempe decreases the PER, weight gain in rats, and digestibility coefficients. Deep-fat frying of tempe reduces the PER by 2.0%.[46]

V. ANTINUTRITIONAL AND/OR TOXIC FACTORS

The conditions established for preparation of tempe provide for vigorous growth of the desired microorganisms and provide very little chance for the growth of unwanted pathogens.[62] Since tempe is a naturally fermented food, it is always possible that there might be contamination of tempe with aflatoxin-producing fungi such as *Aspergillus flavus* and *Aspergillus parasiticus*. The mold species responsible for tempe fermentation do not produce aflatoxin.[39,63,67] Further, there have been no reports on toxins found in foods prepared from soybean. The mold required for tempe fermentation is reported to protect tempe and tempe-like foods against aflatoxin production and aflatoxin-producing molds.[65] The tempe mold *R. oligosporus* hydrolyzes aflatoxin and thus prevents the accumulation of this mycotoxin in tempe.[36,66]

Table 11
PHYTIC ACID AND FLATUS FACTORS OF TEMPE BEFORE AND AFTER FERMENTATION

Component	Fermentation		Ref.
	Before	After	
Phytic Acid (mg/g)	1.23	0.96	54
Flatus Factors (mg/g) (Raffinose family of sugars)	16.50	2.0	59

Table 12
INDONESIAN TEMPE PREPARED FROM DIFFERENT SUBSTRATES

Substrate	Local name	Tempe name
Soybeans	Kedelai	Tempe kedelai
Winged beans	Kecipir	Tempe kecipir
Velvet beans	Benguk	Tempe benguk
Jack beans	Koro pedang	Tempe koro pedang
Pigeon pea	Gude	Tempe gude
Solid waste material from tofu	Gembus	Tempe gembus

The existence of naturally occurring trypsin and chymotrypsin inhibitors in soybeans and their possible physiological roles in nutrition have long been recognized. Cooking of soybeans and subsequent fermentation with *R. oligosporus* decreases trypsin and chymotrypsin inhibitors in tempe.[31] In contrast, Zamora and Veum[61] found higher antritrypsin activity in soybeans fermented with *R. oligosporus* than in unfermented beans. A decrease in flatulence-producing sugars was also reported during tempe fermentation (Table 11). Calloway et al.[67] found that tempe was "essentially" nonflatulence producing when fed to human subjects.

Sudarmadji and Markakis[54] reported that the *R. oligosporus* in tempe produces very strong phytase activity that subsequently hydrolyzes phytic acid in tempe to free inositol and phosphate. Phytic acid decreases the bioavailability of various minerals (Ca, Mg, Zn, and Fe) because of its chelating properties. Raw soybeans contain about 1.4% phytic acid.[54] About 33% of the phytic acid is hydrolyzed during tempe fermentation with *R. oligosporus* (Table 11).

VI. TEMPE AND TEMPE-LIKE FOODS FROM OTHER LEGUMES

A number of dry beans (winged beans, Jack beans, velvet beans, pigeon peas, sweet lupine, yellow peas, broad beans, cowpeas, horse beans, and chickpeas), bean fractions (mung bean starch, solid waste materials from tofu), cereals (wheat and barley), and mixtures of cereals and beans have been successfully substituted for soybeans in the preparation of tempe and tempe-like foods.[8,70] Indonesian people differentiate the types of tempe by adding the local substrate name to the word tempe (Table 12). For example, tempe prepared from winged bean seeds is called *tempe kecipir*.

Table 13
NUTRIENT CONTENT OF WINGED BEAN TEMPE-KECIPIR[69]

Nutrient	Dry seeds	Tempe kecipir
Water (%)	9.5—10.4	58.2
Protein (%)	34.4	14.5—17.5
Fat (%)	16.9	8.0—10.0
Ash (%)	4.2	1.4
Crude fiber (%)	10.7	1.9
Vitamin C (mg/100 g)	3.5	0.0
Thiamin (mg/100 g)	0.3	0.2
Calcium (mg/100g)	468.0	186.0
Iron (mg/100 g)	6.8	2.2
Phosphorus (mg/100 g)	181.8	177.2
Food energy (cal/100 g)	426.1	211.6
Methionine (mg/g N)	58.0	60.0
Cystiene (mg/g N)	54.0	46.0
Lysine (mg/g N)	488.0	294.0
Net Protein Utilization	—	51.0

A. Tempe Kecipir

Gandjar[69] prepared a tempe-like food called tempe kecipir from winged bean seeds. To prepare tempe kecipir, winged bean seeds are washed and boiled in water for 60 min. The seed coats are removed from whole seeds and separated from cotyledons by washing with water. The cotyledons are soaked for 24 hr in excess water, then chopped into small pieces and sterilized for 30 min at 121°C. The sterilized cotyledons are cooled and inoculated with *R. oligosporus* or other *Rhizopus* strains. The inoculated cotyledons are wrapped in a perforated plastic wrap or wilted banana leaves and incubated at 30°C for up to 48 hr to form a tempe cake. Gandjar[68,69] did not detect any beany smell in the 48 hr fermented tempe kecipir.

The nutrient composition of tempe kecipir is shown in Table 13. During fermentation, an increase in the content of the amino nitrogen, total acids, and soluble carbohydrates, and a decrease in total solids was observed.[68] Gandjar[68] reported that acceptable quality of tempe kecipir can be obtained within 48 hr of fermentation.

B. Tempe Gembus

Tempe gembus is prepared by using the solid waste material from tofu factories. This type of tempe is made in central and east Java and consumed as a side dish with rice.[71]

For tempe gembus preparation, solid waste material from tofu production is pressed to remove excess water, steamed, cooled, and inoculated with tempe mold. The inoculated material is wrapped in banana leaves and incubated for 2 days in a dark place. The nutrient composition of tempe gembus is presented in Table 14. The fiber and calcium contents are high in tempe gembus since its substrate (solid waste material from tofu) contains large amounts of insoluble carbohydrates and salts. During tofu preparation, calcium salts are added to precipitate the proteins of soymilk, and this leads to high calcium content in the residue as well as in the tofu. During tempe gembus fermentation, mold hydrolyzes most of the insoluble carbohydrates. Tempe gembus is considered to be inferior in nutritional quality compared to tempe prepared from other substrates.

C. Other Products

Several other legumes, and cereal and legume mixtures, have been used in the prep-

Table 14
COMPOSITION OF TEMPE GEMBUS[72]

Component	Solid waste material (steamed)	Tempe gembus
Moisture (%)	82.6	84.9
Ash (%)	0.5	0.7
Fat (%)	2.7	2.1
Protein (%)	4.0	4.0
Carbohydrate (%)	10.1	8.4
Crude fiber (%)	29.6	30.4
Calcium (mg/100g)	204.0	226.0
Iron (mg/100g)	1.3	1.4
Thiamin (mg/100g)	0.1	0.1

Table 15
ORGANOLEPTIC CHARACTERISTICS OF CHICKPEA, HORSEBEAN, AND SOYBEAN TEMPE[74]

Organoleptic characteristics	Chickpea tempe	Horsebean tempe	Soybean tempe
Appearance	White	White	White
Texture	Firm	Soft	Firm
Color after drying	Light brown	Gray	Brown
Flavor before frying	Strong alcoholic and ester flavored	Weak alcoholic and ester flavored	Bland
Flavor after overnight storage	Bland	Bland	Bland
Flavor after frying	Very agreeable	Agreeable	Agreeable

aration of tempe and tempe-like foods.[70,73,74] In most cases, the preparation method followed is similar to that of the Indonesian traditional method. In India, Bai et al.[73] developed and standardized a process for successful preparation of tempe with a mixture of groundnuts (peanuts) and soybeans. Tempe prepared with a mixture of groundnuts and soybeans results in an increase in fat, soluble solids, and an improved PER in comparison to tempe prepared from soybeans.

Robinson and Kao[74] made tempe-like foods from horse beans, and chickpeas by fermenting 0.2 to 0.4 cm diameter grits with *R. oligosporus.* The organoleptic characteristics of horse bean, chickpea, and soybean tempe are presented in Table 15. Chickpeas can be substituted for soybeans to prepare a tempe with appearance, texture and flavor similar to that of soybean tempe.[74,75] There is an increase in reducing sugars, soluble proteins, and water-soluble vitamins (thiamin, niacin, riboflavin, pyridoxin, cobalamin, pantothenic acid, and ascorbic acid), a decrease in antitryptic activities, and improved PER values in horse bean and chickpea tempe.[75] Aflatoxins have not been detected in horsebean or chickpea tempe.

D. Nutritional Intervention

The Food Reseach Institute, CRDN (Central Research Development for Nutrition) at Bogor, Indonesia, has developed a low-cost, high quality tempe mixture called TFR (tempe, fish, and rice mixture) (Table 16). This mixture is used for preparation of porridge, which is widely used for feeding infants 6 months to 12 months of age. The

Table 16
COMPOSITION AND NUTRITIONAL VALUE OF TEMPE, FISH, AND RICE MIXTURES (TFR)

Ingredient	TFR-I	TFR-II
	Composition	
Tempe (%)	30.0	30.0
Fish (%)	9.0	10.0
Rice (%)	30.0	30.0
Sugar (%)	27.0	25.0
Peanut oil (%)	4.0	5.0
	Nutritive Value	
Protein (%)	27.3	23.7
Fat (%)	9.4	7.8
Carbohydrate (%)	59.6	61.2
Ash (%)	2.3	2.4
Moisture (%)	5.0	4.9
NPU (standardized)	64.0	62.0
NPU (operative)	46.0	51.0

Note: NPU = Net protein utilization.

TFR mixture has also been tested and fed to children suffering from Protein Calorie Malnutrition (PCM). TFR increases body weight and improves the general condition of children who previously suffered from PCM.

Another food mixture consisting of 70% rice and 30% tempe or 30% soybeans has been tested to feed the children (below five years of age) in two different provinces of Java for three months.[76] These feeding studies indicated that the children's acceptance of tempe-rice mixture was better than a rice-soybean mixture. Several other powdered mixtures, including tempe popped rice, dry fish, tempe rice-egg, and a high NPU vegetarian mixture (67% rice and 23% tempe), have been developed in Indonesia to improve the nutritional quality of the diet.

VII. CONCLUSIONS

Tempe is the most popular fermented soybean food of Indonesia. It serves as a major source of protein, vitamins, and minerals. As the population continues to grow, and the prospect of food shortages becomes more acute, people in developing countries must depend increasingly on foods derived from plants, such as tempe, rather than animals to meet their calorie and protein needs in the diet.[77] The future of fermented foods derived from legumes depends upon individual preference. The current concerns of the relationship between animal-derived foods and health is changing the food habits of individuals. This trend may favor the use of more plant-derived foods.[77] A marked increase in the interest of vegetarianism and plant-derived foods has developed in North America and Europe within the last decade. There is an increasing awareness of the low cost and high nutritional quality of foods such as fermented legume foods. Tempe has great potential as a key protein source in underdeveloped countries due to its low cost and easy preparation. In the future, tempe may play a major role in satisfying the protein deficiency problems in underdeveloped countries.

Although *Usar,* a traditional inoculum or starter, has been in use for centuries in preparation of tempe, very little is known about its characteristics and biochemistry. Therefore, it is important to study in detail the microbial and biochemical properties of usar. During tempe preparation, two major problems are usually faced — energy required and losses of solids during preparation. There is a need for research on how to reduce energy input and losses of solids and nutrients in the production of tempe. The development of tempe and tempe-like foods by substituting other beans, and mixtures of cereals and other beans, for soybeans should be continued. The use of tempe as a food supplement and/or weaning food should be examined further, particularly in underdeveloped countries where PCM is prevalent. Since the tempe is prepared by traditional methods, consideration should be given to assure safety of the tempe from food spoilage pathogens by developing pure starter cultures.

REFERENCES

1. Shurtleff, W. and Aoyagi, A., *The Book of Tempeh: A super soyfood from Indonesia.* Harper & Row, New York, 1979.
2. Thio, G. L., *Report of the Royal Tropical Institute,* Amsterdam, 1972.
3. Geerligs, H. C. P., Enige Chinese voedingsmiddeten uit soybean bereid, *Pharm. Weekbl.* 1895.
4. Boorsma, P. A., Scheikunding onderzoek van in Nederlands Indies, inheemse, Voedingsmiddlen and de Soyabean, *Gneesk, Tijuschr. Ned. Indie,* 40, 247, 1900.
5. Heyene, K., *De Nutrige Planten Van Indonesia,* W: Van Heeve. Netherlands, 1913.
6. Jansen, B. C. P., *Mededeelingen van de Burgelijke Geneeskundige Dienst van Nederlandsch Indie* I, 68, 1923.
7. Jansen, B. C. P. and Donath, W. F., Metabolic experiments on rats and digestibility of the proteins of some foodstuffs, *Mededeelingen Burgerlijken Geneeskundigen Dienst Ned. Indie,* 24, 1924.
8. Steinkraus, K. H., Ed., *Handbook of Indigenous Fermented Foods,* Marcel Dekker, New York, 1983, 8.
9. Ebine, H., *Shou-ken* Jpn. 5, 121, 1979.
10. Winarno, F. G., *The Present Status of Soybeans in Indonesia,* FATEMETA, Bogor Agricultural University, 1976.
11. Ko, S. D., and Hesseltine, C. W., Tempeh and related foods, *Econ. Microbiol.,* 4, 115, 1979.
12. Hesseltine, C. W., A Millenium of fungi, food and fermentation, *Mycologia,* 57, 149, 1965.
13. Burkill, I. H., *A Dictionary of the Economic Products of the Malay Pennisula,* Crown agents, London, 1935.
14. Vorderman, A. G., *Geneeskundig Tijkdschrift voor Nederlandsch Indie,* 42, 395, 1902.
15. Stahel, G., Tempe: a tropical staple, *J. N.Y. Bot. Garden,* 47, 1946.
16. Van veen, A. G., and Schaefer, G., The influence of tempeh fungus on the soybean, *Documents Neerlandica et Indonesica de Morbis Tropicis,* 3, 270, 1950.
17. Dupont, A., *Ph.D. thesis,* Fakultas Ilmu Pasti dan Alam, Bandung, 1954.
18. Steinkraus, K. H., Yap, B. H., Van Buren, J. P., Provdenti, M. I., and Hand, D. B., Studies on tempe — an Indonesian fermented food, *Food Res.,* 6, 777, 1960.
19. Hesseltine, C. W., Smith, M., Bradle, B., and Ko, S. D., Investigations of tempeh, an Indonesian food, *Dev. Ind. Microbiol.,* 4, 275, 1963.
20. Hesseltine, C. W., Research on Northern Regional Research Laboratory on fermented foods, in Proc. Conf. Soybean Products for Protein in Human Foods, U.S. Department of Agriculture, ARS-71-72, 1961.
21. Boedyin, K. B., Notes on the mucorales of Indonesia, *Sydowia Ann. Mycol.,* 3, 234, 1958.
22. Saono, S., Brotonegoro, S., Abdulkadir, S., Basuki, T. J., and Badjra, I. G. P., Microbiological studies of tempeh, kecap and tauco, in *ASEAN Project on Soybean and Protein Rich Foods,* Progress Report of Research activities January-May, 1976.
23. Dwidjoseputro, D. and Wolf, F. T., Microbiological studies of Indonesian fermented foodstuffs, *Mycopathol. Mycol. Appl.,* 41, 211, 1970.
24. Yeoh, Q. L. and Merican, Z., Malaysian tempe, presented at the Symp. Indigenous Ferment. Foods, Bangkok, Thailand, 1977.

25. Wang, H. L. and Hesseltine, C. W., Oriental fermented foods, in *Prescott & Dunn's Industrial Microbiology,* 4th ed., Reed, G., Ed., AVI Publishing, Westport, Conn. 1982, 504.
26. Steinkraus, K. H., Van Buren, J. P., Hackler, L. R., and Hand, D. B., A pilot plant process for the production of dehydrated tempe, *Food Technol.,* 19, 63, 1965.
27. Wang, H. L., Swain, E. W., and Hesseltine, C. W., Mass production of *Rhizopus oligosporus* spores and their application in tempe fermentation, *J. Food Sci.,* 40, 168, 1975.
28. Martinelli, A. F. and Hesseltine, C. W., Tempe fermentation: package and tray fermentation, *Food Technol.,* 18, 167, 1964.
29. Wang, H. L., Vespa, J. B., and Hesseltine, C. W., Acid protease production by fungi used in soybean food fermentation, *Appl. Microbiol.,* 27, 906, 1974.
30. Wang, H. L. and Hesseltine, C. W., Wheat tempe, *Cereal Chem.,* 5, 563, 1966.
31. Hesseltine, C. W., Decamargo, R., and Rackis, J. J., A mold inhibitor in soybeans, *Nature,* 220, 1226, 1963.
32. Murata, K., Ikehata, H., and Myamoto, T., Studies on the nutritional value of tempe, *J. Food Sci.,* 32, 580, 1967.
33. Wang, H. L. and Hesseltine, C. W., Studies on the extracellular proteolitic enzymes of *Rhizopus oligosporus, Can. J. Microbiol.,* 11, 727, 1965.
34. Steinkraus, K. H., Microbiology of foods produced in tropical Asia, in *International Symposium on Microbiological Aspects and Fermentation in Tropical Asia* Saono, S. and Winarno, F. G., Eds., Food Technology Development Center (FTDC)-IPB, Bogor, Indonesia, 1984.
35. Liem, T. H. I., Steinkraus, K. H., and Cronk, T. C., Production of vitamin B-12 in tempe, a fermented soybean food, *Appl. Environ. Microbiol.,* 34, 773, 1977.
36. Wang, H. L. and Hesseltine, C. W., Use of microbial cultures: legumes and ceral products, *Food Technol.,* 1, 79, 1981.
37. Bates, C., Lyon, A., Sorenson, S., Keller, B. and Jenkins, S., Utilization of tempe in North America, presented at the Symp. Indigenous Ferment. Foods, Bangkok, Thailand, 1977.
38. Iljas, N., Peng, A. C., and Gould, W. A., Tempe find ways to preserve Indonesian soyfood, *Ohio Rep.,* 55, 22, 1970.
39. Hesseltine, C. W., Smith, M., and Wang, H. L., New fermented cereal products, *Dev. Ind. Microbiol.,* 8, 179, 1967.
40. Smith, A. K., Rackis, J. J., Hesseltine, C. W., Robbins, D. J., and Both, A. N., Tempe: nutritive value in relation to processing, *Cereal Chem.,* 41, 173, 1964.
41. Ching, F. L., Preparation of shrimp-flavored tempe using athermotolerable *Rhizopus* sp. T-3 isolated from Indonesia, *Proc. Orient. Ferment. Foods,* FIRDI, Hsinchu, Taiwan, 1980, 167.
42. Hackler, L. R., Steinkraus, K. H., Van Buren, J. P., and Hand, D. B., Studies on the utilization of tempe protein by weanling rats, *J. Nutr.,* 82, 452, 1964.
43. Murata, K., Ikehata, H., Edani, Y., and Koyanagi, K., Studies on the nutritional value of tempe. II. Rat feeding test with tempe, unfermented soybeans, and tempe supplemented with amino acids. *Agric. Biol. Chem.,* 35, 233, 1971.
44. Stilling, B. R. and Hackler, L. R., Amino acid studies on the effect of fermentation time and heat-processing of tempe, *J. Food Sci.,* 30, 1043, 1965.
45. Steinkraus, K. H., Van Buren, J. P., and Hand, D. B., Studies on tempe — an Indonesian fermented soybean food, in Progress in Meeting Protein Needs of Infants and Preschool Children, Publ. No. 843, National Academy of Sciences, National Research Council, Washington, D.C., 1961.
46. Wang, H. L., Ruttle, D. I., and Hesseltine, S. W., Protein quality of wheat and soybeans after *Rhizopus oligosporus* fermentation, *J. Nutr.,* 96, 109, 1968.
47. Roelofsen, P. A., Tempe bereiding in kyij-gsgevangenschap, *Vakblad Biol.* 26, 146, 1964.
48. Steinkraus, K. H. and Van Veen, A. G., Biochemical nutritional and organoleptic changes occurring during production of traditional foods, in *Global Impacts of Applied Microbiology,* Freitas, Y. M., Fernandez, F., Eds., IBP-UNESCO, University of Bombay, Bombay, India, 1971, 444.
49. Curtis, P. R., Cullen, R. E., and Steinkraus, K. H., Microbiological Synthesis of vitamin B-12 in tempe, presented at the Symp. Indigenous Ferment. Foods, Bangkok, Thailand, 1977.
50. Wagenknecht, A. C., Mattick, L. R., Lewin, L. M., Hand, D. B., and Steinkraus, K. H., Changes in soybean lipids during tempe fermentation, *J. Food Sci.,* 26, 373, 1961.
51. Packett, L. V., Chen, L. H., and Liu, J. T., Antioxidant potential of tempe as compared to tocopherol, *J. Food Sci.,* 36, 798, 1971.
52. Iljas, N., Preservation and Shelf-life Studies of Tempe, Master's thesis, Ohio State University, Columbus, 1969.
53. Ikehata, H., Wakaizumi, M., and Murata, K., Antioxidant and antihemolytic of a new isoflavone, "factor 2" isolated from tempe, *Agric. Biol. Chem.,* 32, 740, 1968.
54. Sudarmadji, S. and Markakis, P., The phytate and phytase of soybean tempe, *J. Sci. Food Agric.,* 28, 381, 1977.

55. Gyorgy, P., The nutritive value of tempe, in Progress in Meeting Protein Needs of Infants and Preschool Children, Publ. No. 843, National Academy of Sciences, National Research Council, Washington, D.C., 1961, 182.
56. Gyorgy, P., Murata, K., and Ikehata, H., Antioxidants isolated from fermented soybeans (tempe), *Nature,* 203, 1964.
57. Gyorgy, P., Murata, K., and Sugimoto, Y., Studies on antioxidant activity of tempe oil, *J. Oil Chem. Soc.,* 51, 377, 1974.
58. Wang, H. L., Swain, E. W., Wallen, L. L., and Hesseltine, C. W., Free fatty acids identified as antitrytic factor in soybeans fermented by *Rhizopus oligosporus, J. Nutr.,* 105, 1351, 1975.
59. Shallenberger, R. S., Hand, D. B., and Steinkraus, K. H., Changes in sucrose, raffinose and stachyose during tempe fermentation, Rep. of 8th Dry Bean Res. Conf., Bellaire, Mich., 1967, 68.
60. Wang, H. L., Ruttle, D. I., and Hesseltine, C. W., Antibacterial compound from a soybean product fermented by *Rhizopus oligosporus* (33930), *Proc. Soc. Exp. Biol. Med.,* 131, 579, 1969.
61. Zamora, R. G. and Veum, J. L., The nutritive value of dehulled soybeans fermented with *Aspergillus oryzae* or *Rhizopus oligosporus* as evaluated by rats, *J. Nutr.,* 109, 1333, 1979.
62. Djurtoft, R. and Jansen, J. S., Tempe from broad beans, cowpeas, barley, wheat, or mixture thereof, presented at the Symp. Indigenous Ferment. Foods, Bangkok, Thailand, 1977.
63. Winarno, F. G., The nutritional potential of fermented foods in Indonesia, in *Traditional Food Fermentation as Industrial Resources in ASCA Countries,* The Indonesian Institute of Science (LIPI), Jakarta, 1981.
64. Van Veen, A. G. and Graham, D. C. W., Fermented peanut presscake, *Cereal Sci. Today,* 13, 96, 1968.
65. Ko, S. D., Self protection of fermented foods against aflatoxin, in Proc. Fourth Int. Congr. Food Sci. Techol.. 1974.
66. Cole, R. J. and Kirksey, J. W., Conversion of aflatoxin B_1 to isomeric hydroxy compounds by *Rhizopus* sp., *J. Agric. Food Chem.,* 20, 1100, 1972.
67. Calloway, D. H., Hickey, C. A., and Murphy, E. L., Reduction of intestinal gas-forming properties of legumes by traditional and experimental food processing methods, *J. Food Sci.,* 36, 251, 1971.
68. Gandjar, I., Fermentation of winged bean seeds-tempe kecipir, presented at the *Symp. Indigenous Ferment. Foods,* Bangkok, Thailand, 1977.
69. Gandjar, I., Fermentation of the winged bean seeds, workshop, Seminar on the Development of the Potential of the Winged Bean, Los Banos, Philippines, 1978, 330.
70. Djurtoft, R. and Nielsen, J. P., Increase in some B-vitamins including B-12 during fermentation of tempe produced from cowpeas or soybeans, *J. Plant Foods,* 5, 135, 1984.
71. Gandjar, I., Tempe gembus, presented at the Symp. Indigenous Ferment. Foods, Bangkok, Thailand, 1977.
72. Gandjar, I. and Slamet, D. S., Tempe gembus hasil fermentasi ampar tahu, *Penelitian Gizi dan Markanan,* 2, 70, 1972.
73. Bai, R. G., Prabha, T. N., Rao, T. N. R., Sreedhara, V. P., and Sreedhara, N., Studies on tempeh. I. Processing and nutritional evaluation of tempe from soybean and groundnut, *J. Food Sci. Technol.,* 12, 135, 1975.
74. Robinson, R. J. and Kao, C., Tempe and miso from chickpea, horse bean, and soybean, *Cereal Chem.,* 54, 1192, 1977.
75. Kao, C. and Robinson, R. J., Nutritional aspects of fermented foods from chickpea, horsebean and soybean, *Cereal Chem.,* 55, 512, 1978.
76. Hermana, Winarno, F. G., Suhadi H., and Nasution, A. H., Pengaruh Konsumsi Bahan Makanan Campuran Dengan Kedelai atau Tempe terhadap Anak Balita Penderita Kurang Kalori Protein, Ph.D. thesis, Institut Pertanian Bogor, Bogor, Indonesia, 1983.
77. Hesseltine, C. W., Future of fermented foods, *J. Process Biochem.,* 16, 18, 1981.

Chapter 7

FERMENTED SOYBEAN MILK AND OTHER FERMENTED LEGUME MILK PRODUCTS

D. R. Rao, S. R. Pulusani, and C. B. Chawan

TABLE OF CONTENTS

I. INTRODUCTION

Soymilk, an aqueous extract of soybeans, is thought to have been developed in China by Liu An King about 164 B.C.[1] Soymilk has been used in China as a spicy hot breakfast soup (tien tou-chiang) or as a warm, sweetened beverage.[1] Yet, the realization of nutritional value of soymilk is only a recent phenomenon. Since the protein quantity and quality of soymilk is roughly comparable to that of cow's milk it has assumed significance in infant and child nutrition in recent years. For example, a soymilk product acceptable to grade school Filipino children has been recently developed.[2]

Although the preparation of soymilk is referred to as far back as 164 B.C.,[1] fermented soymilks and other legume milks have been introduced in recent times. The first published report on the fermentation of soymilk by lactic acid bacteria appeared in 1934, and Kellogg,[3] who patented the procedure, was perhaps the first one to prepare fermented soymilk. However, as early as 1911, a Chinese plant near Paris was reportedly producing fermented soymilk among other soybean products.[1] In the 1940s, Gehrke and Weiser[4,5] published two articles in which they reported successful growth of lactic acid bacteria in soymilk. The first patent on the preparation of yogurt-like product from soymilk was issued in 1963.[6] Preparation of cheese-like products from soymilk was described in the 1960s.[7-10] Hang and Jackson[7] prepared cheese-like products from soymilk using *Streptococcus thermophilus* as a starter culture, the suitability of which was confirmed by Matsuoka et al.[10] in 1967. A semisynthetic sour milk beverage was prepared by Yamanaka et al.[11] using a mixture of soybean protein, amino acids, cow's milk, and a starter culture mix consisting of *S. thermophilus* and *Lactobacillus bulgaricus.* A series of papers was published by Angeles and Marth,[12-15] describing the growth and activity of lactic cultures in soymilk. Recently, Mital and Steinkraus[16] extensively reviewed the fermentation of soymilk by lactic acid bacteria. Also, other reports from India, Japan, Malaysia, and the U.S. appeared in the literature describing the preparation of different fermented legume milks.

No published accounts on the amounts of fermented soymilks (legume milks) consumed, quantities produced, and the economic value are available. Fermented soymilk and legume milk products such as "soyogurt", "tairu", and fermented "toned milk" are routinely consumed in the Orient, especially in China, Japan, Malaysia, and India. Soymilk was probably introduced to the West in the early part of the century. Dr. Julius F. Miller is often credited with popularizing the use of soymilk in both East and West, especially for feeding infants suffering from allergies and/or malnutrition.[1] However, the beany flavor present in the soymilk has been resented by the Westerners. Since the fermentation process often induces better flavors and/or masks disagreeable flavors, experiments were initiated at the Northern Regional Research Laboratory, U.S. Department of Agriculture,[17] to develop an acceptable product from soymilk by fermenting with lactic acid bacteria.

II. PREPARATION

A. Microbiology

Gehrke and Weiser[4,5] conducted one of the earliest studies on the growth and biochemical activities of different microorganisms grown in soymilk. Based on coagulation, proteolysis, acid production, and uniformity of cell size, it was concluded that soymilk serves as an excellent culture medium for the propagation of *L. bulgaricus, Lactobacillus acidophilus, Streptococcus liquefaciens, Streptococcus lactis, Enterobacter aerogenes,* and *Escherichia coli.*[4] In another investigation,[5] a comparative study was made of pH, titrable acidity, volatile acidity, and acetylmethylcarbinol plus diacetyl (flavor compounds) on samples of cow's milk and soymilk inoculated with the

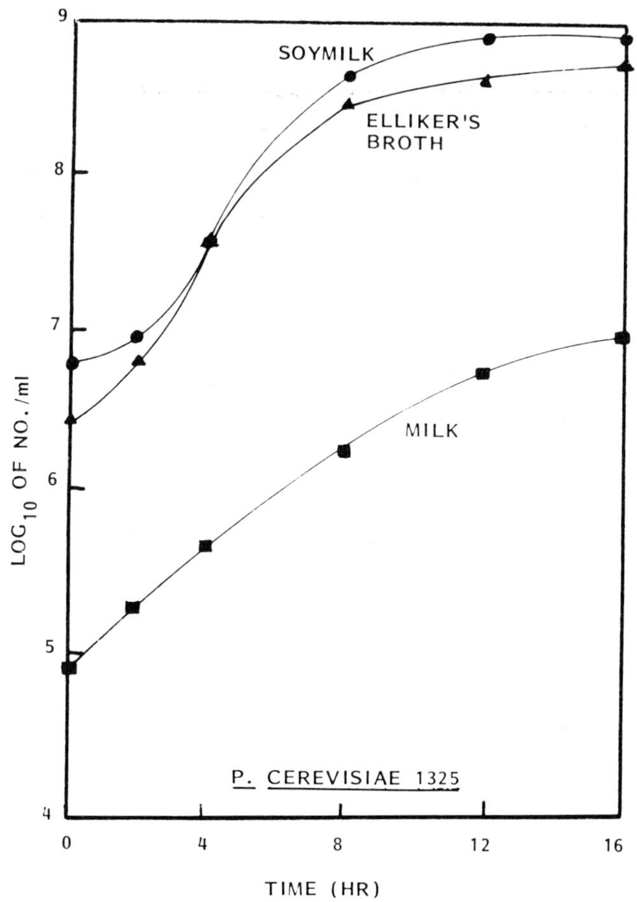

FIGURE 1. Growth of *Pediococcus cerevisiae* in soymilk, milk, and Elliker's broth (Adapted from Angeles, A. G. and Marth, E. H., *J. Milk Food Technol.*, 34, 30, 1971. With permission.)

butter culture organisms, (*S. lactis, Streptococcus citrovorous*, and *Streptococcus paracitrovorous*). Volatile acidity and pH values were comparable, while samples of cultured soymilk contained larger amounts of acetylmethylcarbinol plus diacetyl. Addition of citric acid was reported to have equal stimulatory effect on the production of these flavor compounds in cow's milk and soymilk. However, the lactic acid content was higher in cultured cow's milk than in cultured soymilk.[5]

Yamanaka and Furukawa[18] reported that acid production by *S. thermophilus, Streptococcus fecalis, L. acidophilus, L. bulgaricus,* and *Lactobacillus casei* was higher in soymilk/skim milk combinations than in skim milk alone. Supplementation of the combined milks with glucose stimulates acid production by the above lactic cultures. Sucrose, however, is ineffective in enhancing acid production by the lactic cultures with the exception of *L. acidophilus.* Matsuoka et al.[10] observed that *S. thermophilus* produces more acid in soymilk than *S. lactis* and *L. bulgaricus.* Similar observations have been reported by Kim and Shin.[19] Angeles and Marth[12-15] reported on the growth and activity of lactic acid bacteria in soymilk. They[12] observed that the growth rates of 13 species of lactic acid bacteria in sterile soymilk are generally greater than, or comparable to, those in cow's milk or Elliker's broth. Typical growth rates (*Pediococcus cerevisiae*) in different media are shown in Figure 1. *S. thermophilus, Lactobacillus delbrueckii, Lactobacillus pentosus,* and *Leuconostoc mesenteroides* are able to utilize

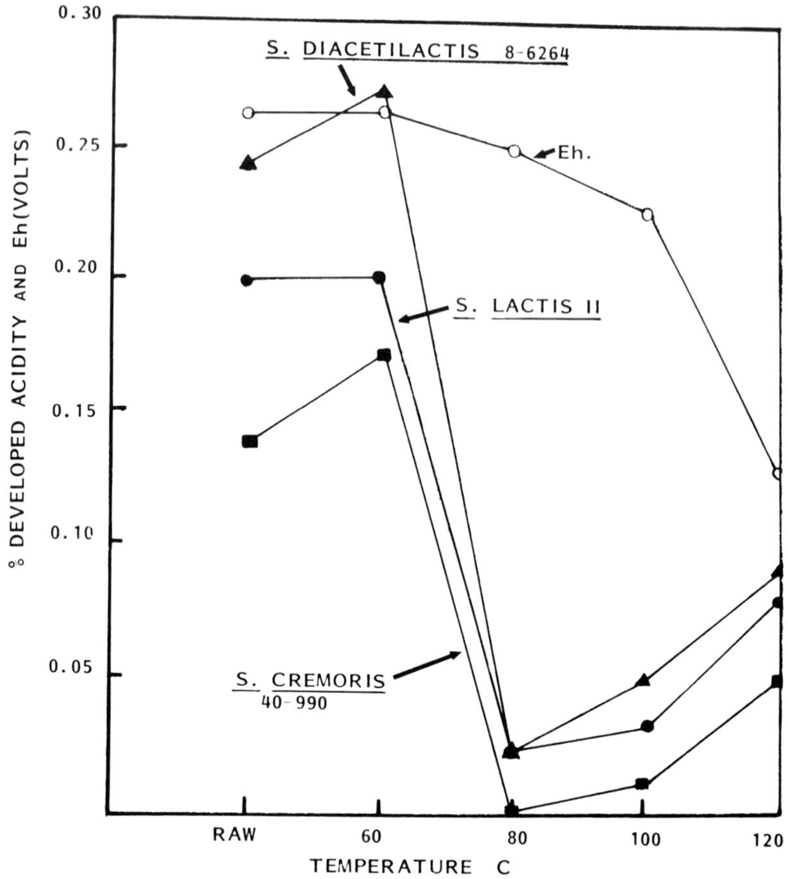

FIGURE 2. Effect of heat treatment of soymilk on subsequent acid production by *Streptococcus* species and on the redox potential of unfermented soymilk. (Adapted from Angeles, A. G. and Marth, E. H., *J. Milk Food Technol.*, 34, 63, 1971. With permission.)

sugars in soymilk resulting in substantial acid production. On the other hand, *S. lactis, Streptococcus cremoris, Streptococcus diacetylactis, L. casei,* and *Lactobacillus helveticus* require the addition of glucose, lactose, or whey powder to soymilk for enhanced acid production, while sucrose addition does not have an effect. The growth rates of *L. bulgaricus* and *L. acidophilus* were not studied. In another study,[13] the effect of heat treatment of soymilk on the growth and activity of lactic acid bacteria in soymilk was studied. Heating soymilk to 60°C for 5 min results in increased acid production by *Streptococcus* and *Leuconostoc* species and reduction in acid production by *P. cerevisiae* and *Lactobacillus* species. Extended heating however, reduces acid production by all the lactic acid bacteria. Minimal acid production occurs in soymilk heated at 80°C from 1 to 60 min. Also, heating at 100°C for short duration results in minimal acid production. However, severe heating at 100 or 120°C progressively improves the quality of soymilk as a substrate for the growth of lactic acid bacteria, and the redox potential of soymilk decreases (Figure 2). The relative amounts of sulfhydryls are the highest in soymilk heated at 80°C and they decrease upon more severe heating. Thus, beneficial effects of severe heating are attributed to expulsion of sulfhydryls and a concomitant decrease in the redox potential of the soymilk.

Angeles and Marth[14,15] also investigated the lipolytic and proteolytic activities of lactic cultures in soymilk. Ten of eleven cultures tested showed lipolytic activity in the agar-well method when triolein was the substrate, but none of the cultures were lipolytic when soybean oil was the substrate. However, *L. casei* and *L. delbrueckii* cause lipolysis in soymilk upon incubation for 14 days. *L. delbrueckii* and *S. thermophilus* are mildly proteolytic in soymilk. In these studies, *L. acidophilus* and *L. bulgaricus* were not included.

Kothari[20] studied the effect of soymilk processing factors on the acid production by four single-strain cheese cultures and two yogurt cultures. *S. cremoris*(R6), *S. diacetilactis*(DRC-H), *Lactic fermentor*(LF), and *L. bulgaricus* (Japan) produce more acid in soymilk prepared from Bragg variety, while *S. thermophilus* and *S. lactis* produce more acid in soymilk prepared from Pb-1 and Hood soybean varieties, respectively. Acid production by cheese and yogurt cultures is significantly higher in soymilk prepared from presoaked beans (maximum between 8 and 12 hr at 8°C) than in soymilk prepared from cooked or blanched beans. For example, the acid production by *S. thermophilus* for soaking, cooking, and blanching is 0.29%, 0.16%, and 0.11%, respectively. Acid production could be enhanced by addition of 1% lactose. Apparently, grinding time influences the rate of acid production by lactic cultures and the optimum grinding time is 7 min. The observations reported by Kothari[20] confirm the conclusions drawn by Angeles and Marth[13] that acid production by lactic cultures is reduced when cultured in heat-treated soymilk (Figure 2).

Wang et al.[21] studied the fermentation of soymilk by eight strains of *L. acidophilus* and four strains of *L. bulgaricus* that are often used in the dairy industry. Four strains of *L. acidophilus* (B-2178, B-2092, B-1910, B-1833) grew well (as measured by pH) in soymilk, and the remaining four strains were able to grow better when the soymilk was supplemented with 4% glucose or lactose. In general, soymilk does not adequately support the growth of *L. bulgaricus* strains. However, *L. bulgaricus* B-548 grows well in soymilk containing glucose or lactose. Supplementation of soymilk with sucrose does not enhance the growth of *L. bulgaricus* since it does not utilize sucrose as a source of energy. Mital and Steinkraus[22] also studied the growth of several selected lactic acid bacteria in soymilk processed by two different methods (Table 1). All the organisms tested exhibit more growth in hot grind soymilk than in the milk prepared from defatted soybeans. These differences in growth rates of lactic acid bacteria in different soymilks are attributed to the loss of some of the "growth factors" during solvent extraction. Many of the cultures tested (except *L. bulgaricus* and *L. fermenti*) were able to utilize sucrose for acid production. Since sucrose is the major fermentable sugar in soymilk, lactic acid bacteria possessing the ability to utilize sucrose can be successfully employed to manufacture fermented products from soymilks.[22] Apparently solvent extraction of soybeans prior to soymilk preparation decreases sucrose content, and exogenous sucrose is necessary for the growth of lactic acid bacteria. In addition to sucrose, there are other fermentable oligosaccharides (raffinose and stachyose) in soymilk.[23] Mital and Steinkraus[24] also examined the ability of several lactic acid bacteria to degrade these oligosaccharides, since some lactobacilli have been previously shown to possess α-galactosidase,[25] an enzyme necessary for hydrolyzing α-1-6 glycosidic bonds in these oligosaccharides. They[24] found that *Lactobacillus buchneri, L. fermenti, Lactobacillus cellobiosis,* and *Lactobacillus plantarum* produce significant amounts of acid in media containing raffinose and stachyose. Further, Mital and Steinkraus[24] demonstrated that the lactic acid producing bacteria also utilize galacto-oligosaccharides present in soymilk.

Kothari[26] studied the effects of addition of simple sugars, skim milk, and lactic acid whey to soymilk on the acid production by several lactic starter cultures in soymilk. He reported that sucrose substantially stimulates acid production by yogurt cultures including *L. bulgaricus* (a strain obtained from Japan), but not by cheese cultures.

Table 1
GROWTH AND ACID PRODUCTION BY LACTIC ACID ORGANISMS IN COW'S MILK AND SOYMILKS[22]

Organism	Cow's milk			Soymilk 1 (hot grind)			Soymilk 2 (defatted beans)		
	Viable counts/mℓ[a]	pH[b]	%TA[b]	Viable counts/mℓ[a]	pH[b]	%TA[b]	Viable counts/mℓ[a]	pH[b]	%TA[b]
Initial range		6.40	0.16—0.18		6.40	0.16—0.18		6.65	0.10—0.12
S. thermophilus	4.30×10^8	4.46	0.62	1.65×10^8	4.56	0.47	1.06×10^8	4.65	0.43
L. acidophilus	1.10×10^8	5.95	0.26	1.74×10^9	4.70	0.43	9.40×10^8	4.90	0.31
L. bulgaricus	1.10×10^9	3.98	0.67	3.40×10^6	6.32	0.21	2.40×10^6	6.28	0.12
L. buchneri	2.39×10^7	6.30	0.18	2.80×10^8	5.48	0.34	1.00×10^8	5.95	0.20
L. cellobiosis	6.68×10^6	6.34	0.18	6.30×10^8	4.80	0.49	6.00×10^7	5.80	0.25
L. fermenti	1.18×10^7	6.35	0.19	4.40×10^8	5.75	0.27	1.82×10^7	6.20	0.14
L. plantarum	2.00×10^7	6.28	0.19	1.84×10^9	4.35	0.49	5.70×10^8	5.25	0.24
S. thermophilus + *L. acidophilus*	2.02×10^8	4.56	0.61	3.40×10^8	4.28	0.63	1.64×10^8	4.20	0.61
S. thermophilus + *L. bulgaricus*	5.60×10^8	4.0	0.71	6.20×10^7	4.26	0.47	7.20×10^7	4.41	0.48
S. thermophilus + *L. buchneri*	1.78×10^8	4.60	0.67	1.46×10^8	4.42	0.42	1.54×10^8	4.34	0.39
S. thermophilus + *L. cellobiosis*	1.70×10^8	4.65	0.69	2.45×10^8	4.50	0.49	8.27×10^7	4.50	0.41
S. thermophilus + *L. fermenti*	1.73×10^8	4.70	0.61	3.40×10^8	4.50	0.44	2.50×10^8	4.48	0.38

Note: Average of duplicate experiments.

[a] The viable counts represent the maximum attained during the 16-hr incubation period.
[b] Titratable acidity and pH were determined after 16 hr.

Cow and buffalo skim milks at 5 and 10% levels are good substitutes for lactose. While it is generally agreed that *L. bulgaricus* does not utilize sucrose, Ariyama[6] reported fermentation of sucrose by *L. bulgaricus.* In 1977, Stern et al.[27] tested the ability of eight strains of *L. acidophilus* (ARS culture collection) to produce acid from fructose, galactose, glucose, mannose, lactose, maltose, sucrose, raffinose, and stachyose. They demonstrated very wide strain variation with respect to the utilization of sugars. For example, the acid production (mℓ of 0.025 *N* NaOH required to neutralize 5.0 mℓ sample) varied from a low of 3.0 to a high of 13.4 when the sucrose was the substrate. Strain NRRL-B-1910 is a superior utilizer of raffinose and stachyose. In contrast, Mital and Steinkraus[24] reported that *L. acidophilus* (ATCC No. 4356) does not utilize raffinose and stachyose.

The growth and activity of four lactic acid bacteria in peanut milk was reported by Beuchat and Nail.[28] *L. bulgaricus* B-1909 and *L. acidophilus* B-1910 (presumably the same used by Stern et al.[27]) grows most rapidly, producing the greatest amount of acid in peanut milk. Maximum acid production (0.38%) occurs between 48 and 72 hr of fermentation in 2% lactose supplemented milk inoculated with *L. bulgaricus* B-1910. For the same period, *L. acidophilus* B-1910 produces 0.53% titrable acidity in peanut milk. Viable cell counts are the highest after 17 hr of fermentation, reaching maximum numbers of 7.3×10^8 cells/mℓ for *L. bulgaricus* B-1909 and 9.9×10^7 cells/mℓ for *L. acidophilus* B-1910; however, after 72 hr, viable cell counts decrease to 1.5×10^6 and 9.3×10^4, respectively. Inoculation with 1:1 combination of these two strains does not result in any higher acid production. They further found that all test strains do grow in peanut milk not fortified with carbohydrates. The growth of *L. bulgaricus* B-548 and *L. acidophilus* B-2092 is not affected by 2% supplemental glucose, sucrose, or lactose; however the growth of the strains B-1909 and B-1910 is stimulated by glucose and lactose. After several transfers of bacteria in peanut milk maintenance medium, the organisms have a shorter lag time once inoculated into peanut milk. Smoot et al.[29] studied the fermentation of soy and winged bean milks by *L. acidophilus, L. bulgaricus, L. plantarum,* or *S. thermophilus.* All four lactic acid bacteria grow well in both test milks with adequate acid production (pH 5.2) to produce a firm gelatinous curd. Maximum cell numbers of 10^8 or greater are attained within 25 to 36 hr of incubation. Recently, the effect of blanching on the soybeans prior to soymilk preparation, and fortification of soymilk with lactose, citrate, and whey on the growth of LF-40 (a lactic culture essentially consisting streptococci), *S. diacetilactis* (DRC 1), *S. thermophilus* (HST), and *L. bulgaricus* (W) was investigated.[30] The results reported by these workers are, in general, in agreement with those of the previous investigators and indicate that cultured whey-soymilk has a slightly weaker consistency, but better flavor, than cultured lactose-enriched soymilk.

B. Processing and Sensory Qualities

Recently Reddy et al.[31] reviewed methods of preparing fermented legumes, including soymilk and other fermented legume milk products. Even though soymilk was produced in the West as early as 1911, the increase in popularity of soymilk and its products in this area is a recent phenomenon. Further, the increase in popularity of soymilk in the West has been limited because of the "objectionable" beany flavor and the flatulence caused by the high content of oligosaccharides present in soybeans. Several approaches have been taken to improve the organoleptic qualities and to reduce the oligasaccharides content of soymilk. These approaches include addition of sugar, oil stabilizer, and fermentation with lactic cultures.

Hang and Jackson[7,8] were among the earliest to prepare lactic fermented soymilk products. They attempted to prepare a product with better palatability, texture, and shelf-life than tofu and conventional soy cheese. Using *S. thermophilus* 101 (Kilzade products, Wisconsin) as the lactic culture, they prepared soy cheese containing less moisture than that produced by calcium sulfate precipitation and possessing better body and texture than cheese produced conventionally. Addition of skim milk (15% dry weight of soymilk/skim milk mixture) and rennet improves the flavor and texture of soy cheese and considerably reduces the time required for the manufacture of curd. Attempts by Shroder and Jackson,[32] however, failed to further improve the flavor by the addition of skim milk at 25, 50, and 75% of total dry weight. Obviously, the dominating heavy flavor in the finished cheese persisted. Mold ripening results in desirable changes in texture, but there is a simultaneous development of bitter flavors. Thus, the suggested efforts to improve the texture and body by increased proteolysis results in the development of bitter and unacceptable flavors. On the other hand, Obara[9] treated the soybean curd obtained by salt precipitation with proteolytic enzymes prior to fermentation with *S. cremoris* and *S. lactis* to obtain a higher rate of ripening and more desirable flavors. A combination of papain bioprase and pronase is more effective when compared to the other proteolytic enzymes. Obara[9] suggested modifications in the traditional cheese-making process to avoid bitter flavors in soy cheeses.

A sour milk beverage consisting of soybean protein, milk, and sucrose was prepared by Yamanaka et al.[11] They used *S. thermophilus* and *L. bulgaricus* as the lactic starters. Addition of an amino acid mixture (alanine, arginine, aspartic acid, methionine, glycine, lysine, and sodium glutamate) masks the beany flavor. Yamanaka and Furukawa[18] produced a harder curd by increasing the proportion of soymilk in skim milk/soymilk combination fermented by different lactic organisms.

In 1974 Wang et al.[21] reported the preparation of yogurt-like product from soymilk. The product prepared by fermenting soymilk with the *L. acidophilus* (B-1910) is white in color, has a sour but not harsh flavor and a smooth custard-like texture. However, whey formation is a problem, indicating the possible need for a stabilizer. When 4% sucrose is added to soymilk prior to fermentation, the yogurt-like product displays a pleasant sour sweetish flavor (Table 2).

Since then many attempts have been made to reduce the beany flavor of soymilk: modifying the preparation of soymilk, lactic fermentation, and addition of flavor compounds. Kanda et al.[17] developed a method in which soybeans are boiled in 0.2% $NaHCO_3$ prior to blending, followed by ultrasonic treatment. A yogurt-like product is obtained by fermenting the soymilk with 1:1 mixture of *L. acidophilus* strains B-1910 and B-2092. The outline of yogurt production from soymilk is shown in Figure 3. Several factors affect the quality of soy yogurt. For example, yogurt made from soymilk sterilized at 90°C has desirable taste, color, and titrable acidity, but a very soft texture. Sterilization of soymilk at 121°C yields yogurt with a darker color and harder texture. A suggested temperature for sterilization of milk is 100°C for either 10 or 20 min. Additionally, yogurt made from soymilk containing 3.6 and 4.5% protein has a more desirable texture than yogurt made from soymilk containing lower levels of protein. Acidity of the final product also decreases as the protein content of the soymilk decreases. Based on both acidity and hardness, Kanda et al.[17] suggested an optimum protein level of 3.6 to 4.5% in soymilk. When gelatin is used, whey separation in yogurt is prevented. Lemon flavor is very complementary to soy yogurt. In taste panel studies, members gave higher quality-rating scores to yogurt made from soymilk than yogurt made from traditional milk (6.0 vs. 4.5 on a 10-point scale where 1 = repulsive and 10 = excellent). Apparently, this yogurt can be stored at 5°C for up to 19 days without significant change in acceptability. The effect of lactic fermentation and addition of fruit flavors on the sensory qualities of yogurt prepared from soybeans has been reported.[22,33] Mital and Steinkraus[22] compared fermented and lactic-fermented soymilks to the flavor acceptability of fermented cow's milk. The fermented soymilks

Table 2
FLAVOR AND ODOR EVALUATION OF *L. ACIDOPHILUS* B-1910 FERMENTED SOYMILK AND CONTROL SOYMILK BY TASTE PANEL

	Sensory response of 14 tasters (%)	
Description	Control milk	Fermented milk
Odor		
Grassy/beany	36	7
Acid/sour	—	29
Raisins	14	57
Flavor		
Grassy/beany	50	7
Acid/sour	14	93
Sweet	64	29
Raisins	14	36
Astringent	21	29

Adapted from Wang, H. L., Kraidej, L., and Hesseltine, C. W., *J. Milk Food Technol.*, 37, 71, 1974.

have a satisfactory gelatinous curd and an acceptable flavor. However, the taste panelists prefer the higher acidity of the fermented cow's milk. Schmidt and Bates[33] prepared yogurt by inoculating soymilk (Illinois process) with a 10% inoculum of commercial yogurt culture (*L. bulgaricus* and *S. thermophilus*). Fruit flavorings were also added. A final pH of 4.7 gave a uniformly textured product. However, increasing the total solids and addition of fruit flavors is necessary to improve the sensory quality of the yogurt made from soymilk. Similar yogurt-like products were produced by Stern et al.[27] and Schmidt et al.[34]

Published accounts indicate that fermented soymilk, a product analogous to "tairu" made from cow's milk may become popular in Malaysia. A method for the preparation of tairu as reported by Ahmad[35] is outlined in Figure 4. Fermented soybean protein foods, as described elsewhere in this book, have been traditional in Japan and the Orient. However, Fukushima[36] pointed out that many investigations have been made in Japan regarding the preparation of cheese and yogurt analogues from vegetable protein foods. While most cheese products may not be commercial, a fermented soymilk drink recently made its way into the Japanese market. The manufacturing process of this fermented soymilk drink is shown in Figure 5. According to available information, the beany flavors in soymilk disappear upon fermentation by lactic acid bacteria and addition of oil, sugar, and stabilizers. The majority of lactic cultures used in the preparation of this Japanese fermented soymilk beverage are *L. casei, L. acidophilus,* and *L. bulgaricus.* In England, Pinthong et al.[37-39] developed a soy-based yogurt product. They were able to produce a firm homogeneous curd with little separation of whey using *L. bulgaricus* as the inoculum for soymilk. However, supplementation with 1% glucose and 0.1% yeast extract is essential for optimum acid production by *L. bulgaricus.* In Kenya, Nout and Williams[40] prepared fresh and fermented soymilk products using dairy whey as a supplement (Figure 6). They reported that 1 kg of dry soybeans

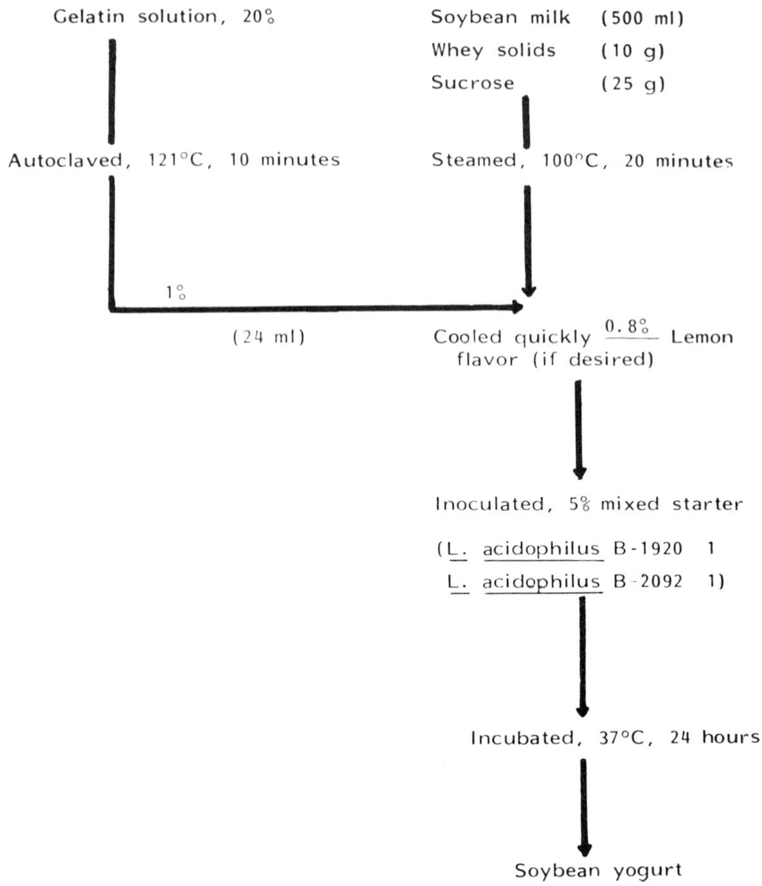

FIGURE 3. Outline for yogurt production from soymilk. (Adapted from Kanda, H., Wang, H. L., Hesseltine, C. W., and Warner, K., *Process Biochem.*, 11, 23, 1976. With permission.)

yields 2 kg of fresh cheese and 1 kg of pressed cheese. Cheese samples stored at 12°C are very acceptable in appearance and color for the first 3 weeks and then start to develop darker color with slime and putrefaction. The cheese has a slight beany flavor with a slightly soapy taste.

At Pantnagar Soy Dairy in India, soy curd resembling "Dahi" has been prepared[41] using a mixture of *L. acidophilus* and *S. thermophilus* as the lactic fermenters. Orlowski et al.[42] studied the effect of formulation and processing variables on the quality of soy yogurt. Preferred samples contain sucrose and cerelose at 3% level added prior to fermentation.

Other fermented products made from soymilk include, soy acidophilus (acidophilus soymilk), soymilk kefir, soymilk piima, soymilk buttermilk, soymilk koumiss, and soy cheeses.[43]

III. NUTRITIONAL VALUE

Information on the nutritional value of fermented soymilk is scanty. Ahmad[35] reported the nutritive composition of "tairu" made from cow's milk and soymilk (Table 3). The protein content in soymilk (3.6%) is higher than that of the cow's milk "tairu" (2%). Carbohydrates (5.9%) are also higher in soymilk "tairu" than in cow's milk

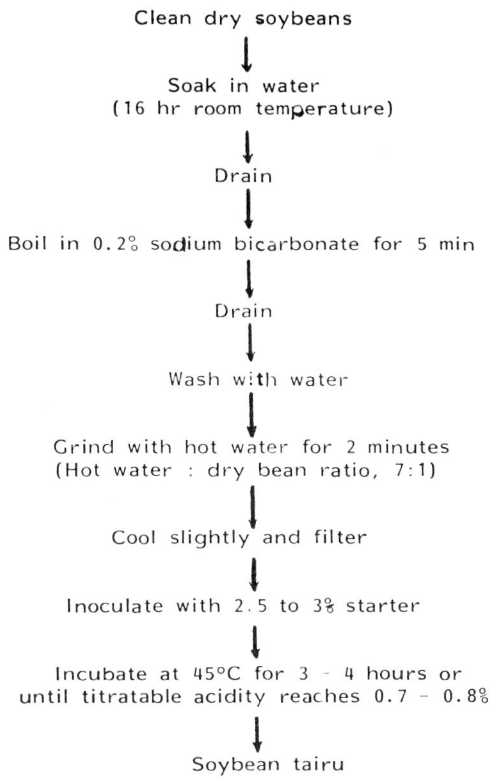

FIGURE 4. Flow sheet for production of malaysian soybean milk tairu. (Adapted from Ahmad, I. H., 1977.)

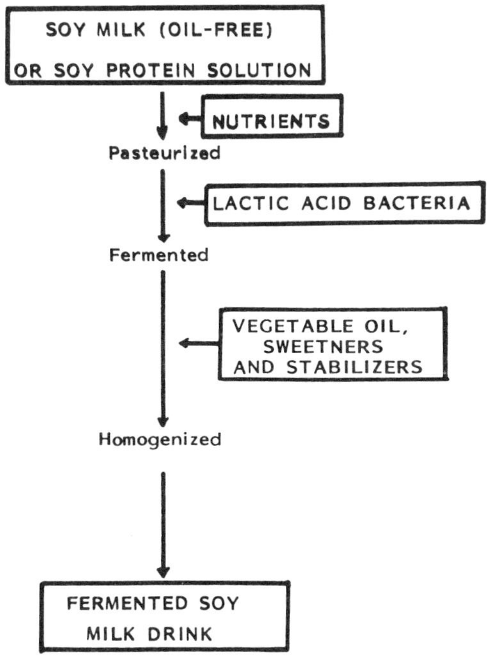

FIGURE 5. Manufacturing process of fermented soymilk drink. (Adapted from Fukushima, D., *J. Am. Oil Chem. Soc.*, 56, 357, 1979. With permission.)

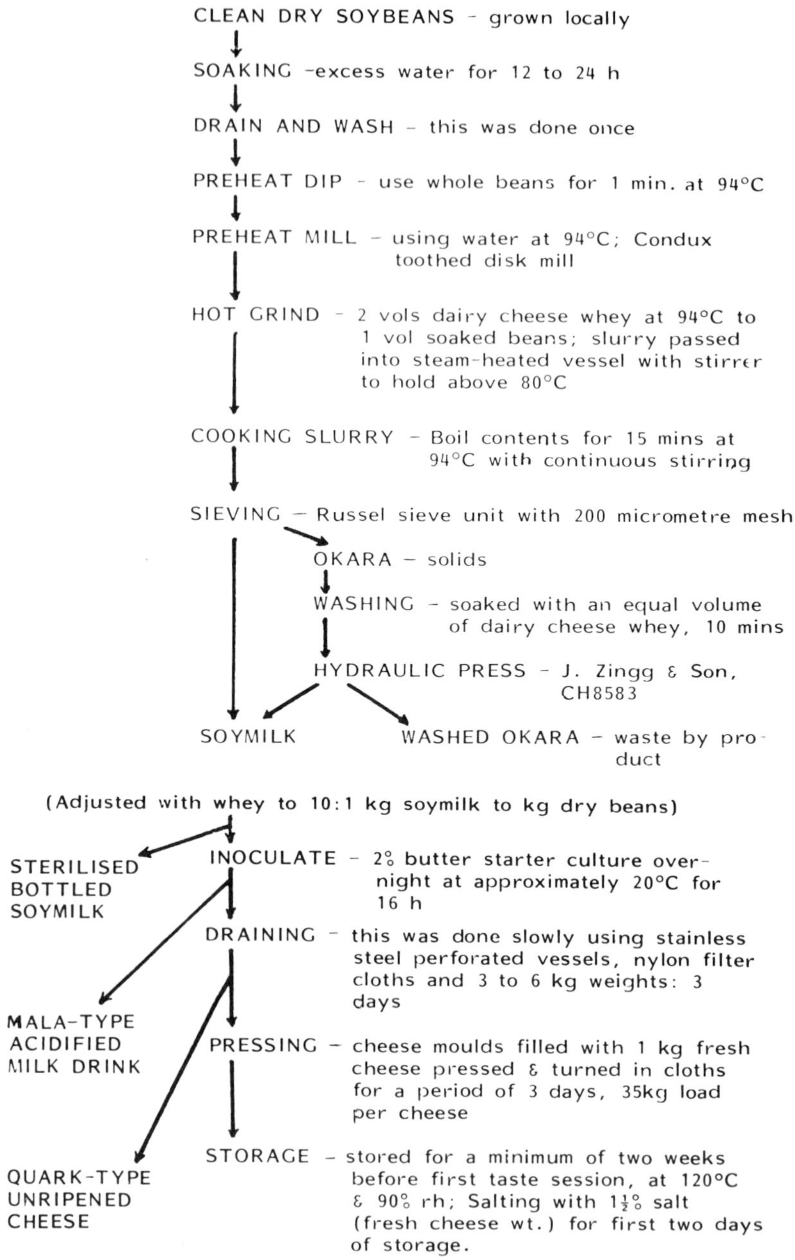

FIGURE 6. Flow sheet of soymilk process and fermented soymilk products. (Adapted from Nout, M. J. R. and Williams, A. I., 1983.)

"tairu" (2.9%). However, calcium and phosphorus are substantially less in soymilk "tairu" than in cow's milk "tairu". Changes in the amino acid composition due to lactic fermentation of soymilk were reported by Rao et al.[44] Upon fermentation of soymilk by a mixture of cultures consisting of *S. thermophilus, L. acidophilus,* and *L. bulgaricus,* the essential amino acid content increases and there are notable increases in lysine, histidine, and phenylalanine concentrations. Also, there is some increase in the concentration of methionine. Further work is needed in this area. Sands and Hankin[45] prepared soy yogurt using a lysine-excreting strain of *L. acidophilus.* They ob-

Table 3
NUTRITIVE COMPOSITION OF COW'S MILK AND SOYMILK TAIRU[35]

Nutrients	Soymilk tairu (per 100g)	Cow milk tairu (per 100g)
Calories	49.1	50.1
Water (g)	88.2	90.5
Fat (g)	1.5	2.9
Protein (g)	3.6	2.0
Carbohydrate (g)	5.9	2.9
Ash (g)	0.81	0.7
Calcium (mg)	95.0	280.0
Phosphorus (mg)	160.0	220.0
Total solids (g)	11.8	9.5
pH	4.5	4.3

served a 14% increase in lysine content upon fermentation. Since methionine is the limiting amino acid in soybean, identification of strains of lactic cultures synthesizing methionine would be of practical significance.

IV. OTHER FERMENTED LEGUME MILK PRODUCTS

The earliest scientist to suggest the preparation of peanut milk was perhaps George Washington Carver. In India, production of toned peanut milk called "Miltone" is a commercial reality. Miltone consists of peanut milk extended with buffalo milk. As early as 1967 a yogurt-like product (Dahi) has been successfully prepared from Miltone using lactic cultures at the Central Food Technological Research Institute (CFTRI), India.[46] According to reports from CFTRI,[46] some 3.6 million liters of Miltone are produced annually and a significant portion of this is utilized for the preparation of yogurt by lactic fermentation. In the U.S., Beuchat and Nail[47] attempted to prepare yogurt-like products from peanut milk. The preparation of fermented peanut milk as described by Beuchat and Nail[47] is shown in Figure 7. Although an acceptable custard-like texture develops in peanut milk fermented with *L. bulgaricus* B-1909 and *L. acidophilus* B-1910, excessive whey makes the products somewhat undesirable as yogurt substitutes. They demonstrated that fermented peanut milk can be substituted for buttermilk in a corn muffin recipe without alteration of sensory characteristics. Bucker et al.[48] tested 19 lactic cultures and reported that 6 cultures could produce lactic acid in peanut milk.

Smoot et al.[29] developed a lactic fermented beverage from winged bean milk. Sensory studies indicated that both American and Oriental taste panels preferred fermented soymilk over the fermented winged bean milk.

Recently a yogurt-like product was prepared from the cowpea and mung bean milks.[49] Data indicate that products simulating yogurt can be prepared from these milks successfully. In these studies it was found that protein extractability from both the legumes is low and yogurt from cowpea and mung bean are ranked slightly inferior to the control yogurt.

V. CONCLUSIONS AND FUTURE RESEARCH

Many attempts have been made in the past few decades to produce fermented soymilk and other fermented legume milk products. A number of lactic acid bacteria grow and produce flavor compounds in legume milks, especially soymilk. In the West, the

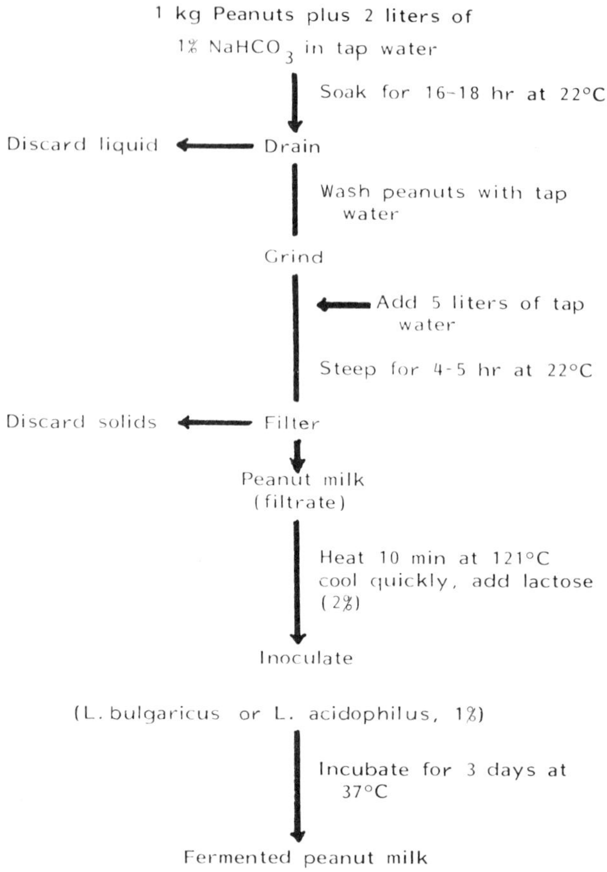

FIGURE 7. Flow diagram describing the preparation of fermented peanut milk. (Adapted from Beuchat, L. R. and Nail, B. J., *J. Food Sci.*, 43, 1109, 1978. With permission.)

popularity of soymilk and its products has been limited because of the "objectionable beany flavor" and, to some extent, the flatulence caused by the high content of oligosaccharides present in soybeans. Lactic fermentation of soymilk is a promising approach in alleviating these problems. However, further research is needed in improving the flavor of the lactic fermented soymilk and other legume milk products. Basic information on the production of flavor compounds, breakdown of oligosaccharides and related microbiology is necessary to solve these problems. Another area of future research should be in the nutritional value of fermented soymilk and other legume milk products. There are indications that lactic fermentation of soymilk improves the amino acid profile. Clearly more work is needed in this area. Other legume milks should be studied for their potential for lactic fermentation since in many parts of the world different legumes, especially cowpeas, are extensively consumed.

REFERENCES

1. Shurtleff, W., Soyfoods pioneers, Dr. Harry Miller: Taking soymilk around the world, *Soyfoods*, 1, 28, 1981.
2. Steinkraus, K. H., David, L. T., Ramos, L. J., and Banzon, J., Development of flavored soymilks and soy/coconut milks for the Philippine market, *Philipp. Agric.*, LII, 268, 1968.
3. Kellog, J. H., Method of making acidophilus milk, U.S. Patent 1, 982, 941, 1934.
4. Gehrke, C. W. and Weiser, H. H., Comparative studies on growth and biochemical features of microorganisms grown in cow's and soybean milk, *Food Res.*, 12, 360, 1947.
5. Gehrke, C. W. and Weiser, H. H., A comparative study of the biochemical activity of *Streptococcus lactis, Streptococcus citrovorus,* and *Streptococcus paracitrovorus* when grown in cow's milk and soybean milk, *J. Dairy Sci.*, 31, 213, 1948.
6. Ariyama, H., Process for the manufacture of synthetic yogurt from soybean, U.S. Patent 3,096,177, 1963.
7. Hang, Y. D. and Jackson, H., Preparation of soybean cheese using lactic starter organisms. I. General Characteristics of the finished cheese, *Food Technol.*, 21, 95, 1967.
8. Hang, Y. D. and Jackson H., Preparation of soybean cheese using lactic starter organisms. II. Effects of addition of rennet extract and skim milk, *Food Technol.*, 21, 97, 1967.
9. Obara, T., Basic investigations on the development of foods from enzymatically treated soybean protein concentrates to increase use of United States Soybeans in Japan, U.S. Department of Agriculture, Final Tech. Rep. PL 480, Project UR-A11-(40) 26, 1968.
10. Matsuoka, H., Sasago, K., and Sekiguchi, M., Manufacturing of a cheese-like product from soybean milk, *J. Food Sci. Technol.*, Tokyo, 15, 103, 1967.
11. Yamanaka, Y., Okamura, O., and Hasegawa, Y., Method of preparing a sour milk beverage, U.S. Patent 3,535,177, 1970.
12. Angeles, A. G. and Marth, E. H., Growth and activity of lactic acid bacteria in soymilk, *J. Milk Food Technol.*, 34, 30, 1971.
13. Angeles, A. G. and Marth E. H., Growth and activity of lactic acid bacteria in soy milk. II. Heat treatment of soymilk and culture activity, *J. Milk Food Technol.*, 34, 63, 1971.
14. Angeles, A. G. and Marth, E. H., Growth and activity of lactic acid bacteria in soymilk. III. Lipolytic activity, *J. Milk Food Technol.*, 34, 69, 1971.
15. Angeles, A. G. and Marth, E. H., Growth and activity of lactic acid bacteria in soymilk. IV. Proteolytic activity, *J. Milk Food Technol.*, 34, 123, 1971.
16. Mital, B. K. and Steinkraus, K. H., Fermentation of soymilk by lactic acid bacteria — a review, *J. Food Protection*, 42, 895, 1979.
17. Kanda, H., Wang, H. L., Hesseltine, C. W., and Warner, K., Yogurt production by lactobacillus fermentation of soybean milk, *Process Biochem.*, 11, 23, 1976.
18. Yamanaka, Y. and Furukawa, N., Studies on utilization of soybean protein for food manufacturing. II. Influence of soymilk added to skim milk on the acidity and the hardiness of curd produced by lactic bacteria for dairy use, *J. Food Sci. Technol.* (Tokyo), 17, 456, 1970.
19. Kim, C. S. and Shin, S. H., Studies on preparation of cheese-like product from soybean milk, *Korean J. Food Sci. Technol.*, 3, 57, 1971.
20. Kothari, S. L., Studies on fermented soymilk products. II. Effect of soymilk processing factors on the acid production by lactic acid bacteria in soymilk, *Indian J. Microbiol.*, 13, 109, 1973.
21. Wang, H. L., Kraidej, L., and Hesseltine, C. W., Lactic acid fermentation of soybean milk, *J. Milk Food Technol.*, 37, 71, 1974.
22. Mital, B. K. and Steinkraus, K. H., Growth of lactic acid bacteria in soymilks, *J. Food Sci.*, 39, 1018, 1974.
23. Kawamura, S., Review of PL-480 work on soybean carbohydrates, Proc. Int. Conf. on Soybean Protein Foods, Peoria, Ill. ARS-71-35, U.S. Department of Agriculture, 1967.
24. Mital, B. K. and Steinkraus, K. H., Utilization of oligosaccharides by lactic acid bacteria during fermentation of soymilk, *J. Food Sci.*, 40, 114, 1975.
25. Mital, B. K., Shallenberger, R. S., and Steinkraus, K. H., α-galactosidase activity of lactobacilli, *Appl. Microbiol.*, 26, 783, 1973.
26. Kothari, S. L., Studies on fermented soymilk products. III. Activity of cheese and yogurt cultures in soymilk and fortified soymilk, *Indian J. Microbiol.*, 15, 18, 1975.
27. Stern, N. J., Hesseltine, C. W., Wang, H. L., and Konishi, F., *Lactobacillus acidophilus* utilization of sugars and production of a fermented soybean product, *Can. Inst. Food Sci. Technol. J.*, 10, 197, 1977.
28. Beuchat, L. R. and Nail, B. J., Fermentation of peanut milk with *Lactobacillus bulgaricus* and *Lactobacillus acidophilus, J. Food Sci.*, 43, 1109, 1978.
29. Smoot, L. A., Garcia, V. V., and Pierson, M. D., Fermentation of soy and winged bean milks by lactic acid bacteria, *J. Food Prot.*, 42, 839, 1979.

30. Patel, A. A. and Gupta, S. K., Fermentation of blanched-bean soymilk with lactic cultures, *J. Food Prot.*, 45, 620, 1982.
31. Reddy, N. R., Pierson, M. D., Sathe, S. K., and Salunkhe, D. K., Legume-based fermented foods: their preparation and nutritional quality, *CRC Critical Rev. Food Sci. Nutr.*, 17, 335, 1982.
32. Schroder, D. J. and Jackson, H., Preparation of soybean cheese using lactic starter organisms. 3. Effects of mold ripening and increasing concentrations of skim milk solids, *J. Food Sci.*, 36, 22, 1971.
33. Schmidt, R. H. and Bates, R. P., Sensory acceptability of fruit flavored oilseed milk formulations, *Proc. Fla. State Hort. Soc.*, 89, 217, 1976.
34. Schmidt, R. H., Mathews, R. F., and Davidson, S. M., Sensory acceptance of tomato salad type yogurt systems from oilseed/dairy combinations, *Proc. Fla. State Hort. Soc.*, 90, 201. 1977.
35. Ahmad, I. H., Malaysian tairu, presented at the Symp. Indigenous Ferment. Foods, Bangkok, Thailand, 1977.
36. Fukushima, D., Fermented vegetable (soybean) protein and related foods of Japan and China, *J. Am. Oil Chem. Soc.*, 56, 357, 1979.
37. Pinthong, R., Macrae, R., and Rothwell, J., The development of soya-based yogurt. I. Acid production of lactic acid bacteria, *J. Food. Technol*, 15, 647, 1980.
38. Pinthong, R., Macrae, R., and Rothwell, J., The development of a soya-based yogurt. II. Sensory evaluation and analysis of volatiles, *J. Food Technol*, 15, 653, 1980.
39. Pinthong, R., Macrae, R., and Dick, J., The development of a soya-based yogurt. III. Analysis of oligosaccharides, *J. Food Technol.*, 15, 661, 1980.
40. Nout, M. J. R. and Williams, A. I., The use of dairy whey in the preparation of fresh and fermented soymilk products: preliminary investigations, *The Production Preservation and Processing of Food*, Vol 1, McLoughlin, J. V. and Mckenna, B. M., Eds., Boole Press, Dublin, 1983, 135.
41. Singh, S., Potential of soy protein improving Indian diet, in Int. Soya Protein Food Conf. Proc., Singapore, January 25 to 27, 1978.
42. Orlowski, J. K., Nelson, A. I., and Wei, L. S., Effect of formulation and processing variables on the quality of soybean yogurt, in World Soybean Res. Conf. II, March 26 to 29, North Carolina State University, 1979.
43. Shurtleff, W. and Aoyagi, A., Tofu and Soymilk Production, *The Book of Tofu, II*, New-Age Foods Study Center, Lafayette, Calif., 1979, 242.
44. Rao, D. R. and Pulusani, S. R., Amino acid composition and nutritional implications of skim milk and soy milk fermented by various lactic cultures, Institute of Food Technologists 82 Program, 42nd Ann. IFT Meet., Las Vegas, June 22 to 25, 1982.
45. Sands, D. C. and Hankin, L. Fortification of foods by fermentation with lysine-excreting mutants of lactobacilli, *J. Agric. Food Chem.*, 24, 1004, 1976.
46. Swaminathan, M. and Parpia, H. A. B., Milk substitutes based on oil seeds and nuts, *World Rev. Nutr. Diet*, 8, 184, 1967.
47. Beuchat, L. R. and Nail, B. J., Fermentation of peanut milk with *Lactobacillus bulgaricus* and *Lactobacillus acidophilus*, *J. Food Sci.*, 43, 1109, 1978.
48. Bucker, E. R., Jr., Mitchell, J. H., Jr., and Johnson, M. B., Lactic fermentation of peanut milk, *J. Food Sci.*, 44, 1934, 1979.
49. Rao, D. R., Chawan, C. B., and Pulusani, S. R., Preparation of yogurt-like product from cowpea and mung beans, 81st Ann. Meet., Southern Assoc. of Agric. Scientists, Nashville, Tenn., February, 1984.

Chapter 8

ONCOM (FERMENTED PEANUT PRESS CAKE)

L. R. Beuchat

TABLE OF CONTENTS

I. INTRODUCTION

Fermented peanut press cake (oncom, ontjom, lontjom) is a very popular product in Indonesia.[1,2] Although it has been consumed for centuries, biomodifications which contribute to nutritional and organoleptic changes during fermentation are far from being understood. However, interested researchers at the U.S. Department of Agriculture in Peoria, Ill., the New York State Agricultural Experiment Station, Cornell University, Geneva, N.Y., the Tropical Products Institute in England, and in our laboratory at the Agricultural Experiment Station, University of Georgia, have explored some of the changes which occur as a result of mold growth and aging of oncom. This chapter presents a description of the general procedure for preparing oncom and summarizes information describing the complex changes resulting from fermentation.

II. FERMENTATION PROCESS

A flow sheet illustrating a general procedure for preparing oncom is shown in Figure 1. After oil has been extracted from peanuts, the press cake, called boongkil, is broken up and soaked in water for about 1 day.[3] Press cake from industrial operations is low in residual oil content, while village products contain considerable amounts, e.g., up to 20%. Oil which rises to the surface of the soak water is removed, and the washed press cake is then drained and gently pressed to remove excess water. The press cake is combined with coarsely sieved cassava press cake or residue from soybeans from which soybean milk has been prepared and steamed for 1 to 1.5 hr. The cooled press cake is then inoculated by mixing with powdered oncom from a previous batch and transferred to a mold to form a flat cake about 2 to 3 cm in thickness. The molds (10 × 20 cm) are constructed of bamboo wood. Banana or other suitable leaves are often used to wrap the press cake for a 1- to 2-day fermentation period. This procedure retards the loss of moisture but does not establish anaerobic conditions. Oxygen is necessary to induce desirable growth patterns of the molds used for the fermentation. The optimum temperature for development of characteristic flavor, texture, aroma, and color is 25 to 30°C.

The mold of choice for preparing oncom is *Neurospora intermedia (N. sitophila).* This mold produces orange to orange-pink conidiospores, thus resulting in a product with a very attractive colored surface. *Rhizopus oligosporus,* the mold traditionally used to ferment soybeans in the preparation of tempe, is also used to make oncom. This mold produces black spores, resulting in ash-grey colored oncom. Both molds are readily available as inocula from earlier preparations of oncom. *Neurospora* is reported to be a very common mold in woody material under tropical rain forest conditions shortly after sterilization or pasteurization has occurred, such as in the burning of forest land.[4] Presumably, the Indonesian villager would have little problem when independently initiating a new batch of oncom.

The addition of a readily utilizable source of starch, such as cassava, potato, or potato peels, appears to be very beneficial in the fermentation process.[5,6] Cassava has also been reported to promote growth of *Rhizopus* on peanuts.[7] Sodium chloride, as well as some other ionizable salts, promotes the release of mycelium-bound proteinase and therefore increases proteinase activity.[8] Salt may exert an osmotic or ionic effect on mycelia, whereby extracellular enzymes are more readily freed to act upon peanut press cake constituents.[9]

The flavor of oncom has been described as fruity and somewhat alcoholic; however, the fried product takes on a mince-meat or almond flavor.[7] The accumulation of free amino acids and peptides in fermented peanut press cake results in the meaty flavor characteristic of other fermented oilseed products, including miso and tempe. Flavor

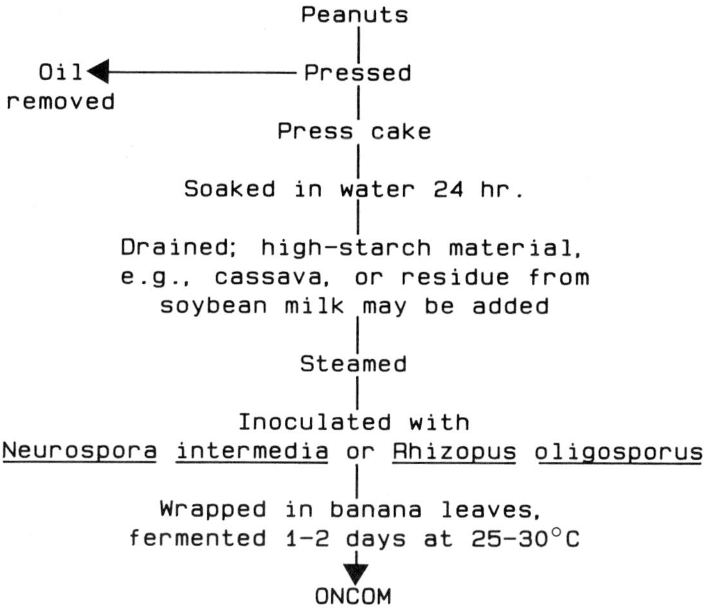

FIGURE 1. Flow sheet for preparation of oncom.

FIGURE 2. Oncom.

development in oncom is in part attributable to lipase activity of molds. While data have not been published correlating the presence or type of particular fatty acids, or their reaction products to flavor and aroma of oncom, it is well known that flavor development in cocoa and fermented dairy products is highly dependent on lipase activity of a large number of microorganisms. For example, a commercially available lipase from *Rhizopus delemar,* a mold showing strong lipolytic activity in peanuts,[9] is said to enhance flavors of dairy products.

Oncom (Figure 2) may be fried in oil or it may be roasted, covered with boiling water, and seasoned with salt and sugar before eating.[3] It may also be roasted, cut into pieces and covered with a ginger sauce before being consumed.[7]

III. CHANGES IN COMPOSITION

According to van Veen et al.,[6] unfermented peanut press cake from the village and cottage industry in western Java contains 10 to 16% moisture, 6 to 20% oil, 38 to 51% crude protein, 14 to 20% carbohydrate, 4 to 6% ash, and 5 to 8% crude fiber. Oncom (fermented peanut press cake) averages 70% moisture, 3 to 9% oil, 20 to 30% crude protein, about 4% carbohydrate, 1% ash, and 2% fiber. Winarno,[10] on the other hand, stated that fresh oncom has a moisture content of 57%; protein, 13%; fat, 6%; and carbohydrates, 22%. Variation in proximate composition exists largely because of different procedures used to prepare oncom.

A. Carbohydrates

The carbohydrate content of full-fat peanut kernels is low (12 to 18%) compared to lipid and protein. Removal of oil by solvent extraction or expulsion processes results in up to a two-fold increase in remaining constituents, including carbohydrates. Cellulose, simple sugars, and oligosaccharides compose the largest proportion of the carbohydrate fraction in peanut press cake. Reducing substances generally decrease as a result of fungal utilization of low molecular weight reducing sugars as sources of energy. Organic acids, which may result from breakdown of simple sugars, contribute to the darkening of product color and to the development of undesirable flavors.

Although the content of flatulence-causing sugars, raffinose and stachyose, is considerably lower in peanuts than in beans (*Phaseolus* sp.), and often present in only trace quantities, hydrolysis of these and other oligosaccharides during fermentation undoubtedly contributes to increased digestibility and decreased sweetness. Worthington and Beuchat[11] studied the α-galactosidase activity of ten molds, most of which are traditionally or commercially used to ferment oilseeds and grains. *Neurospora intermedia* was demonstrated to essentially eliminate sucrose, raffinose, and stachyose during the first 21 hr of peanut fermentation (see Table 1). They concluded that utilization of flatulence-causing sugars may contribute to increased digestibility of oncom as suggested by others.[12]

Rhizopus oligosporus is somewhat less effective in reducing the sucrose, raffinose, and stachyose contents during peanut fermentation.[11] The mold was reported to utilize small amounts of stachyose, but only after 68 hr of fermentation. Raffinose and sucrose were not utilized. These data are consistent with those in a report by Sorenson and Hesseltine[13] in which *R. oligosporus* was shown not to utilize raffinose and stachyose as sole sources of carbon. Shallenberger et al.,[14] on the other hand, reported a decrease in stachyose and sucrose content without apparent changes in raffinose content in soybeans fermented over a 72-hr period with a *Rhizopus* mold.

Molds used in other traditional fermented foods have also been shown to have strong α-galactosidase activity when grown on peanut substrates. These include *Aspergillus oryzae*, used on soy sauce and miso preparation, and *Monascus purpureus*, used to ferment rice in the preparation of ang-kak or red rice.[2] *Mucor hiemalis* utilizes sucrose, but not stachyose or raffinose, whereas *Actinomucor elegans*, the principle mold used to prepare sufu, is capable of utilizing stachyose only after an extended fermentation of peanuts. An increased or enhanced digestibility of raw peanuts after fermentation by molds has been attributed in part to hydrolysis of raffinose and stachyose during fermentation processes.

B. Lipids

Both *N. intermedia* and *R. oligosporus* are active lipase producers, hydrolyzing triglycerides to yield free fatty acids which accumulate to various levels, depending upon fermentation conditions. Beuchat and Worthington[9] studied changes in total lipids,

Table 1
SUCROSE, RAFFINOSE, AND STACHYOSE CONTENTS OF PEANUTS FERMENTED WITH TRADITIONAL FOOD FUNGI

Fungus	Ferment. time (hr)	Contents (g/100 g of sample)		
		Sucrose	Raffinose	Stachyose
Control 1[a]		6.9	0.15	0.65
Control 2[b]		5.3	0.10	0.60
Control 3[c]		5.2	0.10	0.47
Aspergillus niger	21	0.29	0.04	0.30
	44	0.01	0.02	T[d]
	68	0.08	ND[e]	T
	98	0.08	ND	ND
Aspergillus oryzae	21	0.07	0.01	0.24
	44	0.04	T	T
	68	T	ND	ND
	98	0.05	ND	ND
Neurospora intermedia	21	0.03	ND	0.03
	44	0.02	ND	T
	68	T	ND	T
	98	0.06	T	0.02
Rhizopus delemar	21	0.02	0.04	0.35
	44	0.51	ND	T
	68	0.15	ND	ND
	98	0.06	ND	ND
Rhizopus arrhizus	21	5.9	0.11	0.47
	44	6.6	0.18	0.41
	68	7.7	0.30	0.32
	98	9.3	0.61	0.19
Rhizopus oligosporus	21	5.8	0.08	0.45
	44	6.8	0.12	0.53
	68	6.7	0.17	0.62
	98	6.5	0.27	0.39
Monascus purpureus	21	5.4	0.08	0.48
	44	5.7	0.16	0.37
	68	7.8	0.20	0.09
	98	9.1	0.06	T
Mucor hiemalis	21	0.53	0.13	0.51
	44	0.22	0.09	0.58
	68	0.34	0.13	0.60
	98	0.20	0.24	0.61
Actinomucor elegans	21	5.6	0.10	0.40
	44	5.9	0.10	0.44
	68	6.7	0.13	0.39
	98	6.7	0.28	0.34

[a] Unfermented sample, not autoclaved, freeze-dried without incubation.
[b] Unfermented sample, autoclaved, freeze-dried without incubation.
[c] Unfermented sample, autoclaved, incubated 98 hr at 28°C before freeze-drying.
[d] T, trace (less than 0.01 g/100 g of sample).
[e] ND, not detected.

Table 2
LIPID AND FATTY ACID COMPOSITION OF PEANUTS FERMENTED WITH FUNGI

Fungus	Temp (°C)	Time (hr)	Lipid (%)[b]	Fatty acids (%)[a] 16:0	18:0	18:1	18:2	20:0	20:1	22:0	24:0
Control		0	50.8	10.6	2.7	51.2	28.3	1.4	1.3	2.8	1.6
Control	28	99	50.5	10.6	2.8	50.9	28.5	1.4	1.4	2.8	1.6
Control	21	116	50.5	10.7	2.8	51.1	28.2	1.4	1.3	2.9	1.4
N. intermedia	28	99	52.1	10.3	2.7	50.3	28.3	1.4	1.4	3.0	1.6
N. intermedia	21	116	51.6	10.5	2.7	50.7	28.2	1.3	1.3	2.7	1.6
A. oryzae	28	90	53.3	10.3	2.3	50.9	28.4	1.4	1.3	2.8	1.6
A. niger	28	90	52.5	10.1	2.7	50.7	28.4	1.4	1.4	2.8	1.7
R. oligosporus	28	90	48.3	9.4	2.5	51.6	28.7	1.4	1.3	2.8	1.7
R. delemar	28	74	51.4	9.8	2.4	52.1	29.1	1.2	1.2	2.4	1.4

[a] Percentage of total fatty acid in peanut lipid; average of four determinations.
[b] Percentage of peanut, dry weight basis; average of two determinations.

total fatty acids, and free fatty acids in peanuts fermented with the oncom molds, as well as *A. oryzae, Aspergillus niger*, and *Rhizopus delemar,* for incubation times ranging to 116 hr (Table 2). Alkali titration of extracts of the fermented peanuts indicated that lipolytic activity was essentially linear for *Rhizopus* spp. throughout fermentation, while an initial lag of about 40 hr was required before increases in free fatty acids were noted in peanuts inoculated with *N. intermedia.*

The distribution of fatty acids in peanuts changes somewhat upon fermentation. With the exception of slightly lower percentages of palmitic (16:0) and slightly higher percentages of oleic (18:1) and linoleic (18:2) acids in peanuts fermented with *R. oligosporus* and *R. delemar*, the remaining major fatty acids (stearic [18:0], eicosanoic [20:0], eicosenoic [20:1], docosanoic [22:0], and tetracosanoic [24:0]) in lipids of peanuts fermented with *N. intermedia, A. oryzae, A. niger, R. oligosporus*, and *R. delemar* were essentially unchanged from the controls.

The free fatty acid fractions of fermented peanuts, in contrast, have been shown to contain significantly higher levels of saturated fatty acids, particularly palmitic and stearic acids, and lower levels of linoleic acid than did the total lipid extracts.[9] Oleic acid was essentially unchanged. Differences in free fatty acid distribution are those which would be expected from the action of 1,3-lipases, since saturated acids are located primarily in the 1,3 positions and linoleic acid is in the 2 position of peanut triglycerides.[15] Oleic acid is approximately equally distributed among the three positions.

No evidence of preferential utilization of any of the free fatty acids was detected in peanuts fermented with *N. intermedia* or *R. oligosporus.*[9] This supports the observations of Wagenknecht et al.,[16] who reported that *R. oryzae* did not preferentially utilize those fatty acids of soybeans which are in common with peanuts. The mold did, however, utilize 40% of the linolenic acid, a fatty acid not found in significant quantity in peanuts. Murata et al.[17] likewise reported decreases in linolenic acid content of tempe (soybean) cultured with *R. oligosporus*; increases in percentages of palmitic and oleic acids, and decreases in stearic and linoleic acids, were also reported.

Fermentation of solvent-defatted peanut flour results in fatty acid profiles quite different from those observed for full-fat peanuts. If one initiates the fermentation using flour containing less than 1% lipid and then measures relative percentages of fatty acids in the final product, profiles reflect the presence of both fungal and peanut fatty acids.[15] Flour fermented with Ascomycetes (*N. intermedia* and *A. oryzae*) showed increases in linoleic, palmitic, and stearic acids, accompanied by decreases in oleic, eicosanoic, eicosenoic, docosanoic, and tetracosanoic acids. Data tend to reflect changes away from normal percentages of acids in peanut lipids and toward those reported to be normally present in Ascomycetes mycelia.[18] In peanut flour fermented with genera of Phycomycetes (*R. oligosporus, M. hiemalis* and *A. elegans*), increases in palmitic, stearic, and eicosanoic acids were accompanied by decreases in oleic and tetracosanoic acids. Thus, the lower the initial lipid content of peanut press cake, the more likely the change in fatty acid profile of oncom will be.

Antioxidants such as jenistin, diadzein and 6,7,4′-trihydroxyisoflavone are in a bound inactive form in soybeans, but active in tempe.[19] Studies have shown that tempe can prevent lipid oxidation in tocopherol-stripped corn oil.[20] However, it is not known if naturally occurring antioxidants in peanuts are released by *R. oligosporus* and *N. intermedia* during the preparation of oncom.

C. Protein

Research relating changes in protein after fermentation of peanut press cake and other peanut substrates indicates that protein content is little altered, but solubility is greatly increased. Van Veen et al.[6] reported that the crude protein (Kjeldahl $N \times 6.25$) in press cake was elevated slightly while true protein (precipitation with 5% trichloracetic acid) decreased from 94% of the crude protein in press cake to 74% of crude protein in oncom. An increase in the percentage of crude protein has been demonstrated in peanut flour fermented with the oncom molds, *N. intermedia* and *R. oligosporus,* as well as *M. hiemalis, A. elegans,* and *A. oryzae.*[5] It was suggested that there was not an actual increase in protein weight as a result of fermentation, but rather a loss of nonprotein volatiles during the fermentation process, thus accounting for proportionate increases in unaltered peanut constituents. Ash was also noted to increase after fermentation.

Electrophoretic patterns of proteins and free amino acid composition of peanut materials fermented with *N. intermedia, R. oligosporus,* and other molds, have been studied. Beuchat et al.[21] observed that large molecular weight globulins such as arachin were hydrolyzed to smaller components during fermentation (Figure 3). The extent of peptide formation varied with the mold under study, but only small changes in the total amino acid profile occurred. The percentage of amino acids as free, and the proportions of specific amino acids within this fraction, varied greatly among the ferments as well as between ferments and nonfermented controls. The amount of total amino acids which were free in peanuts fermented with *N. intermedia* was 8.67%; *R. oligosporus* had somewhat less protease activity, resulting in only 2.99% of the total amino acids as free.

As a result of protein breakdown during fermentation, the nitrogen solubility profiles of peanut proteins at various pH values change drastically. For example, the percentage of nitrogen soluble at pH 4 to 6 has been demonstrated to increase from less than 5% in unfermented peanuts to greater than 20% in peanuts fermented with *N. intermedia* and *R. oligosporus.*[22] Increased nitrogen solubility in peanut miso has also been shown to result from protease production by *R. oligosporus.*[23]

The influence of pH, temperature, and time on protease activity of *N. intermedia* on peanut protein was studied by Beuchat and Basha.[24] Production of protease on solid and liquid media was monitored over a 4-day period. Maximal activity of protease extracted from 4-day-old cultures occurred at pH 6.5 when unfractionated peanut protein substrate was used. Media containing more than 0.2 *M* sodium chloride resulted in decreased protease and mycelium production, while cassava starch was without effect at concentrations up to 1.4%. The addition of up to 0.3 *M* sucrose to growth media greatly stimulated protease production and mycelial growth. Maximal proteo-

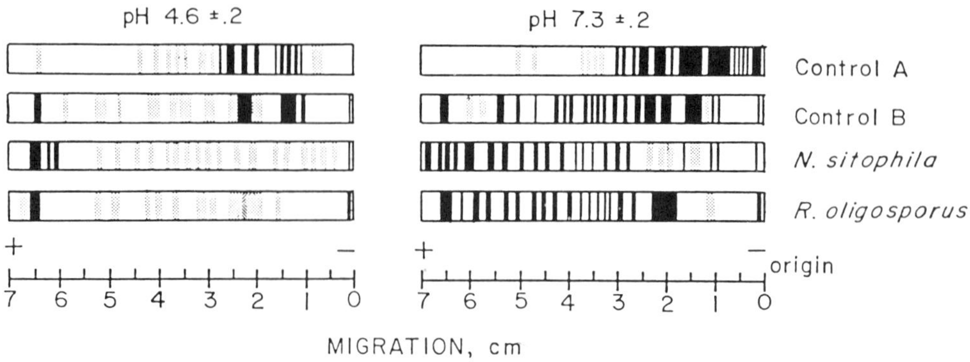

FIGURE 3. Polyacrylamide gel patterns of control peanut proteins and peanut meal proteins fermented with *N. intermedia (N. sitophila)* and *R. oligosporus*. Extraction was at pH 4.6 and 7.3; control A was not autoclaved, not fermented; control B was autoclaved, not fermented. (Adapted from Beuchat, L. R., Young, C. T., and Cherry, J. P., *Can. Inst. Food Sci. Technol. J.*, 8, 40, 1975.)

lytic activity was observed in extracts from mycelium cultured in media adjusted from pH 6.0 to 7.5. Activity was greater when soluble peanut protein was used as a substrate, compared to unfractionated or globulin protein substrates.

Wang[8] and Wang et al.[25] observed that mold proteinase and peptidase production and activity in peanut substrates are greatly affected by pH, temperature, moisture, and ionizable salts. Thus, control of these parameters during the 1- to 2-day period for oncom fermentation is critical for producing a product with desirable sensory qualities.

IV. NUTRITIVE VALUE

The protein efficiency ratios (PER) of fermented peanuts are not increased over properly heated nonfermented materials.[5,6,26] On the other hand, increased digestibility of fermented peanuts and soybeans has been attributed to various hydrolytic activities associated with *N. intermedia* and *R. oligosporus*.[11,27-29]

The E/T (g of essential amino acids per g of nitrogen) and A/E (mg of each individual essential amino acid per g of total essential amino acids) ratios, in addition to protein scores, calculated for free amino acid fractions of peanut meal fermented with *N. intermedia, R. oligosporus,* and seven other molds, have been reported.[21] The E/T ratio indicates the proportion of total nitrogen which is in the form of essential amino acids. With the exception of *M. purpureus,* E/T ratios of free amino acid fractions of ferments were improved over controls. Noting significant hydrolysis of proteins during fermentation, the authors suggested that specific flavor sensations in oncom may be attributable to free amino acids. For example, leucine produces a weak meaty flavor plus a must flavor, glycine has a sweet taste and produces a cooling sensation, sodium glutamate enhances flavor, and sulfur-containing amino acids such as methionine and cysteine, when heated, produce pyrogenic flavors. The fermentation process undoubtedly contributes to the enhancement of these flavor characteristics in peanut press cake.

In contrast to reports that thiamin levels are decreased or unchanged during oilseed and grain fermentations,[6,26,30] thiamin was shown to increase significantly in peanut flour fermented with *N. intermedia, R. oligosporus, A. oryzae,* and *A. elegans.*[5] In the latter study, riboflavin was shown to increase in peanut flour fermented with these molds as well as *M. hiemalis,* which agrees with an earlier report on oncom prepared using *N. intermedia.*[6] Niacin is also reported to increase in peanut flour fermented with *N. intermedia*;[5] however, pantothenate levels were unchanged after fermentation.

Van Veen and Schaefer[31] speculated on the beneficial effects of tempe on prisoners-of-war suffering from dysentery. Certain strains of *R. oligosporus* are known to produce an antibacterial compound especially active against some Gram-positive organisms.[32] This antibacterial activity may be significant for those people whose diets are often nutritionally inadequate. The suggested suppression of intestinal bacterial due to the consumption of tempe by young men[33] may well have been due to the presence of antimetabolites secreted by *R. oligosporus*. Production of microbial antimetabolites by *R. oligosporus* in fermented peanuts has not been reported; however, since soybeans and peanuts provide similar nutrient conditions, it can be speculated that oncom also contains by-products of the mold used for fermentation which may have antimicrobial activity.

The release of bound trypsin inhibitors from soybeans has been reported by Wang et al.[34] It has been noted that there may be more than one trypsin-chymotrypsin inhibitor in peanuts and that the inhibitor(s) may depend on the variety and age of the peanut.[35] It is possible, therefore, that some of the inconsistencies reported on the nutritional value of fermented peanut substrates exist due to the presence, absence, degree of availability, and quantity of naturally occurring inhibitors. Variation may also be dependent in part upon the physiological characteristics associated with *N. intermedia* and *R. oligosporus*.

Degradation of phytic acid, myo-inositol 1,2,3,4,5,6-hexakis (dihydrogen phosphate), by *N. intermedia*, *R. oligosporus*, and a *Neurospora* strain isolated from Indonesian oncom was studied by Fardiaz and Markakis.[36] The phytic acid content of uninoculated peanut press cake was 1.36% on a dry weight basis, but decreased rapidly during a 72-hr fermentation period. Oncom prepared with *R. oligosporus* contained the least amount of phytic acid (0.05%), while oncom prepared with the Indonesian strain of *Neurospora* contained the greatest amount (0.70%). Penta-, tetra-, tri-, di-, and mono-phosphates of inositol, as well as inorganic phosphate and inositol, were found in the fermenting cake. The authors noted that since little is known about the effect of different inositol phosphates on the bioavailability of essential minerals, it is difficult to fully appreciate the significance of the partial dephosphorylation of phytic acid in human nutrition.

V. SUMMARY

In summary, the nutritive value of protein and lipids appear to be changed only slightly, if at all, during fermentation of peanut press cake. Increases in riboflavin, niacin, and possibly thiamin, in the finished product may exist. Enzymatic hydrolysis of some peanut constituents definitely enhances edibility and, presumably, the digestibility of peanut press cake. Flavors developed during the fermentation process provide variety to otherwise bland and perhaps monotonous diets. The per capita consumption of oncom seems destined to increase in the future, particularly in light of increasingly significant cost advantages in producing food protein in plant vs. animal origin.

REFERENCES

1. Beuchat, L. R., Flavor chemistry of fermented peanuts, *Ind. Eng. Chem. Prod. Res. Dev.*, 21, 533, 1982.
2. Beuchat, L. R., Fungal fermentation of peanut press cake, *Econ. Bot.*, 30, 227, 1976.
3. Hesseltine, C. W. and Wang, H. L., Traditional fermented foods, *Biotechnol. Bioeng.*, 9, 275, 1967.
4. Stanton, W. R., Microbially produced foods in the tropics, *Proceedings of the International Symposium on the Conversion and Manufacture of Foodstuffs by Microorganisms*, Saikon Publishing, Kyota, Japan, 1971, 133.

5. Quinn, M. R., Beuchat, L. R., Miller, J., Young, C. T., and Worthington, R. E., Fungal fermentation of peanut flour: effects on chemical composition and nutritive value. *J. Food Sci.*, 40, 470, 1975.
6. Van Veen, A. G., Graham, D. C. W., and Steinkraus, K. H., Fermented peanut press cake, *Cereal Sci. Today*, 13, 96, 1968.
7. Hesseltine, C. W., A millennium of fungi, food and fermentation, *Mycologia*, 57, 149, 1965.
8. Wang, H. L., Release of proteinase from mycelium of *Mucor hiemalis*, *J. Bacteriol.*, 93, 1794, 1967.
9. Beuchat, L. R. and Worthington, R. E., Changes in the lipid content of fermented peanuts, *J. Agric. Food Chem.*, 22, 509, 1974.
10. Winarno, F. G., Fermented vegetable protein and related foods of Southeast Asia with special reference to Indonesia, *J. Am. Oil Chem. Soc.*, 56, 363, 1979.
11. Worthington, R. E. and Beuchat, L. R., α-Galactosidase activity of fungi on intestinal gas-forming peanut oligosaccharides, *J. Agric. Food Chem.*, 22, 1063, 1974.
12. Steinkraus, K. H., Lee, C. Y., and Buck, P. A., Soybean fermentation by the ontjom mold *Neurospora*, *Food Technol.*, 19, 1301, 1965.
13. Sorenson, W. G. and Hesseltine, C. W., Carbon and nitrogen utilization by *Rhizopus oligosporus*, *Mycologia*, 58, 681, 1966.
14. Shallenberger, R. S., Hand, D. B., and Steinkraus, K. H., Changes in sucrose, raffinose, and stachyose during tempeh fermentation, Rep. Eight Day Bean Res. Conf., ARS-74-71, 68, 1967.
15. Brockerhoff, H. and Yurkowski, M., Stereospecific analyses of several vegetable fats, *J. Lipid Res.*, 7, 62, 1966.
16. Wagenknecht, A. C., Mattick, L. R., Lewin, L. M., Hand, D. B., and Steinkraus, K. H., Changes in soybean lipids during tempeh fermentation, *J. Food Sci.*, 26, 373, 1961.
17. Murata, K., Ikehata, H., and Miyamoto, T., Studies on the nutritional value of tempeh, *J. Food Sci.*, 32, 580, 1967.
18. Shaw, R., The polyunsaturated fatty acids of microorganisms, *Adv. Lipid Res.*, 4, 107, 1966.
19. Gyorgy, P., Murata, K., and Idehata, H., Antioxidants isolated from fermented soybeans, *Nature*, 203, 870, 1964.
20. Packett, L. V., Chen, L. H., and Liu, J. Y., Antioxidant potential of tempeh as compared to tocopherol, *J. Food Sci.*, 36, 798, 1971.
21. Beuchat, L. R., Young, C. T., and Cherry, J. P., Electrophoretic patterns and free amino acid composition of peanut meal fermented with fungi, *Can. Inst. Food Sci. Technol. J.*, 8, 40, 1975.
22. Quinn, M. R. and Beuchat, L. R., Functional property changes resulting from fungal fermentation of peanut flour, *J. Food Sci.*, 40, 475, 1975.
23. Shieh, Y.-S. C., Beuchat, L. R., Worthington, R. E., and Phillips, R. D., Physical and chemical changes in fermented peanut and soybean pastes containing kojis prepared using *Aspergillus oryzae* and *Rhizopus oligosporus*, *J. Food Sci.*, 47, 523, 1982.
24. Beuchat, L. R. and Basha, S. M. M., Protease production by the ontjom fungus, *Neurospora sitophila*, *Eur. J. Appl. Microbiol.*, 2, 195, 1976.
25. Wang, H. L., Vespa, J. B., and Hesseltine, C. W., Acid protease production by fungi used in soybean food fermentation, *Appl. Microbiol.*, 27, 907, 1974.
26. Van Veen, A. G. and Steinkraus, K. H., Nutritive value and wholesomeness of fermented foods, *J. Agric. Food Chem.*, 18, 576, 1970.
27. Smith, A. K., Rackis, J. J., Hesseltine, C. W., Smith, M., Robbins, D. J., and Booth, A. N., Tempeh: Nutritive value in relation to processing, *Cereal Chem.*, 41, 1973, 1964.
28. Steinkraus, K. H., Research on traditional Oriental and Indian fermented foods, Special Rep. No. 16, New York Agricultural Experiment Station, Cornell University, Geneva, 10, 1973.
29. Steinkraus, K. H., Van Buren, J. P., Hackler, L. R., and Hand, D. B., A pilot plant process for the production of dehydrated tempeh, *Food Technol.*, 19, 63, 1965.
30. Roelofsen, P. A. and Talens, A., Changes in some B vitamins during molding of soybeans by *Rhizopus oryzae* in the production of tempeh kedelee, *J. Food Sci.*, 29, 224, 1964.
31. Van Veen, A. G. and Schaefer, G., The influence of the tempeh fungus on the soya bean, *Doc. Neerl. Indones. Morbis Trop.*, 2, 270, 1950.
32. Wang, H. L., Ruttle, D. I., and Hesseltine, C. W., Antibacterial compound from a soybean product fermented by *Rhizopus oligosporus* (33930), *Proc. Soc. Exp. Biol. Med.*, 131, 579, 1969.
33. Calloway, D. H., Hickey, C. A., and Murphy, E. L., Reduction of intestinal gas-forming properties of legumes by traditional and experimental food processing methods, *J. Food Sci.*, 36, 251, 1971.
34. Wang, H. L., Vespa, J. B., and Hesseltine, C.W., Release of bound trypsin inhibitors in soybeans by *Rhizopus oligosporus*, *J. Nutr.*, 102, 1495, 1972.
35. Birk, Y. and Gertler, A., Chemistry and biology of proteinase inhibitors from soybeans and groundnuts, Proc. Intl. Res. Conf. on Proteinase Inhibitors, 142, 1970.
36. Fardiaz, D. and Markakis, P., Degradation of phytic acid in oncom (fermented peanut press cake), *J. Food Sci.*, 45, 523, 1981.

Chapter 9

IDLI

N. R. Reddy, M. D. Pierson, and D. K. Salunkhe

TABLE OF CONTENTS

I. INTRODUCTION

Idli is a naturally fermented food of India, primarily popular in southern and other parts of India, and to some extent, in Sri Lanka.[1,2] It is mainly served as a breakfast food along with chutney (a relish usually containing fresh coconut and spices) and/or sambhar (a thin, spiced soup prepared from red gram cotyledons, vegetables, tamarind juice, salt, spices, coconut paste, and herbs). Its appeal as a breakfast food is due to its textural and organoleptic attributes.[1-4] Further, idli serves as an important source of protein, vitamins, and calories in the diet of people. Often, it is given to typhoid patients in the hospitals, to patients who have digestive disorders, young children, and infants, because of its improved digestibility and nutritional quality.[5]

According to Ramakrishnan,[5] idli originated approximately in 1100 A.D. Although it has been produced and consumed by people in south India for centuries, some of the modifications and organoleptic attributes which occur during fermentation are not clearly understood. For example, how methionine is synthesized and increased during fermentation is not known. Some of the research work related to idli fermentation and nutritional quality has been carried out in the U.S. at places such as New York Agricultural Experiment Station, at Cornell University, in Ithaca, and Utah State University, in Logan. This chapter presents information on the method of preparation, nutrient content, nutritional value, and antinutritional aspects of idli. Preparation of idli from other legumes is also presented.

II. METHOD OF PREPARATION

A. Ingredients

The main ingredients used in the idli preparation are polished white rice (*Oryza sativa* L.) and black gram (*Phaseolus mungo* L.) cotyledons (Figure 1). Black gram plays a major role in idli fermentation as a source of microorganisms and as a fermenting substrate.[6,48,49] The proportion of these ingredients in the idli preparation may vary and depends upon the availability and cost of the raw ingredients. Generally, rice is cheaper than black gram cotyledons, so, the product made with higher proportions of rice costs less. Several researchers have used different proportions of black gram cotyledons to rice, i.e., from 4:1 to 1:4 w/w for making idli. For instance, Khandwala et al.[7] and Rao[8] used four parts of black gram cotyledons to one part rice; Joseph et al.[9] used one part black gram cotyledons to two parts of rice; Mukherjee et al.,[10] Padhye and Salunkhe,[11] and Reddy and Salunkhe[12,13] employed one part black gram cotyledons to one part rice; and Lewis et al.,[14] Desikachar et al.,[15] and Steinkraus et al.[16] used proportions of rice to black gram cotyledons of 2:1 to 4:1, and reported a preference for 2:1 and 3:1 over 4:1, where a starchy rice taste predominated. In India, several other ingredients are added to the idli preparation to enhance taste and flavor. For example, idli usually consists of three parts rice to one part black gram cotyledons, plus salt added to taste.[17] Kanchipuram idli (a regional idli type) is made from one part of rice to one part of black gram cotyledons, plus a number of other ingredients such as cashew nuts, butter, chili pepper, ginger, tumeric, and cumin seeds to modify the taste and flavor.

Substitutions can be made in the preparation of idli ingredients. Kodri (*Paspalum scrobiculatum*) can be substituted for rice and dehulled soybeans (*Glycine max* L.) or green gram (*Phaseolus aureus* L.) for black gram.[5] However, use of these substitutions for rice or black gram may give marked changes in the product texture and flavor.

B. Preparation of Idli

The flow diagram for idli preparation is presented in Figure 2. The ingredients, black

FIGURE 1. Black gram whole seeds and cotyledons.

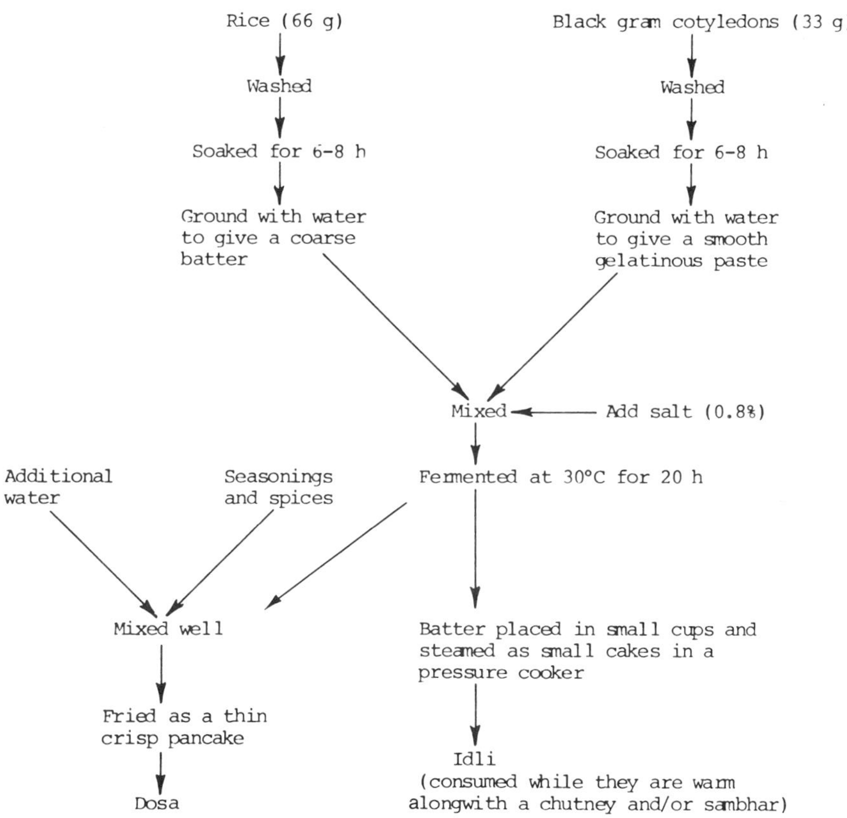

FIGURE 2. Flow sheet for preparation of idli.

gram cotyledons and polished white rice, are washed several times with tap water to remove adhering dirt and dust particles, and surface microorganisms, soaked in water for 6 to 8 hr separately, ground for a couple of minutes in a blender (in villages soaked black gram and rice are ground in stone mortars to obtain a coarse to smooth batter), and mixed together with salt (0.8 to 1.0% w/w). The combined batter is then allowed

FIGURE 3. Steamed idli. (From Reddy, N. R., Salunkhe, D. K., and Sathe, S. K., *CRC Crit. Rev. Food Sci. Nutr.*, 16, 49, 1982. With permission.)

to ferment for 20 to 24 hr or longer at 30°C depending upon the proportion of ingredients used and temperature.[11-13,16,18] Traditionally in India, the combined black gram and rice batter is fermented overnight in a warm place.[48,49] To avoid contamination and keep insects away, the batter is fermented in closed containers. The fermented batter is placed in small cups, or in a special pan having cups, and steamed as idli in a pressure cooker (Figure 3).

The amount of water added to the ingredient mixture can also vary from 1.5 to 2.2 times over the dry weight of the ingredients.[7,9,14,15,19] The batter should be thick for idli (i.e., about 1.5 times of water) and much more fluid for dosa. A range of added water from 2.0 to 2.2 times the initial dry weights of ingredients will provide the desired viscosity for dosa. To prepare dosa, additional seasonings are added to the batter and fried as a thin crisp pancake. Sometimes, the fermented idli batter is also used with proper addition of spices to prepare hopper (called "appam" or "uttappa" in south India), a crepelike pancake.[20]

The microorganisms involved in the fermentation of black gram and rice batter have been isolated and identified by various investigators (Table 1). These microorganisms have been found to be naturally present in black gram cotyledons and rice[5] (Table 2). Acidification and leavening are the two important changes that occur during the fermentation of black gram and rice batter (Table 3). According to Mukherjee et al.,[10]

Table 1
ISOLATION AND IDENTIFICATION OF MICROORGANISMS INVOLVED IN IDLI FERMENTATION BY VARIOUS INVESTIGATORS

Bacteria	Yeasts	Ref.
Lactobacillus delbrueckii	*Odium lactis*	46, 48, 49
Lactobacillus lactis	*Torulopsis holmii*	
Streptococcus lactis	*Candida parapsilosis*	21
	Candida peliculosa	
Leuconostoc mesenteroides		10
Streptococcus faecalis	*Torulopsis candida*	22, 48, 49
Pediococcus cerevisiae	*Trichosporon pullulans*	
Lactobacillus fermenti	Yeasts	18
Bacillus spp.		36

Table 2
BACTERIAL FLORA OF RAW INGREDIENTS (RICE AND BLACK GRAM COTYLEDONS) AND FERMENTED BATTER OF BLACK GRAM AND RICE[5]

Bacterial flora	Rice	Black gram cotyledons	Fermented batter of black gram and rice
Total counts per g	2.25×10^4	1.27×10^4	126×10^9
Individual counts as % of total counts			
Leuconostoc mesenteroids	33.3	69.2	42.9
Lactobacillus fermenti	16.6	—	14.3
Lactobacillus delbrueckii	8.3	7.1	14.3
Lactobacillus lactis	—	—	—
Streptococcus faecalis	8.3	7.1	14.3
Pediococcus acidilactici	16.5	—	—
Bacillus sp.	16.6	14.9	—
Yeasts (counts/g)	—	—	1050.0

Note: Unfermented batter of black gram and rice contained total counts of 1.4×10^9 per g.

the acid and gas required for leavening action are produced exclusively by activity of heterofermentative lactic acid bacteria, *Leuconostoc mesenteroides,* although there is a sequential change in the bacterial flora. Gas and acid produced by the bacteria contribute leavening and flavor to the product, respectively. Further, the acid produced during fermentation inhibits the growth of food poisoning and food spoilage organisms. Idli can also be produced by joint action of yeasts in the mixture.[21-23,48,49] In India, sour buttermilk or bakers yeast is sometimes added to enhance the fermentation.

C. Physico-Chemical and Microbiological Changes

Desikachar et al.,[15] Mukherjee et al.,[10] Steinkraus et al.,[16] Ramchandran,[24] Akolkar,[25] Ramakrishnan,[18] and Padhye and Salunkhe[11] observed an increase in batter volume, soluble acids and solids, and soluble crude protein and a decrease in soluble

Table 3
PHYSICO-CHEMICAL CHANGES OCCURRING IN THE BLACK GRAM AND RICE BLEND FERMENTATION[10]

Fermentation time (hr)	Batter volume (mℓ)	pH	Total acidity (g lactic acid per g dry blend)
0[a]	505	6.0	0.77
4	505	5.7	1.20
8	645	5.2	2.23
12	705	4.9	2.50
16[b]	745	4.7	2.71
20	745	4.5	3.14
24	525	—	3.14
48	380	4.4	3.57

[a] Black gram cotyledons washed several times in water and soaked in distilled water for 8 hr, ground and mixed with rice semolina (30 mesh) and 1.0% salt.
[b] Batter had general appearance and aroma of idli.

sugars and pH as a result of fermentation. The pH dropped from the initial value of 6.0 to 4.3 to 4.7 during 20 to 22 hr fermentation. The total bacterial count ranged from 10^3 to 10^5 per g and rose to 10^8 to 10^{11} per g at the end of 20 to 22 hr fermentation.[10,11,16] Mukherjee et al.[10] studied the sequence of microorganisms that develop during soaking and fermentation of black gram and rice batter at 30°C. They reported that *Leuconostoc mesenteroides* and *Streptococcus faecalis,* respectively, multiplied from 3.3×10^8 and 1.9×10^8 cells per g in the initial black gram and rice batter to 1.5×10^9 cells per g in the 23 hr fermented batter. On the basis of predominance of these microorganisms, they concluded that *Leuconostoc mesenteroides* is essential for leavening of batter and also responsible along with *Streptococcus faecalis* for acid production in idli, dosa, and related fermentations. This was later confirmed.[18,25,48,49]

D. Organoleptic and Textural Characteristics of Idli

The overnight or 20 to 24 hr fermented batter is steamed to obtain the traditional idli (Figure 3). Idli has a soft and spongy (porous crumb) texture and pleasant sour flavor. The air voids and fine network (spongy texture) of the idli can be seen in Figure 4. The soft spongy texture observed in the leavened steamed idli or puddings made out of black gram is due to the presence of two components, namely surface active protein (globulin) and an arabinogalactan (polysaccharide), in black gram.[26-31] The mucilaginous principle of black gram, first noted by Kadkol et al.[32] was later isolated and characterized by Susheelamma and Rao[26-28] and identified to be arabinogalactan. The ratio of arabinose to galactose was noted to be 3:2 by Susheelamma and Rao.[28] The surface active protein (globulin) has been resolved into two homogenous constituents with high foam forming activity, having a molecular weight of about 100,000 daltons each.[27] These two polypeptides are rich in acidic as well as basic amino acids but low in sulfur containing amino acids. Susheelamma and Rao[28] reported that the arabinogalactan has a molecular weight of about 140,000 daltons. They further stated that arabinogalactan is highly viscogenic around pH 5.0 to 7.0. Arabinogalactan stabilizes the network of foam formed by the surface active protein and is essential for raising of the rice/black gram batters and for porosity of the steamed products.[29] Further, arabinogalactan also helps to hold the leavening gases and prevents disruption of the

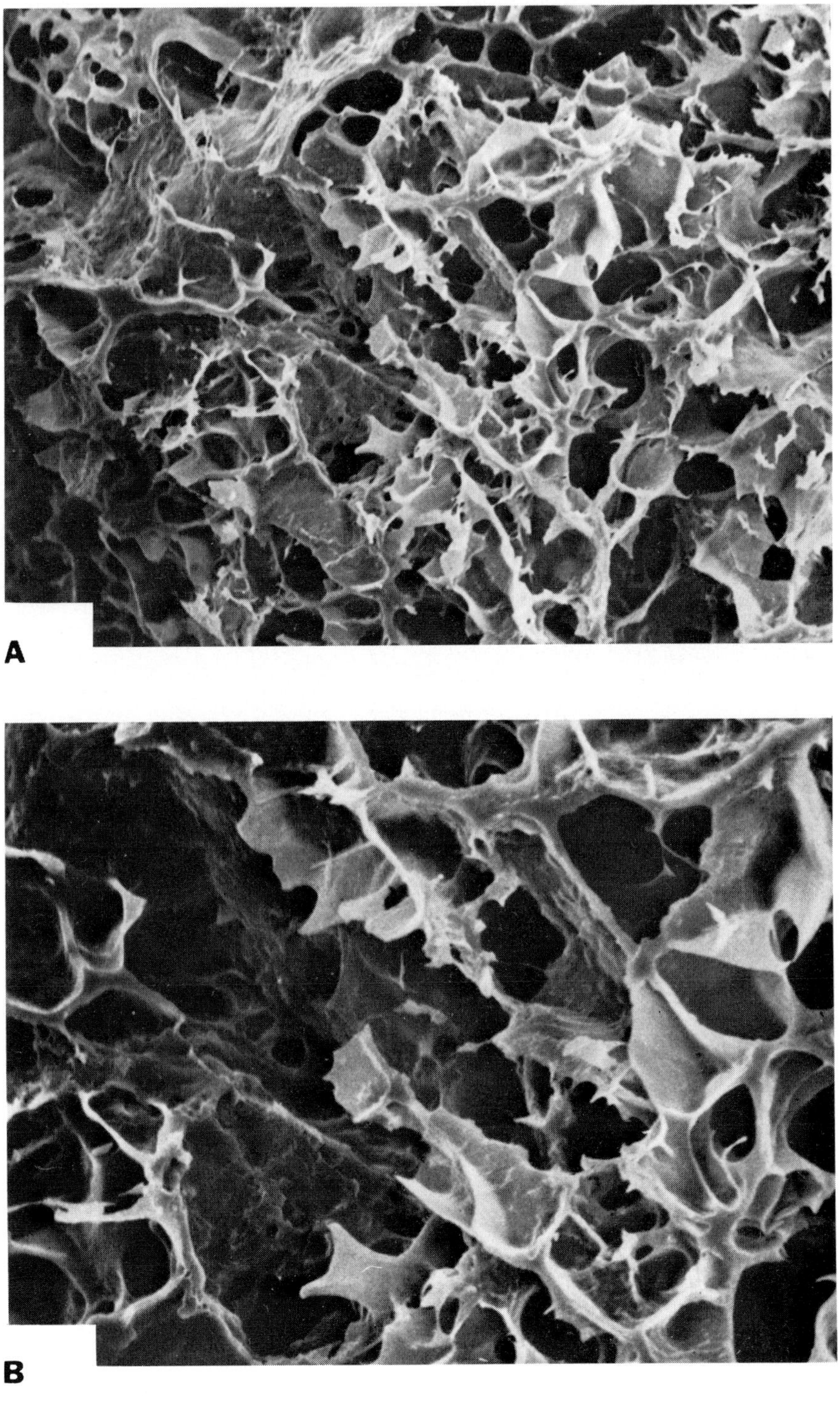

FIGURE 4. Scanning electron photomicrograph showing fine structural network (spongy type texture of idli). (A) Magnification × 500. (B) Magnification × 1000. (From Reddy, N. R., Salunkhe, D. K., and Sathe, S. K., *CRC Crit. Rev. Food Sci. Nutr.*, 16, 49, 1982. With permission.)

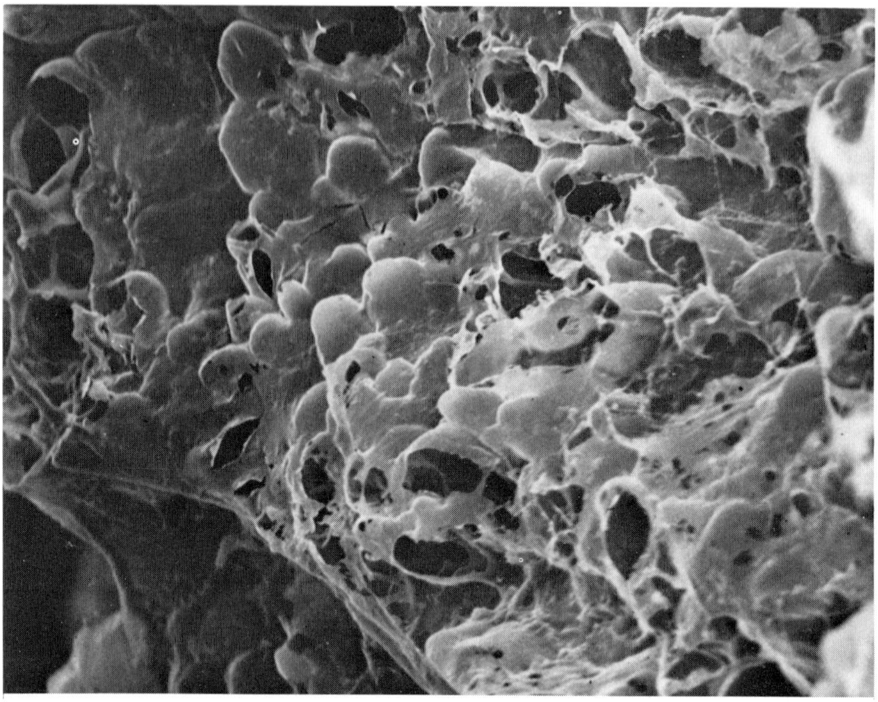

FIGURE 5. Scanning electron photomicrograph of steamed idli, showing ruptures at the weak surfaces. (Magnification × 1000.) (From Reddy, N. R., Salunkhe, D. K., and Sathe, S. K., *CRC Crit. Rev. Food Sci. Nutr.*, 16, 49, 1982. With permission.)

protein-foam (spongy texture) by heat at cooking temperatures. However, a certain degree of disruption of the protein network does occur during steaming. Such disruption may be due to excessive stress developed by the steam which may result in disruption of vulnerable (weak) protein surfaces (Figure 5).

Proportions of black gram and rice in the batters of 1:1 to 1:2 appear to be optimal for achieving a balance between surface activity and viscosity and obtaining products with low bulk density and satisfactory texture.[29] Some of the steps in processing will reduce the surface activity and viscosity of black gram flours. For example, the germination process will decrease surface activity of proteins and viscosity of arabinogalactan as the germination progresses.[30] However, overnight fermentation does not adversely affect the functional properties of surface active protein and arabinogalactan of the black gram.

III. NUTRITIONAL COMPOSITION AND QUALITY OF IDLI

Nutritional composition and quality of different types of idli prepared by using different proportions of rice to black gram are presented in Tables 4, 5, and 6. Data related to soaking losses of nutrients are very limited. The results are somewhat contradictory in regard to changes in vitamin and amino acid content, and protein quality. These differences could be due in part to utilization of different proportions of rice to black gram in preparation of idli.

Rao[8] reported an increase in choline and folic acid contents as a result of fermentation (Table 4). Ramchandran[24] has also reported an increase in thiamin, riboflavin, and niacin during fermentation of idli batter. Rajalakshmi and Vanaja[33] found a 2- to 3-fold increase in thiamin and riboflavin content of the fermented over the unfer-

Table 4
NUTRITIONAL QUALITY OF IDLI[8]

	Unfermented		Fermented and steamed idli
PER (protein efficiency ratio)	2.28		2.55
Methionine (g%)	0.129		0.153
Choline (g%)	0.324		0.434
Folic acid (μg%)	42.4		67.3
Vitamin B-12 (ng%)	3.9		3.7
RBC regeneration	No improvement		Improvement
Hemoglobin regeneration		No change	
Serum protein regeneration		No change	
Liver protein regeneration		No change	

Note: Idli prepared by using four parts black gram cotyledons and one part rice.

Table 5
NUTRITIONAL QUALITY OF IDLI[16,33]

	Unfermented	Fermented
Protein (%)	13.9	14.0
Methionine (g/g N)	0.123	0.136
Cysteine (g/g N)	0.024	0.022
PER	1.50	2.00
Total phosphorus (mg/100g)	262.0	241.0
Phytic acid (mg/100g)	174.0	113.0
Thiamin (mg/100g)	0.21	0.58
Riboflavin (mg/100g)	0.25	0.54

Note: Prepared by using one part black gram and two parts rice.

mented samples (Table 5). When black gram and rice were in a ratio of 1:1, Van Veen et al.[34] found a decrease in riboflavin content during fermentation. Ananthachar and Desikachar[35] concluded that thiamin content of idli was not altered by fermentation, possibly because of different technique of fermentation employed by them. They used a yeast and *Lactobacillus* in the fermentation. Shah[36] recently isolated a strain of *Bacillus* spp. from rice/black gram batter using a medium containing cobalt. This microorganism is reported to be responsible for increase of vitamin B_{12} content in idli (Table 7). Changes in the vitamin content during fermentation are of considerable importance from the standpoint of nutrition, particularly since cereal based diets are often deficient in B-group vitamins.[37]

Several workers found increases in the amino acid content, especially methionine during black gram and rice blend fermentation, irrespective of the proportions of black gram and rice used (Tables 4, 5, and 6). Rao[8] reported that the methionine content of idli was 18.6% greater than that of the unfermented batter (Table 4). An increase of 10.6%[16] and 60.0%[11] of methionine in fermented batters over unfermented ones has been reported, although different proportions of black gram to rice were used by these investigators. Van Veen et al.[34] reported a slight increase in methionine content after 24 hr fermentation, using a proportion of rice to black gram of 1:1 and an initial soaking of 8 hr at room temperature. The increase in methionine content may be related to a decrease in inorganic sulfur during fermentation, since Reddy and Salunkhe[12] observed a loss of 77.8% of inorganic sulfur during black gram and rice blend fermentation. Whether such relationship exists remains to be resolved however. An increase

Table 6
NUTRITIONAL QUALITY OF IDLI[11-13,34]

	Unfermented	Fermented	Fermented and steamed idli
Total phosphorus (mg/g)	3.7	3.7	3.7
Inorganic phosphorus (mg/g)	0.4	1.1	1.4
Phytic acid phosphorus (mg/g)	2.5	1.5	1.5
Calcium (mg/g)	0.4	0.4	0.4
Magnesium (mg/g)	1.4	1.4	1.4
Zinc (μg/g)	23.5	23.4	23.4
Iron (μg/g)	46.9	46.4	46.4
Inorganic sulfur (mg/g)	1.3	0.3	0.3
Oligosaccharides			
Sucrose (mg/g)	8.3	5.6	5.1
Raffinose (mg/g)	ND	ND	ND
Stachyose (mg/g)	3.1	2.9	2.4
Verbascose (mg/g)	15.5	12.7	10.3
Amino acids (mg/100g)			
Lysine	4.8	6.2	—
Cysteine	0.2	Trace	—
Arginine	6.9	8.9	—
Tryptophan	3.6	3.3	—
Methionine	1.0	1.6	—
Predicted BV	22.0	40.0	—
PER	1.99	1.84	—

Note: Prepared by using one part black gram and one part rice. ND denotes not detectable.

Table 7
PHYSICAL CHARACTERISTICS AND VITAMIN B_{12} CONTENT OF NATURALLY FERMENTED AND EXPERIMENTALLY FERMENTED RICE/BLACK GRAM BATTER[36]

	Batter characteristics			
		pH of the batter		
	% increase in volume	Initial	Final	Vitamin B_{12}[a] content (μg/100 g batter)
Natural fermentation				
Unfermented batter	0	6.4	6.4	0.16
Fermented batter	80	6.4	4.5	0.73
Experimental fermentation				
Unfermented batter	0	6.4	6.4	0.16
Batter fermented with *Bacillus* spp.	30	6.4	6.0	1.40

[a] Vitamin B_{12} was estimated by microbiological method using *L. leichmanii* ATCC 7830.

in methionine during fermentation is of nutritional significance, since most legumes are deficient in sulfur-containing amino acids. This deficiency restricts the biological value of bean proteins. As a result of fermentation, Padhye and Salunkhe[11] observed a significant increase in predicted biological value (Table 6). Fermentation also improves the PER of idli over the unfermented mixture (Tables 4 and 5). In contrast,

Table 8
BACTERIAL FLORA OF RAW INGREDIENTS (RICE AND SOYBEAN COTYLEDONS) AND FERMENTED BATTER OF SOYBEAN AND RICE[5]

Bacterial flora	Rice	Soybean cotyledons	Fermented batter of soybean and rice
Total counts per g	2.25×10^4	29.5×10^4	11.0×10^9
Individual counts as % of total counts			
Leuconostoc mesenteroides	33.3	40.0	29.1
Lactobacillus fermenti	16.6	—	14.5
Lactobacillus delbrueckii	8.3	—	18.2
Lactobacillus lactis	—	7.1	10.9
Lactobacillus buchneri	—	—	7.3
Pediococcus acidilactici	16.5	20.0	—
Streptococcus faecalis	8.3	20.0	10.9
Bacillus spp.	16.6	—	—
Yeasts (counts/g)	—	—	1560.0

Note: Unfermented batter of soybean and rice contained total counts of 0.04×10^9 per g.

Van Veen et al.[34] found a decrease in PER of idli (Table 6). Khandwala et al.[7] reported that the digestibility of idli was the same as that of unfermented mixture. Khandwala[38] and Rao[8] found that the feeding of idli to rats resulted in a better regenerating effect on red blood cells and in protecting liver than the same mixture before fermentation (Table 4). Rajalakshmi and Vanaja[33] reported improvements in weight gain per gram protein intake, and increases in the liver contents of thiamin, riboflavin, xanthine oxidase, and succinic dehydrogenase in rats fed fermented idli.

IV. ANTINUTRITIONAL FACTORS IN IDLI

The fermentation process appears to have a significant effect on elimination and/or reduction of antinutrients (phytic acid and enzyme inhibitors) and the flatulence problem. A reduction of 35.0 to 40.0% of phytic acid was reported by Rajalakshmi and Vanaja[33] and Reddy and Salunkhe[12] during idli preparation (Tables 5 and 6). Fermentation and steaming also result in appreciable reduction (33.8%) of oligosaccharides (verbascose, stachyose, and raffinose) in idli (Table 6). The final product, idli, contains 1.0% verbascose and 0.2% stachyose and produces low flatus (8.4 mℓ H_2) in rats, when fed to rats at a 50% level in the basal diet.[13,39] Chymotrypsin inhibiting activity can be eliminated completely by fermentation and steaming during the preparation of idli, but not the trypsin inhibiting activity.[11] No hemagglutinating activity is found in the final product, idli.[19]

V. PREPARATION OF IDLI FROM OTHER LEGUMES

Substitution of black gram by oilseeds (soybeans and peanuts)[16,18,25,31,40] and dry beans of *Phaseolus vulgaris* L. (Great Northern beans)[43,44] in preparation of idli have been recently reported. Ramakrishnan et al.[40] prepared a soy idli by using two parts of rice and one part of soybean cotyledons. The microorganisms involved in soy idli fermentation have been recently isolated and identified (Table 8). These microorganisms are naturally present in the raw ingredients. Ramakrishnan et al.[40] and Akolkar[25] in-

Table 9
NUTRITIONAL COMPOSITION OF NATURALLY FERMENTED SOY IDLI[25,40]

	Unfermented	Fermented
PER	0.9	1.2
Amino nitrogen (%)	0.2	0.3
Total sugars (%)	57.2	51.0
Free sugars (%)	0.9	3.2
Total phosphorus (mg/100g)	376.0	376.0
Inorganic phosphorus (mg/100g)	63.3	161.3
Phytic acid (mg/100g)	662.0	425.5
Vitamins		
Niacin (mg/100g)	2.25	3.12
Thiamin (mg/100g)	0.35	0.57
Riboflavin (mg/100g)	0.34	1.19
Vitamin B_{12} (mg/100g)	0.48	2.20

Note: Idli prepared by using two parts rice and one part soybean cotyledons.

oculated each of the microorganisms (presented in Table 8) into sterilized soy idli batter and found the following fermentation for 16 hr:

- *L. mesenteroides, L. fermenti, L. buchneri, L. lactis, S. faecalis* decrease pH.
- *L. mesenteroides* and *L. fermenti* increase the batter volume following incubation.
- *L. delbrueckii* increases thiamin content.
- *L. fermenti* and *L. delbrueckii* increase riboflavin content.
- *L. fermenti, L. lactis, S. faecalis* increase niacin content.
- *L. mesenteroides, L. fermenti, L. delbrueckii, L. buchneri, L. lactis,* and *S. faecalis* are involved in the carbohydrate breakdown.
- *L. mesenteroides, L. fermenti, L. delbrueckii, L. buchneri, L. lactis,* and *S. faecalis* are involved in the protein hydrolysis.
- *Bacillus* spp. increases vitamin B_{12} content and decreases phytic acid content.

They further found that natural fermentation by soy idli was associated with an increase in amino nitrogen (206%), free sugars (208%), niacin (52%), riboflavin (100%), and vitamin B_{12} (358%) and a decrease in phytic acid (64%) (Table 9). Hemagglutinating and trypsin inhibiting activities and flatulence-causing sugars were also reported to be reduced significantly during soy idli fermentation.[5,25,40-42] Rao[42] recently isolated and identified a strain of *L. mesenteroides* (HA) from soy idli batter using an enrichment culture technique. This strain differs in its characteristics from the strain *L. mesenteroides* isolated from normal idli batter, (prepared with black gram and rice). Further Rao[42] found that the newly isolated strain *L. mesenteroides* (HA) hydrolyzes hemagglutinins from soybeans, red kidney beans, navy beans, and black beans. This strain secretes a mixture of enzymes (protease, 4-aspartyl glycosylamine amido hydrolase, β-N-acetyl glucosaminidase, and α-D-mannosidase) which are reported to be involved in the hemagglutinin hydrolysis. When this organism was used along with *L. delbrueckii, L. fermenti,* and *Bacillus* spp. to ferment sterilized soy idli batter, the idli obtained was found to have the same characteristics as idli prepared by natural fermentation and at the same time found to be completely free from hemagglutinin. Feeding of soy idli along with basal diet ingredients to rats increased the protein efficiency ratio, growth, and liver and bone status more than the unfermented or fer-

Table 10
FOOD INTAKE, BODY WEIGHT GAIN, AND LIVER COMPOSITION IN RATS FED SOY IDLI WITH AND WITHOUT FERMENTATION[25]

	Diets			
	20% Casein	Unfermented soy idli	Fermented soy idli	Fermented soy idli + lime and green leafy vegetables
Protein content of the diet (%)	20.0	19.5	19.5	19.9
Initial body weight (g)	47.0 ± 1.9	46.0 ± 1.4	46.0 ± 1.9	47.0 ± 2.1
Food intake (g)	222.0 ± 4.0	117.0 ± 2.0	177.0 ± 2.0	222.0 ± 3.0
Gain in body weight (g)	73.0 ± 3.5	19.0 ± 0.9	40.0 ± 1.7	70.0 ± 2.0
Gain per 100 cal (g)	8.1 ± 0.3	4.6 ± 0.2	6.6 ± 0.2	9.2 ± 0.2
Calcium content of femur (mg)	43.4 ± 0.8	17.5 ± 0.6	26.7 ± 0.4	41.6 ± 0.9
Composition of liver		Values per 100 g liver		
Protein (g)	20.8 ± 0.5	14.3 ± 0.5	18.3 ± 0.4	20.1 ± 0.3
Vitamin A (mg)	10.1 ± 0.3	4.9 ± 0.2	6.2 ± 0.2	10.6 ± 0.4
Thiamin (mg)	1.1 ± 0.1	0.5 ± 0.0	1.5 ± 0.1	1.1 ± 0.1
Riboflavin (mg)	3.4 ± 0.1	1.6 ± 0.1	3.1 ± 0.2	3.3 ± 0.3
Niacin (mg)	2.6 ± 0.1	1.4 ± 0.1	2.9 ± 0.1	3.1 ± 0.4

Note: Values are means ± S. E. Age at start, 3 weeks; 9 rats per group; period of treatment, 4 weeks. Energy value of the diets used, 343 to 411 cal/100 g.

mented soy idli along.[5,25] Ramakrishnan[18] also conducted studies at Trivandrum Hospital (Kerala, India) where soy idli was used as a dietary supplement for children with kwashiorkor and marasmus. He found: (1) soy idli was well tolerated by children, and there was a disappearance of edema in kwashiorkor, and (2) weight gain was observed in children with both kawashiorkor and marasmus (after edema disappeared).

Sathe and Salunkhe[43] found that whole Great Northern beans could be substituted for black gram cotyledons in the preparation of idli. Incorporation of whole beans may offer an advantage of fiber inclusion in idli, which lacks in its conventional preparation. In a separate study these investigators[44] found that the Great Northern beans contain an arabinogalactan (arabinose to galactose ratio of 2:1.7) necessary for the texture of idli. These data suggest that beans other than black gram may have the necessary components needed for the texture of fermented steamed products.

The diet of many children in India are deficient in calories, protein, calcium, and vitamins (vitamins A and C).[45] These problems of undernutrition and protein deficiency can be overcome by the use of fermented foods prepared with various combinations of cereals and legumes. Further, incorporation of calcium mixtures ($CaCO_3$, CaO, and $Ca(OH)_2$) and leafy vegetables into fermented foods may increase the calcium and vitamin (A and C) content in the diet and eliminate these undernutrition and protein deficiency problems. Leafy vegetables are excellent sources of riboflavin, vitamin C, and carotene.[5] Akolkar[25] reported that incorporation of lime and leafy vegetables into fermented soy idli improves the nutritional status of rat (Table 10) and may also increase the nutritional status of children fed such fortified diets.

VI. CONCLUSIONS

Idli is a legume based fermented food that is prepared by steaming a fermented batter of black gram and rice. Idli can be prepared by using various proportions of black gram and rice. It is mainly used as a breakfast food in India. Idli appears to make an important contribution to the diet as a source of protein, calories, and vitamins. This product has enhanced nutritive value over the raw ingredients as evidenced by vitamin and methionine synthesis and hydrolysis of antinutritional factors during fermentation. Still, we do not know how methionine is increased and synthesized. Further research is needed regarding the increase of methionine content during idli fermentation, by which pathway methionine is synthesized, and identification and isolation of microorganisms responsible for methionine synthesis. Idli can be produced easily and used as a dietary supplement in developing countries where protein calorie malnutrition and kwashiorkor are prevalent. Other legumes such as soybeans and Great Northern beans could be substituted for black gram in the preparation of idli, which may provide a meaningful outlet for these underutilized legumes.

REFERENCES

1. Steinkraus, K. H., Ed., *Handbook of Indigenous Fermented Foods,* Marcel Dekker, New York, 1983, 131.
2. Reddy, N. R., Sathe, S. K., Pierson, M. D., and Salunkhe, D. K., Idli, an Indian fermented food: a review, *J. Food Quality,* 5, 89, 1981.
3. Reddy, N. R., Salunkhe, D. K., and Sathe, S. K., Biochemistry of black gram *(Phaseolus mungo* L.): a review, *CRC Crit. Rev. Food Sci. Nutri.,* 16, 49, 1982.
4. Reddy, N. R., Pierson, M. D., Sathe, S. K., and Salunkhe, D. K., Legume-based fermented foods: their preparation and nutritional quality, *CRC Crit. Rev. Food Sci. Nutri.,* 17, 335, 1982.
5. Ramakrishnan, C. V., Studies on Indian fermented foods, *Baroda J. Nutr., (India,)* 6, 1, 1979.
6. Radhakrishnamurthy, R., Desikachar, H. S. R., Srinivasan, M., and Subramanyan, V., Studies on idli fermentation. II. Relative participation of black gram flour and rice semolina in the fermentation, *J. Sci. Ind. Res.(India),* 20C, 342, 1961.
7. Khandwala, P. K., Ambegaokar, S. D., Patel, S. M., and Rao, M. V. R., Studies on fermented foods. I. Nutritive value of idli, *J. Sci. Ind. Res. (India),* 20C, 275, 1962.
8. Rao, M. V. R., Some observations on fermented foods, in Progress in Meeting Protein Needs of Infants and Preschool Children, Publ. No. 843, National Academy of Sciences, National Research Council, Washington, D. C., 1961, 291.
9. Joseph, K. M., Rao, M. N., Indiramma, K., Swaminathan, M., and Subramanyan, V., Studies on the nutritive value of idli fortified with Indian multipurpose food, *J. Sci. Ind. Res. (India),* 20C, 269, 1961.
10. Mukherjee, S. K., Albury, M. N., Pederson, C. S., van Veen, A. G., and Steinkraus, K. H., Role of *Leuconostoc mesenteroides* in leavening the batter of idli, a fermented food of India, *Appl. Microbiol.,* 13, 227, 1965.
11. Padhye, V. W. and Salunkhe, D. K., Biochemical studies on black gram (*Phaseolus mungo* L.). III. Fermentation of the black gram and rice blend and its influence on the in vitro digestibility of the proteins, *J. Food Biochem.,* 2, 327, 1978.
12. Reddy, N. R. and Salunkhe, D. K., Effect of fermentation on phytate phosphorus, and mineral content in black gram, rice, and black gram and rice blends, *J. Food Sci.,* 45, 1708, 1980.
13. Reddy, N. R. and Salunkhe, D. K., Changes in oligosaccharides during germination and cooking of black gram and fermentation of black gram/rice blends, *Cereal Chem.,* 57, 356, 1980.
14. Lewis, Y. S., Johar, D. S., and Subramanyan, V., Studies on process simplification in the preparation of a fermented type of foodstuff idli, *Cent. Food Technol. Res. Inst. Bull. (Mysore, India),* 4(10), 257, 1955.
15. Desikachar, H. S. R., Radhakrishnamurthy, R., Rao, G. R., Kadkol, S. B., Srinivasan, M., and Subramanyan, V., Studies on idli fermentation. I. Some accompanying changes in the batter, *J. Sci. Ind. Res. (India),* 19C, 168, 1960.
16. Steinkraus, K. H., van Veen, A. G., and Thiebeau, D. B., Studies on idli — an Indian fermented black gram-rice food, *Food Technol.,* 21, 110, 1967.
17. Aykroyd, W. R. and Doughty, J., Legumes in Human Nutrition, Food and Agricultural Organization Nutri. Stud., No. 19, Food and Agricultural Organization, Rome, 1964, 59.
18. Ramakrishnan, C. V., Study of Indian fermented foods from legumes and production of similar fermented foods from soybeans. 4th Annu. Res. Rep., Maharaja Sayajirao University of Baroda, Baroda, India, 1977.

19. Reddy, N. R., Effects of Cooking and Germination of Black Gram (*Phaseolus mungo* L.) and Fermentation of Black Gram and Rice *(Oryza sativa* L.). Blend on Phytate, α-Amylase Inhibitors, Phytohemagglutinins, and Flatulence Producing Factors, Ph.D. dissertation, Utah State University, Logan, 1981.
20. Rau, S. R., *The Cooking of India,* Time-Life Books, New York, 1969, 208.
21. Pandalai, N. G. and Kurup, P. V., Isolation of organisms from idli paste, *Ind. J. Microbiol.,* 3, 155, 1963.
22. Batra, L. R. and Millner, P. D., Some Asian fermented foods and beverages and associated fungi, *Mycologia,* 66, 942, 1974.
23. Batra, L. R. and Millner, P. D., Asian fermented foods and beverages, in *Developments in Industrial Microbiology,* Underkofler, L. A., Ed., American Institute of Biological Sciences, Washington, D.C., 1976, 117.
24. Ramachandran, K., Formulation and Evaluation of Low Cost Balanced Meals Based on Locally Available Foods for Children in Post Weaning Period, Ph.D. dissertation, Maharaja Sayajirao University of Baroda, Baroda, India, 1968.
25. Akolkar, P. N., Studies on Soyidli Fermentation, Ph.D. dissertation, Maharaja Sayajirao University of Baroda, Baroda, India, 1977.
26. Susheelamma, N. and Rao, M. V. L., Surface active principles in black gram (*Phaseolus mungo* L.) and their role in the texture of leavened foods containing the legume, *J. Sci. Food Agric.,* 25, 665, 1974.
27. Susheelamma, N. and Rao, M. V. L., Purification and characterization of the surface active proteins of black gram (*Phaseolus mungo* L.), *Intern, J. Pep. Prot. Res.,* 12, 93, 1978.
28. Susheelamma, N. and Rao, M. V. L., Isolation and characterization of the arabinogalactan from black gram (*Phaseolus mungo* L.), *J. Agric. Food Chem.,* 26, 1434, 1978.
29. Susheelamma, N. and Rao, M. V. L., Functional role of the arabinogalactan of black gram (*Phaseolus mungo* L.) in the texture of leavened foods (steamed puddings), *J. Food Sci.,* 44, 1309, 1979.
30. Susheelamma, N. and Rao, M. V. L., Effect of simple processing on the properties of protein and polysaccharide from black gram, *J. Food Technol.,* 14, 463, 1979.
31. Susheelamma, N. and Rao, M. V. L., Potentialities of oil seed flours and proteins for replacing black gram components in the textue of leavened foods, *J. Amer. Oil Chem. Soc.,* 57, 212, 1980.
32. Kadkol, S. B., Desikachar, H. S. R., and Srinivasan, M., The mucilagenous principles of black gram (*Phaseolus mungo* L.), *J. Sci. Ind. Res. (India),* 20C, 252, 1961.
33. Rajalakshmi, R. and Vanaja, K., Chemical and biological evaluation of the effects of fermentation on the nutritive value of foods prepared from rice and grams, *Br. J. Nutr.,* 21, 467, 1967.
34. van Veen, A. G., Hackler, L. R., Steinkraus, K. H., and Mukherjee, S. K., Nutritive quality of idli, a fermented food of India, *J. Food Sci.,* 32, 339, 1967.
35. Ananthachar, T. K. and Desikachar, H. S. R., Effect of fermentation on the nutritive value of idli, *J. Sci. Ind. Res. (India),* 21C, 191, 1962.
36. Shah, V., unpublished data, 1978.
37. Rajalakshmi, R., Nanavaty, K., and Gumazha, A., Effect of cooking procedures on the free and total niacin content of certain foodstuffs, *Ind. J. Nutr. Diet.,* 1, 276, 1964.
38. Khandwala, P. K., Studies on Fermented Foodstuffs, Master's thesis, Bombay University, Bombay, India, 1959.
39. Reddy, N. R., Salunkhe, D. K., and Sharma, R. P., Flatulence in rats following ingestion of cooked and germinated black gram and a fermented product of black gram and rice blend, *J. Food Sci.,* 45, 1161, 1980.
40. Ramakrishnan, C. V., Parekh, L. J., Akolkar, P. N., Rao, G. S., and Bhandari, S., Studies on soyidli fermentation, *J. Plant Foods,* 2, 15, 1976.
41. Agrawal, A. K., Pattern of Various Carbohydrates in Soybean in Various Processings, Master's thesis, Maharaja Sayajirao University of Baroda, Baroda, India, 1976.
42. Rao, G. S., Studies on Fermented Foods with Reference to Hemagglutinin Hydrolyzing Bacteria Isolated from Rice-Soyidli Batter, Ph.D. dissertation, Maharaja Sayajirao University of Baroda, Baroda, India, 1978.
43. Sathe, S. K. and Salunkhe, D. K., Fermentation of the Great Northern bean (*Phaseolus vulgaris* L.) and rice blends, *J. Food Sci.,* 46, 1374, 1981.
44. Sathe, S. K. and Salunkhe, D. K., Isolation and partial characterization of an arabinogalactan from the Great Northern bean (*Phaseolus vulgaris* L.), *J. Food Sci.,* 46, 1276, 1981.
45. Rajalakshmi, R., Preschool child malnutrition: pattern, prevalence, and prevention, *Baroda J. Nutr.,* 3, 1, 1976.
46. Lewis, Y. S. and Johar, D. S., Microorganisms in fermenting grain mashes used for food preparation, *Cent. Food Technol. Res. Inst. Bull. (Mysore, India),* 2, 288, 1953.
47. Ramakrishnan, C. V. and Parekh, L. J., Studies on Indian fermented foods, (unpublished report), 1982.

48. Venkatasubbaiah, P., Dwarakanath, C. T., and Murthy, V. S., Microbiological and physico-chemical changes in idli batter during fermentation, *J. Food Sci. Technol. (India),* 21, 59, 1984.
49. Yajurvedi, R. P., Microbiology of idli fermentation, *Indian Food Packer,* 34, 33, 1980.

Chapter 10

DHOKLA AND KHAMAN

B. B. Desai and D. K. Salunkhe

TABLE OF CONTENTS

I. INTRODUCTION

Dhokla and khaman are similar to idli in appearance and texture; however, the two former fermented foods are prepared somewhat differently than idli. While idli is prepared and consumed throughout India, dhokla and khaman are generally limited to the state of Gujarat (northern India) and parts of neighboring states. The origin of dhokla and khaman is not known.[1] Dhokla and khaman are used mainly as breakfast food and sometimes as supplements to lunch. These foods serve as important sources of protein, calories, and vitamins in the diets of people. The popularity of these fermented foods is due to the organoleptic qualities (texture and taste), nutritional quality and improved digestibility.[1-3] Ramakrishnan and his co-workers[1,4] at M. S. University of Baroda, Baroda, Gujarat, India have carried out extensive research on these fermented foods. They isolated and identified the bacteria involved in the fermentation of dhokla and khaman and studied the nutritional quality of these foods. This chapter presents a description of general procedure for preparing dhokla and khaman and summarizes the information available on the nutrient content, nutritional quality, and acceptability of these foods.

II. PREPARATION OF DHOKLA AND KHAMAN

A. Ingredients

The ingredients used in the preparation of dhokla are polished white rice and bengal gram cotyledons (Figure 1). However, both of these ingredients can be substituted for i.e., coarsely ground meal of wheat (suji), maize, or kodri (*Paspalum scorbiculatum*) for rice, and soybean cotyledons, split peas, red gram (*Cajanus cajan*) cotyledons, or moth bean (*Phaseolus aconitifolitus* Jacq) cotyledons for bengal gram. Some additional seasonings and chopped green leafy vegetables are added to the fermented batter before steaming for taste. The bengal gram cotyledons are used in the preparation of khaman (Figure 2).

B. Preparation

The flow diagrams for preparation of dhokla and khaman are presented in Figures 3 and 4.

1. Dhokla

The ingredients, bengal gram cotyledons and polished white rice, are washed with water to remove adhering dust and surface microorganisms, and soaked in water for 5 to 10 hr separately. The soaked ingredients are ground in a stone mortar separately to obtain a fine paste and mixed together with salt (approximately 1.0% w/w). The combined batter is incubated in a warm place (30 to 32°C) for overnight (12 to 14 hr). Chopped green fenugreek leaves are added to the fermented batter for taste. The fermented batter is poured into greased cups and steam heated in a larger pan for 10 to 15 min. The product is cooled, cut into small pieces (diamond shape) (Figure 1), and seasoned with cracked mustard seeds in groundnut oil.

2. Khaman

The bengal gram cotyledons are washed with water to remove adhering dust and surface microorganisms and soaked in water for 5 to 10 hr. The soaked bengal gram cotyledons are ground in a stone mortar to obtain a coarse batter. Salt is added to the coarsely ground batter at a level of 1.0% w/w, mixed well, and fermented overnight (12 to 14 hr) in a warm place (30 to 32°C). The fermented batter is poured into greased cups and steam heated in a larger pan for 10 min. The steamed khaman is cooled and cut into small pieces (Figure 2) and consumed with soup or chutney.

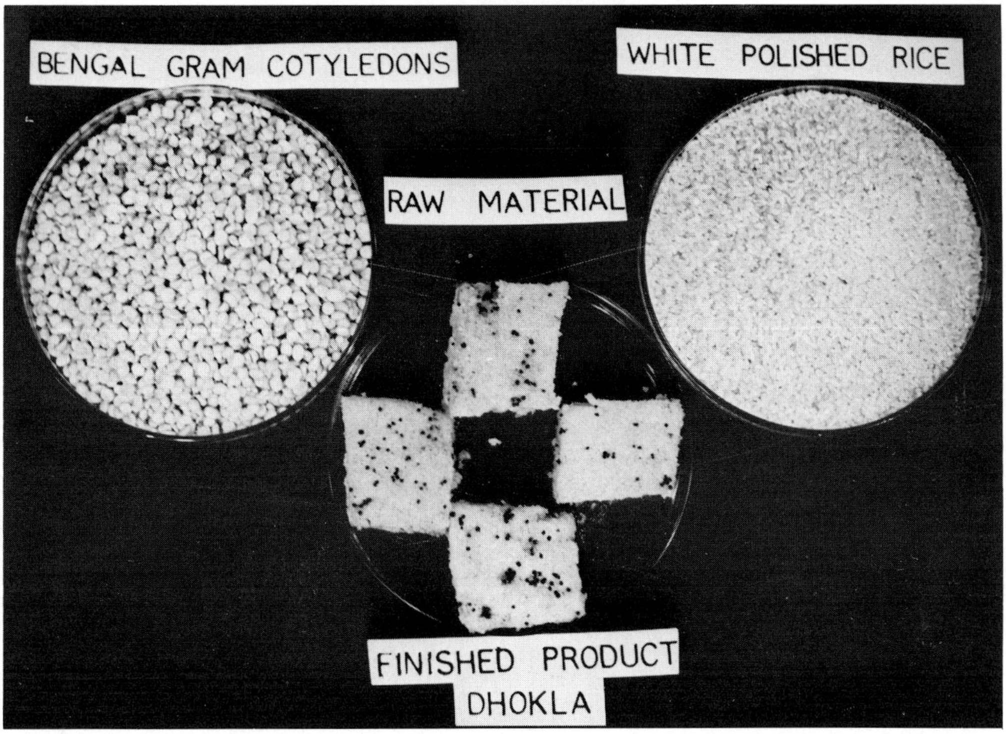

FIGURE 1. Raw material and the finished product, dhokla.

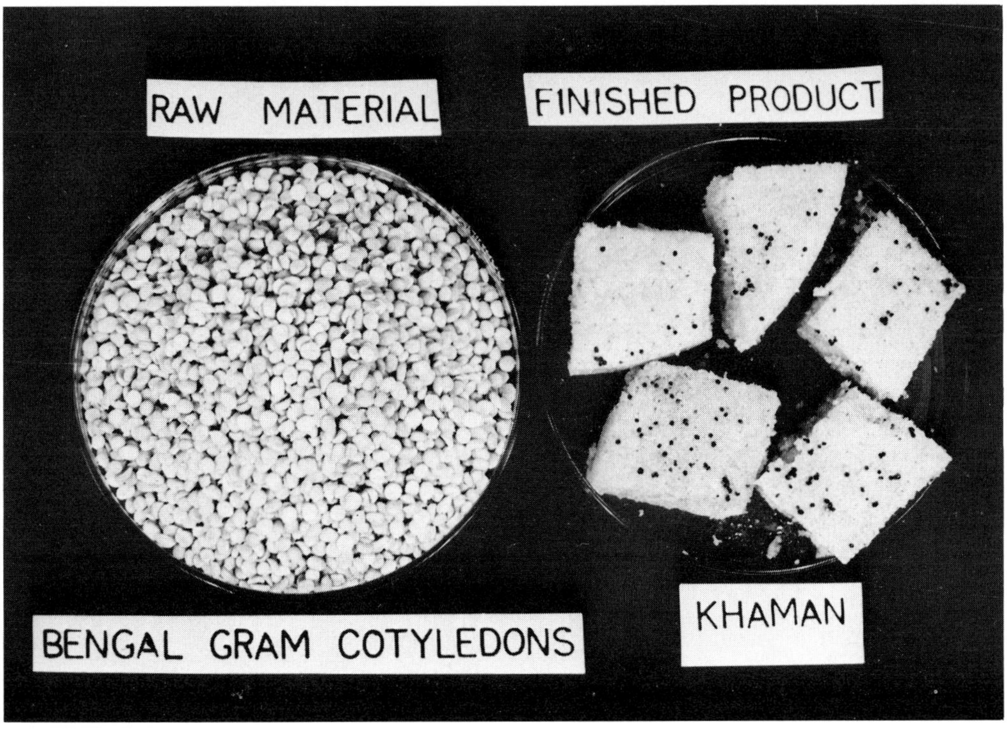

FIGURE 2. Raw material and the finished product, khaman.

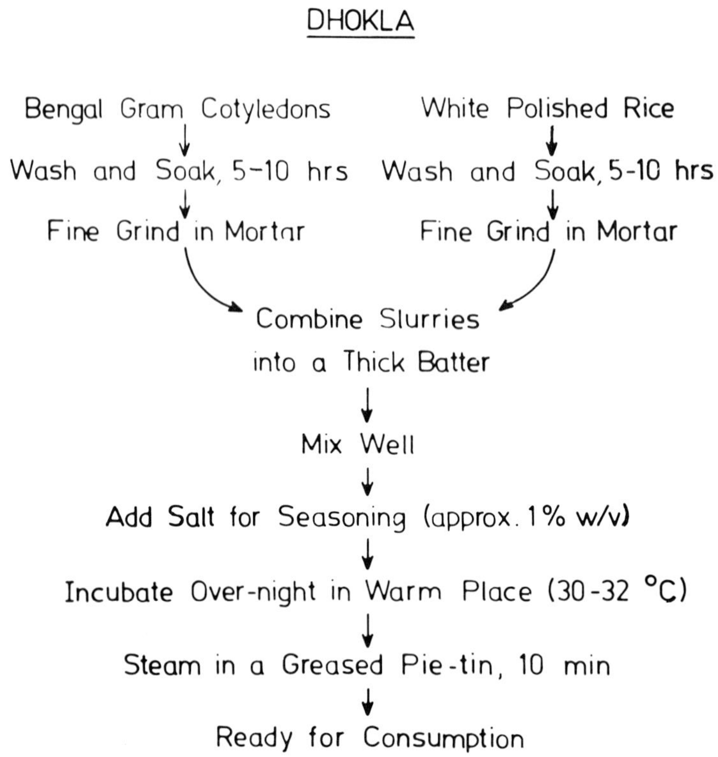

FIGURE 3. Flow diagram for the preparation of dhokla.

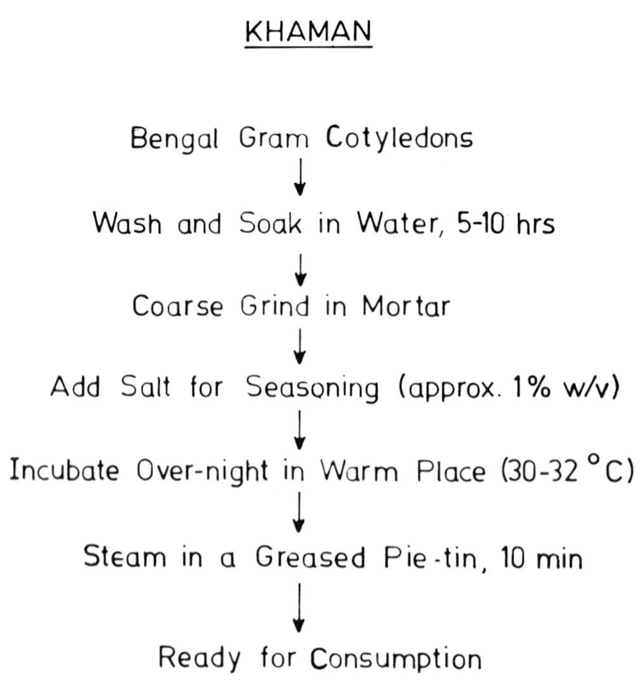

FIGURE 4. Flow diagram for the preparation of khaman.

Table 1
INGREDIENTS USED IN THE PREPARATION OF DHOKLA AND KHAMAN[4]

Ingredients	Weight (g)	Standard measure
	Dhokla	
Cream of wheat (*suji*)	100.0	1 cup
Bengal gram flour	50.0	1/2 cup
Salt	6.5	1 teaspoon
Chopped fenugreek leaves	30.0	1/4 cup
Groundnut oil	4.0 to 5.0	1 teaspoon
Mustard seeds	1.0	1/6 teaspoon
Water	150.0	1 1/2 cup
	Khaman	
Bengal gram cotyledons	100.0	1 cup
Salt	6.0	1 teaspoon
Groundnut oil	4.0 to 5.0	1 teaspoon
Water	150.0	1 1/2 cup

Table 2
BACTERIAL FLORA IN THE RAW INGREDIENTS USED FOR PREPARATION OF DHOKLA AND KHAMAN[1]

Bacterial flora	Bengal gram cotyledons	Bengal gram flour	Soybean cotyledons	Rice	Wheat
Total count ($\times 10^4$/g)	0.19	11.1	29.5	2.25	2.55
Individual counts (% of total counts)					
Leuconostoc mesenteroides	40	45	40	33.3	23.1
Lactobacillus fermenti	20	10	—	16.6	—
Lactobacillus delbrueckii	—	—	—	8.3	—
Lactobacillus lactis	10	5	7.1	—	—
Streptococcus faecalis	—	15	20	8.3	30.6
Pediococcus acidilactici	10	15	20	16.5	30.6
Bacillus spp.	20	10	—	16.6	23.2

Ramakrishnan[4] used various ingredients for preparation of acceptable quality of dhokla and kahman (Table 1).

C. Microorganisms Involved

Researchers at the M. S. University of Baroda, India, recently isolated and identified the microorganisms in the raw ingredients (Table 2) as well as in the fermented batters of dhokla and khaman (Table 3). The total bacterial counts increased from 2×10^7 per g to 3×10^8 per g during fermentation of dhokla and from 3×10^6 per g to 3×10^{10} per g during fermentation of khaman (Table 3). The most predominant bacteria involved in the fermentation of dhokla and khaman are *Leuconostoc mesenteroides* and *Lactobacillus fermenti* (Table 3). Several other microorganisms were also found in dhokla and khaman batters. Rajalakshmi and Vanaja[2] reported that the *Lactobacillus* sp. was the predominant organism in khaman. The bacterial species which were present in the raw ingredients in appreciable amounts could not be detected in the fermented batters of dhokla and khaman. On the other hand, species like *L. mesenteroides* and *L. fermenti* increased during fermentation.

Table 3
BACTERIAL FLORA OF FERMENTED BATTERS OF DHOKLA AND KHAMAN[1]

Bacteria present	Dhokla	Khaman
Total count (× 10^9/g)		
Unfermented	0.02	0.003
Fermented	0.36	33.0
Individual bacterial count as % of total counts		
Leuconostoc mesenteroides	44.4	54.0
Lactobacillus fermenti	44.4	29.0
Lactobacillus lactis	—	9.7
Streptococcus faecalis	11.1	—
Pediococcus acidilactici	—	3.2
Bacillus spp.	—	3.2

D. Physico-Chemical and Organoleptic Characteristics

Several researchers[2-6] observed an increase in batter volume (leavening), amino nitrogen, soluble acids, and a decrease in total sugars and pH as a result of fermentation of dhokla and kahman batters. Acidification and leavening (change in batter volume due to gas production) are the two important changes that occur during fermentation of dhokla and khaman batters. These two characteristics are reported to be responsible for the sour taste and porous texture of products (Figures 1 and 2), dhokla and khaman. The sour taste and porous texture make dhokla and khaman more appealing to the people of Gujarat. The other desirable changes that occur during fermentation of dhokla and khaman include a partial breakdown of protein, starch, and some of the antinutritional factors.

Lakshmi[6] conducted studies by inoculating a mixture of microorganisms (*L. mesenteroides* (HA), *Lactobacillus delbrueckii, L. fermenti,* and *Bacillus* spp.) isolated from idli into sterilized batters of dhokla and khaman. She fermented the inoculated dhokla and khaman batters overnight at 30 to 32°C. She further reported that the products, dhokla and khaman, prepared from the batters inoculated with idli organisms compared well with the products prepared from the naturally fermented batters in terms of taste, texture, and acceptability (Table 4).

III. NUTRITIONAL QUALITY

Unfermented legumes such as bengal gram cotyledons are generally considered hard to digest, but the fermented foods prepared from them, dhokla and khaman, are suitable for young children and patients with digestion disorders.[1] The changes brought about during fermentation of dhokla and khaman should result in increased digestibility and greater protein value. A similar increase in carbohydrate digestibility can also be expected in these foods as a result of fermentation.

A. Nutritive Value

The nutritional composition of dhokla and khaman is presented in Tables 5, 6, 7, and 8. Rajalakshmi and Vanaja[2] studied the effect of fermentation on the nutrient composition and nutritional value of khaman. Fermentation increased the thiamin (49%) and riboflavin (188%) content, and decreased phytate (36%) in khaman (Table 5). They studied the nutritive value of khaman by feeding weanling rats with diets containing either unfermented khaman or fermented khaman. Weanling rats fed on fermented khaman were reported to be superior to those fed on unfermented khaman with regard to weight gain, PER, nitrogen retention, thiamin and riboflavin contents

Table 4
CHANGES IN PHYSICAL AND CHEMICAL COMPOSITION OF DHOKLA AND KHAMAN DURING EXPERIMENTAL FERMENTATION

	Dhokla	Khaman
Percent increase in volume	51	55
pH after fermentation	5.1	5.4
Percent of unfermented values		
Amino nitrogen	129	219
Free sugar	408	676
Thiamin	158	136
Riboflavin	213	300
Niacin	173	190
Inorganic P	288	548
Sensory evaluation (mean score)		
Taste	3	2
Texture	2	1
Color	1	1

Note: A batter of finely ground flours at 15 lb psi for 10 min was prepared with sterile water and inoculated with *L. mesenteroides* (HA), *L. fermenti*, *L. delbrueckii*, and *Bacillus* spp. at a concentration of 8, 8, 16, and 4×10^6/g of autoclaved batter, respectively.

[a] 3= Very good, 2 = good, 1 = acceptable, 0 = not acceptable.

Table 5
NUTRITIONAL COMPOSITION OF FERMENTED AND UNFERMENTED KHAMAN[2]

Component	Unfermented khaman	Fermented khaman
Protein (%)	21.9 (21.8 to 22.0)	22.0 (21.9 to 22.4)
Thiamin (mg%)	0.53 (0.49 to 0.54)	0.79 (0.77 to 0.84)
Riboflavin (mg%)	0.26 (0.24 to 0.28)	0.75 (0.70 to 0.80)
Total phosphorus (mg%)	322 (311 to 333)	270 (259 to 282)
Phytate phosphorus (mg%)	211 (193 to 215)	135 (114 to 156)

Note: Values are means of 4 to 7 determinations with range shown in parentheses.

Table 6
NUTRITIVE VALUE OF FERMENTED AND UNFERMENTED KHAMAN FED TO RATS[2]

	Unfermented khaman	Fermented khaman
Initial body weight (g)	45	45
Food intake (g per 4 weeks)	158	156
Weight gain (g per 4 weeks)	37	45
Protein intake (g per 4 weeks)	29	31
Protein efficiency ratio (weight gain/g protein eaten)	1.3 ± 0.02	1.5 ± 0.05
% of absorbed N retained	74 ± 1.4	82 ± 0.9
Hemoglobin (g/100 mℓ)		
Initial	11.7 ± 0.55	10.9 ± 0.37
Final	14.8 ± 0.36	15.6 ± 0.16
Liver composition		
Protein (g/100g)	15.8 ± 0.37	16.3 ± 0.32
Thiamin (mg/100g)	0.94 ± 0.04	1.21 ± 0.08
Riboflavin (mg/100g)	2.68 ± 0.14	3.66 ± 0.27
Glutathione (mg/100g)	0.23 ± 0.05	0.39 ± 0.11
Xanthine oxidase[a]	6.8 ± 0.46	11.0 ± 1.16
Succinic dehydrogenase[b]	11.1 ± 0.54	25.8 ± 2.4

Note: Weanling albino rats were used, number of animals per group was 9 to 10, period of treatment, 4 weeks. For determination of glutathione, number of animals in each group was 5. Mean values are given with their standard errors.

[a] Enzyme units calculated by dividing 2500 by the time (sec) for reduction of methylene blue.

[b] Moles of 2,3,4-triphenyl tetrazolium chloride reduced per g wet tissue per hour at 37°C under assay condition.

Table 7
PHYSICAL AND NUTRITIONAL COMPOSITION OF UNFERMENTED, FERMENTED, AND STEAMED KHAMAN[1,4]

Component	Unfermented	Fermented	Fermented and steamed
% increase in volume	—	40.0	—
pH	6.3	4.9	
Amino nitrogen (mg%)	178.0	462.5	369.0
Total sugars (%)	37.8	31.3	31.1
Free sugars (%)	0.4	1.1	1.2
Total phosphorus (mg%)	292.0	292.0	292.0
Phytate phosphorus (mg%)	124.0	61.0	52.0
Inorganic phosphorus (mg%)	57.0	141.0	154.5
Niacin (mg%)	4.8	7.6	7.5
Thiamin (mg%)	0.3	0.3	0.3
Riboflavin (mg%)	0.2	0.2	0.2
Trypsin inhibitor (TIU × 10^3/100g)	57.5	45.5	39.0

Table 8
PHYSICAL AND NUTRITIONAL COMPOSITION OF UNFERMENTED, FERMENTED AND STEAMED DHOKLA

Component	Unfermented	Fermented	Fermented and steamed
% Increase in volume	—	31.0	—
pH	6.3	5.3	—
Amino nitrogen (mg%)	53.0	91.0	85.0
Total sugars (%)	44.4	33.3	33.0
Free sugars (%)	1.1	2.1	1.9
Total phosphorus (mg%)	429.0	429.0	429.0
Phytate phosphorus (mg%)	60.0	6.5	1.8
Inorganic phosphorus (mg%)	62.5	165.5	180.5
Thiamin (mg%)	0.6	0.6	0.5
Riboflavin (mg%)	0.27	0.29	0.26
Hemagglutinins (HU × 10^5/100 g)	4.0	4.3	0.3
Trypsin inhibitor (TIU × 10^3/100 g)	175.0	188.0	169.0

Note: Dhokla was prepared by using wheat semolina and bengal gram flour.

of the liver, activities of liver xanthine oxidase and succinic dehydrogenase, and hemoglobin content of blood (Table 6). Recently Ramakrishnan[4] investigated the nutrient composition of khaman during fermentation. He reported that the fermentation of khaman is associated with an increase in amino nitrogen (159%), free sugars (162%), and niacin (58%), and a decrease in phytate phosphorus (51%) and trypsin inhibitors (31%) (Table 7).

The effect of fermentation on the nutritional composition of dhokla was recently studied.[4] Fermentation of dhokla was associated with an increase in amino nitrogen (72%), free sugars (90%), and riboflavin (7%), and a decrease in phytate phosphorus (90%) (Table 8). Ramachandran[5] reported an increase in thiamin, riboflavin, and niacin content during the fermentation of dhokla.

The normal human diets in India are deficient in calories, protein, essential minerals (calcium, iron, zinc, etc.), and vitamins (A and C).[7] Incorporation of additional supplements such as green leafy vegetables and lime mixtures to fermented foods may increase their nutrient content without causing any adverse effects, such as loss of vitamins or availability of essential minerals. Green leafy vegetables are excellent sources of riboflavin, ascorbic acid, and β-carotene. Lime is the cheapest source of calcium available in India. Incorporation of lime at lower levels into fermented foods such as dhokla does not destroy vitamins (Table 9). One can add these supplements to fermented batters before steaming without changing the sour taste and texture.[1] Rajalakshmi and Ramachandran[8] reported that the addition of lime to fermented foods improves the nutritional and bone composition of test rats (Table 10). Ramachandran[5] observed an improvement in skeletal development in preschool children following a diet containing lime, i.e., dhokla containing lime.

A series of field studies on the growth and nutritional status of young children fed fermented foods as a supplementary meal was conducted in the state of Gujarat from 1964 to 1970.[5,8-12] These studies indicate that fermented foods such as dhokla have several beneficial effects on the health of young children by providing significant amounts of nutrients, calories, and vitamins, which are lacking in their normal diet. Recently Ramakrishnan and Parekh[13] reported that the fermented foods dhokla and khaman could be used as supplementary foods to prevent weight loss in poor pregnant women in the first trimester and in the treatment of young children suffering from protein calorie malnutrition, kwashiorkor, and marasmus.

Table 9
EFFECTS OF INCORPORATION OF LIME INTO DHOKLA BATTER ON THIAMIN AND RIBOFLAVIN CONTENTS[5]

Amount of lime powder added (mg/100g)	pH	Thiamin (g/100g)[a]	% Loss	Riboflavin (g/100g)[a]	% Loss
0	5.0	422 (410 to 434)	—	430 (410 to 450)	—
500	6.3	402 (398 to 406)	4.7	420 (410 to 430)	2.3
1500	8.5	326 (301 to 350)	22.0	260 (210 to 290)	39.0

[a] On dry weight basis. Values are means with ranges shown in parentheses.

Table 10
BONE COMPOSITION IN RATS FED DHOKLA, WITH OR WITHOUT LIME[9]

	Dhokla		Lime-treated dhokla		Dhokla with standard calcium salts	
	Tibia	Femur	Tibia	Femur	Tibia	Femur
Length (cm)	2.90	2.88	3.10	3.07	3.00	2.97
Wet weight (mg)	177	285	206	388	200	310
Dry weight (mg)	121	149	146	185	138	175
Fat free dry weight (mg)	111	137	136	173	129	164
Ash (mg)	57	60	67	89	64	79
Calcium (mg)	18.1	22.1	26.5	34.4	24.3	30.4
Calcium (mg/g fat-free dry weight)	164	159	194	198	184	180

Note: Period of treatment = 4 weeks; number of animals per group = 8; values are means per bone.

B. Antinutritional Factors

Cereals and legumes used in the preparation of dhokla and khaman contain several antinutritional factors, namely enzyme inhibitors (α-amylase, chymotrypsin, and trypsin inhibitors), hemagglutinins, tannins, phytates, and oligosaccharides of the raffinose family of sugars.[1] Some of these antinutritional factors are hydrolyzed completely or partially reduced during fermentation by bacteria and their enzymes. Ramakrishnan and his co-workers isolated strains of bacteria capable of hydrolyzing various antinutritional factors in legumes. For example, Rao[14] isolated a strain, *L. mesenteroides* (HA), from idli fermented batter, which has the capability of hydrolyzing hemagglutinins from several legumes.

IV. ACCEPTABILITY OF DHOKLA AND KHAMAN

Dhokla and khaman are well accepted by young children, adults, and patients with digestive disorders because of their organoleptic attributes, nutritional quality, and improved digestibility.[1] Ramakrishnan[4] produced an acceptable quality of dhokla by substituting soybean cotyledons for bengal gram, and wheat semolina for rice. Soy dhokla was well accepted and tolerated by young children. He investigated the effect of soy dhokla on the nutritional status of young children. He reported that feeding young children soy dhokla greatly improved their skeletal and nutritional status compared to those on controls. This study suggests that the further substitution of other unexploited legumes in the preparation of dhokla, khaman, and related foods should be investigated in improving the acceptability and nutritional quality of these foods.

V. CONCLUSIONS

Dhokla and khaman are two important fermented foods of India. These foods serve as important sources of protein, calories, and vitamins. The biochemical and microbiological changes brought about during fermentation of these foods are not clearly understood. Further research is needed on dhokla and khaman regarding protein quality and digestibility, toxicological aspects, mineral content, and bioavailability. Several studies have indicated that dhokla and khaman can be used as a supplementary diet to treat young children suffering from protein calorie malnutrition, kwashiorkor, and marasmus, patients with digestive disorders, and to improve the nutritional status of pregnant women. However, the role of dhokla and khaman in treatment of these nutritional problems is not clearly understood. Substitution of other underutilized legumes for bengal gram in the preparation of dhokla and khaman should be investigated.

REFERENCES

1. Ramakrishnan, C. V., Studies on Indian fermented foods, *Baroda J. Nutr.*, 6, 1, 1979.
2. Rajalakshmi, R. and Vanaja, K. V., Chemical and biological evaluation of effects of fermentation on the nutritive value of foods prepared from rice and grams, *Br. J. Nutr.*, 21, 467, 1967.
3. Steinkraus, K. H., Ed., *Handbook of Indigenous Fermented Foods*, Marcel Dekker, New York, 1983, 131.
4. Ramakrishnan, C. V., Study of Indian fermented foods from legumes and production of similar fermented foods from soybeans, (4th Annu. Res.) Report, Maharaja Sayajirao University of Baroda, Baroda, India, 1977.
5. Ramachandran, K., Formulation and Evaluation of Low Cost Balanced Meals Based on Locally Available Foods for Children in Postweaning Period, Ph.D. dissertation, Maharaja Sayajirao University of Baroda, Baroda, India, 1968.
6. Lakshmi, I., Studies on Fermented Foods, Master's thesis, Maharaja Sayajirao University of Baroda, Baroda, India, 1978.
7. Rajalakshmi, R., Pre-school child malnutrition: pattern, prevalance and prevention, *Baroda J. Nutr. (India)*, 3, 1, 1976.
8. Rajalakshmi, R. and Ramachandran, K., Calcium incorporation in foodstuffs, in *Nutrition and Health*, Proc. VIII. Int. Congr. Nutr., (Hamburg), Kuhnau, J., Ed., Pergamon Press, Elmsford, N.Y., 1967, II, 298.
9. Chandrasekharan, K. N., Nutritional Studies on Children of School Going Age, Ph.D. dissertation, Maharaja Sayajirao University of Baroda, Baroda, India, 1968.
10. Sail, S. S., Nutritional Studies on School Children, Ph.D. dissertation, Maharaja Sayajirao University of Baroda, Baroda, India, 1970.
11. Rajalakshmi, R., Formulation of Low Cost Meals for Postweaning Children in the Age Group of 2 to 5 Years and Their Chemical and Biological Evaluation, Am. P. L.-480 Res. Project Report, Department of Biochemistry, Maharaja Sayajirao University of Baroda, Baroda, India, 1971.
12. Shah, D. G., Nutritional Studies on Preschool Children, Ph.D. dissertation, Maharaja Sayajirao University of Baroda, Baroda, India, 1973.
13. Ramakrishnan, C. V. and Parekh, L. J., Studies on Indian Fermented Foods, (unpublished report), 1983.
14. Rao, G. S., Studies on Fermented Foods with Reference to Hemagglutinin Hydrolyzing Bacteria Isolated from Rice-Soyidli Batter, Ph.D. dissertation, Maharaja Sayajirao University of Baroda, Baroda, India, 1978.

Chapter 11

DAWADAWA

S. A. Odunfa

TABLE OF CONTENTS

I. INTRODUCTION

Dawadawa is the Hausa name for fermented African locust bean (*Parkia biglobosa*). It is the most important food condiment in the entire savanna region of West and Central Africa. Some of the countries where dawadawa is important include the northern areas of Nigeria, Ghana, Togo, Benin, Chad, Sierra Leone, Upper Volta, Gambia, Cameroon, Ivory Coast, Guinea, Mali, Senegal, and the semi-desert country of Niger.[1] Fermented foods constitute a significant part of the diet of many West Africans.[2]

Dawadawa is also known as *iru* in Yorubaland of Nigeria, *ogiri-igala* in Iboland, *kpalugu* among the Kusasis and Dagombas of northern Ghana, *kinda* in Sierra Leone, and *netetou* or *soumbara* in Gambia. Dawadawa has undoubtedly been produced in West Africa for centuries, but as with many other traditional arts, there are no written records of its origin. The history of dawadawa can be traced back several centuries to when locust bean trees were traditionally first planted around communities. A Scottish surgeon, Mungo Park encountered locust bean trees when he explored the Niger Basin from 1795 to 1799. In 1804 to 1806 he described this tree in his "Travels in the Interior Districts of Africa".[3] In fact, the locust bean tree was subsequently named after him by Robert Brown in 1826.[4]

The taxonomy of locust bean trees has been in a state of flux until recently. Because of this, various reports in the literature have referred to this same tree as *Parkia filicoidea, P. bicolor,* and *P. clappertoniana.* The currently accepted nomenclature is *P. biglobosa* with *P. clappertoniana* as the accepted synonym.[4,5] The related species *P. filicoidea* is indigenous to the forests of East and Central Africa. The *P. bicolor* (*Igba-odo* yor.) is found in the forest regions of West Africa; the fruits of it are not eaten by humans, but they serve as food for forest animals.[4] The locust bean tree is planted mainly because of the value of its fruit. These fruit provide a constant source of valuable protein in the dry season (from January to March). The locust bean tree is also used for medicinal purposes[6] and as a source of mouthwash to relieve toothaches.[7] The bean husks (seed coats) are used with indigo dye to improve the luster of fabrics, while the tree bark yields a red tannin for dying leather.[7]

A. Quantity Produced

More than 100 million inhabitants of West Africa use dawadawa as a foodstuff. Despite its wide use as a food, reliable figures for the production of dawadawa are difficult to obtain. Most of the beans are collected by individuals from their farms or surrounding parkland. Cobley and Steel[8] estimated that about 200,000 tons of beans are gathered each year in northern Nigeria alone. In addition, large quantities are produced in the savanna regions of Oyo and Kwara states in southwestern Nigeria. Some of the beans produced in northern Nigeria are sold to the Yorubas in southwestern Nigeria. In southwestern Nigeria, there is a shortage of locust beans because half of the area is rain forest.

Each locust bean tree yields about 25 to 52 kg of pods from which 6 to 13 kg of beans may be obtained.[9,10] About 250,000 tons of locust beans are produced, from which 170,000 tons of dawadawa is made.

B. Consumption Patterns

Dawadawa is a strong smelling product and is normally used as a soup or stew flavoring. Low-income families use dawadawa as a low-cost meat substitute and they generously add it to soups or stews and sorghum or millet-based dumplings and porridges. The Ekitis of the Ondo state of Nigeria advertised this food "here is dawadawa, the premium meat of the stew". As the population of Africa increases, there will be more demand for animal protein sources as well as other protein sources such as dawadawa or similar vegetable protein products. Stews and soups are essential features of the diets of West Africans. The use of stew or soup flavors is widespread in West Africa. Consumer surveys in Nigeria show that flavoring materials constitute an essential ingredient in the preparation of local soups or stews.[11]

The quantity of dawadawa consumed varies with the country and within the country. The average per capita per day consumption of dawadawa in Togo and Ghana is 4 and 2 g, respectively.[12,13] The Yorubas of southwestern Nigeria consume 10 g per day per person.[14] Overall, consumption estimated for parts of Nigeria range from 1 to 17 g per person per day.[14-16]

Although dawadawa is used as a flavoring, it contributes to the calorie and protein intake. Simmons[17] reported that the average daily per capita cal intake of dawadawa is 32.6 cal out of the 2300 total cal intake in Zaria province. The daily per capita intake of protein is higher from dawadawa than from poultry, fish, eggs, and milk, but less than from beef. Nicol[16] reported that the mean daily protein intake from dawadawa in Bero-Okuta of southwestern Nigeria is 3.1% of the total protein intake. Dawadawa constitutes a prominent percentage of the protein intake amoung the low-income rural population because it is easily available. In Ibadan, Nigeria, dawadawa is the second cheapest source of protein after groundnuts and dried fish.[18] Ogunbunmi and Bassir[19] reported that dawadawa contributes appreciable amount of protein to the diet of Nigerians. In Zaria province, 100 g dawadawa costs about 15 pence (23 cents) compared to animal protein, which is 30 pence (45 cents) per 100 g.[17]

II. METHOD OF PREPARATION

The processing of locust beans into dawadawa is carried out exclusively by women as a specialized trade and commercial activity, hence dawadawa is almost always a purchased food.

A. Raw Materials

The mature pods of the African locust bean (*Parkia biglobosa*) occur in large bunches. Each pod may vary between 12 and 30 cm in length. The mature pod contains a yellow dry powdery pulp (*dorowa* Hau.) in which a number of dark brown or black seeds are embedded. The seeds are removed from the pods by pounding the pods gently in a mortar and sieving them to remove the powdery yellow pulp. Most dawadawa producers obtain the dried seeds (Figure 1A) from the local markets.

In place of locust bean seeds, local varieties of soybeans are used as a substitute to make dawadawa in the Benue and Plateau states of Nigeria. In the Zaria province of Nigeria, hemp seeds (*Hibiscus cannabinus*) and red sorrel seeds (*Hibiscus sabdariffa*) are used to prepare dawadawa baso (another type of dawadawa).[17,20]

B. Commercial Preparation of Dawadawa

The preparation of dawadawa is a traditional family art practiced as a rural cottage industry. A flow diagram for dawadawa preparation is presented in Figure 2. Dried locust beans (after removing all adhering pulp) are boiled in water in a covered earthenware or metallic pot for up to 24 hr to soften the seed coat or testa (Figure 3A). Additional water is added as the original water is evaporated. Dawadawa baso may also be added to the beans during boiling to enhance cooking. The boiled seeds are put in a mortar and pressed by foot (Figure 3B) to remove the softened testa.[21] Alterna-

FIGURE 1. Top, locust beans, bottom, dawadawa.

tively, they may be pounded with a pestle. Sand or other abrasive material such as wood ash is added to aid in removal of testa.[17,22] Other methods include rubbing the seeds against the walls of a basket[23] or rubbing them between the palms.[24] The cotyledons are then washed thoroughly and the testa removed using baskets or special sieves such as earthenware perforated calabash or pots (Figure 3C). This requires large quantities of water and in some places, it may be carried out near a stream. Hard unswollen beans are removed during washing.

The washed bean cotyledons are boiled again for 1 to 2 hr in a metallic pot. A softening agent locally called *kuru* (Yor.) or *kanwa* (Hau.) may be added during the second boiling to aid in softening of bean cotyledons, if a softer variety of dawadawa is desired. Kuru contains sunflower seed, whereas kanwa is a native potash rock salt containing mostly potassium carbonate and bicarbonate.

After the second boiling, hot bean cotyledons are drained through raffia sieves or baskets and spread on clean calabash trays in layers about 4 cm in depth. In some areas, earthenware pots are placed in a shallow heap or in a hole in the ground and lined with leaves in place of calabash trays. Some of the yellow mealy pulp, called dorowa (Hau.), may be sprinkled over cooked bean cotyledons at the start of fermentation.[17] On rare occasions, millet flour and wood ash are sprinkled on the cotyledons. Dorowa has a high fermentable carbohydrate content (19% reducing sugars, 9% non-

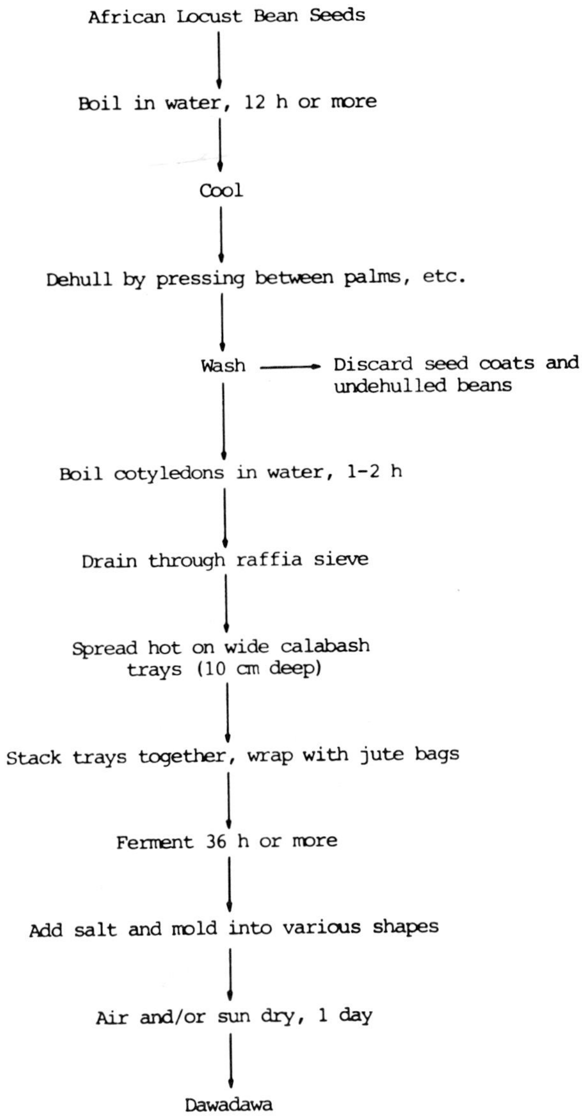

FIGURE 2. Flow sheet for preparation of dawadawa.

reducing sugars, and 36% other carbohydrates).[25] Presumably, this pulp supplies readily available sugars that stimulate the rapid growth of the microorganisms involved in the fermentation. The pulp may also be a source of inoculum.

About 2 or 3 calabash trays containing cooked bean cotyledons and pulp are stacked together and wrapped with 2 or 3 layers of jute sacks or coarse cotton cloth to provide a warm humid atmosphere (Figure 3D). The fermenting beans may be covered with leaves of trees such as baobab, silk cotton, and shea butter tree.[24] The hot seed cotyledons also provide the moisture that increases the initial humidity for fermentation. The cotyledons are left to ferment for 36 hr or longer, depending on local practice. In southeastern Nigeria, the Yorubas allow 2 days for fermentation, whereas in Ghana, fermentation may be 4 days.[26]

The temperature increases during fermentation from the ambient of 25 — 30°C to 45 — 50°C.[21] Campbell-Platt[27] reported a fermentation temperature of 37.5°C after

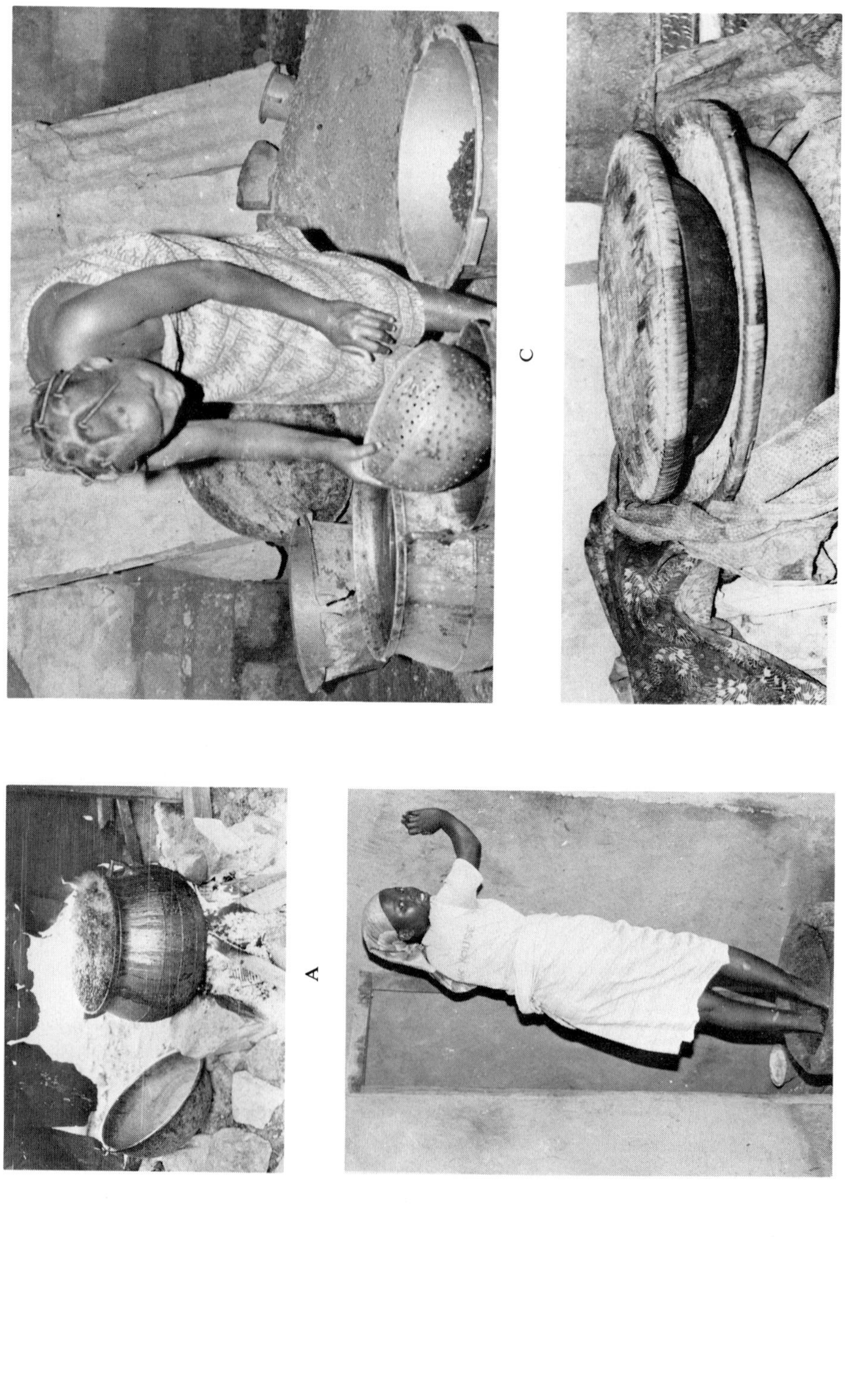

FIGURE 3. Steps in preparation of dawadawa. (A) Boiling of beans. (B) Foot dehulling of beans in a mortar. (C) Separation of cotyledons from hulls with a calabash sieve. (D) Incubation trays containing cooked locust bean cotyledons with cloth used for wrapping.

48 hr and 39°C after 36 hr when a starter culture was used. Fermentation is terminated by removing the jute cover from the trays, exposing the fermenting beans to the air.

A variety of bacteria grows in the fermenting beans during fermentation and produces a mucilaginous substance that covers and links the individual bean cotyledons. The fermented beans have a strongly proteolytic and slightly ammonia-like smell. During the 3-day fermentation, bean cotyledons change in color from light brown to dark brown and become softer.

Two kinds of dawadawa are produced by the Yorubas of southwestern Nigeria. The normal type, *Woro* (Yor.), has distinctly discernible cotyledons and a moisture content of 56%. The softer variety, *Pete* (Yor.), has 65% moisture and contains partly mashed bean cotyledons. The softer variety is prepared by boiling bean cotyledons with potash rock salt.

About 70 kg of fresh dawadawa is obtained from 100 kg locust bean seeds.

To make soybean dawadawa, the soybeans are first fried to a brown color. The fried beans are ground to remove the seed coat or testa. The dehulled soybeans are cooked in water for 3 hr. The cooked beans are drained using a calabash sieve and spread in a basket lined with leaves. Previously fermented soybean dawadawa is added to the basket, the beans mixed, and the basket covered with the same leaves used for lining the basket. The covered baskets are placed in a warm place for 2 to 3 days for fermentation. The fermented beans are sun-dried and pounded to a fine powder.

C. Postfermentation Treatment

The postfermentation treatment of dawadawa varies with local practice. The Yorubas of southwestern Nigeria add salt to the fermented beans. The salted beans are molded into balls that are about 3 cm in diameter. Balls of dawadawa are arranged in calabash trays (Figure 4) for marketing. The calabash trays are covered wth flat raffia trays to keep away flies and other insects. The freshly prepared dawadawa must be sold within 3 days, otherwise it becomes infested with maggots. Refrigeration is not normally available to the dawadawa producers.

The most common postfermentation treatment involves drying the dawadawa. Sun drying yields a stable dark brown to black product. The color is from polyphenol oxidation. Dried dawadawa is usually stored in earthenware pots for up to a year. The Hausas of Nigeria dry dawadawa in the sun for a day or two, subsequently pound it, and make it into flat cakes. Dried root powder of a shrub (*Grewia mollis*) may be added to the cakes, presumably as a binder or preservative.[17] The moisture content of dried dawadawa cake purchased in Zaria is 12%.[28] The extent of drying varies with the locality, and the moisture content may vary between 9 and 19%.[27]

The Mossis of Upper Volta and the Frafras of northern Ghana mold the partially dried dawadawa into round balls (about 5 cm diameter).[27] The Kusasis of northeast Ghana and the Dagartis of southwest Ghana pound the fermented beans into a paste and make them into round balls or pyramids.

The salt that is added to the fermented dawadawa before sundrying serves only as a temporary preservative. The use of salt as a preservative could be more effective if the right concentration were used. A salt concentration of 5% (w/w) inhibits both the growth of *Bacillus subtilis* and its proteolytic activity.[29] Addition of salt also reduces the total volatile nitrogen,[27] hence, salt could be an effective ingredient to prevent further enzymatic activities and fermentation. Salt could also safeguard against food pathogens such as clostridium and salmonella and halt the development of undesirable volatile amines, which give a putrefactive odor to fermented foods.[30]

FIGURE 4. Calabash trays with dawadawa balls displayed in a market in Yorubaland.

D. Upgrading Production Technology

Like other African fermented foods, the production of dawadawa has remained a traditional family art practiced in homes with rudimentary utensils. Consequently, the production of dawadawa has not increased substantially. Its declining popularity, especially among the growing urban population has led to a rapid increase in the import of foreign soup flavors.

In order to increase the production of dawadawa, it is necessary to modernize production techniques and to optimize processing conditions. This will result in a more efficient process and improve and protect shelf-life.

Studies have been conducted on reducing the cooking time. We have reduced the cooking time by soaking the beans in water for 12 hr before cooking in a pressure cooker for 30 min at 1.05 kg/cm^2. However, it takes more time to remove the seed coats from beans cooked in this manner. In contrast, Osinowo[31] cooked beans in a pressure cooker for 90 min without presoaking, and found that the seed coats were easily removed from the cooked beans.

The optimum time, temperature, and relative humidity for dawadawa fermentation were studied by Adewuyi.[32] She assessed the extent of dawadawa fermentation by measuring aroma development, total aerobic plate count, pH, total soluble amino acids, and moisture content. The optimum temperatures and times for dawadawa fermentation were 35°C and 36 hr, and 40°C and 48 hr. The optimum relative humidity (RH) level for fermentation was 51% at 35°C and 71% at 40°C. These studies show that the time for producing dawadawa can be reduced by 50% (from the traditional 4 to 6 days, to 2 days). While the traditional method takes about 48 hr for preparing the beans and 48 to 96 hr for fermentation, the method developed in our laboratory would require only 12 hr for preparing the beans and 36 hr for fermentation. This method saves time, conserves energy, and would lead to increased production.

III. MICROBIOLOGICAL AND PHYSICO-CHEMICAL CHANGES DURING FERMENTATION

Dawadawa fermentation is a solid-state fermentation. This is quite unlike most other

Table 1
PHYSICAL AND MICROBIOLOGICAL CHANGES DURING FERMENTATION OF *PARKIA* SEEDS[21]

Ferment. time (hr)	Moisture content (%)	Temp. (°C)	pH	Plate count	
				Aerobic	Anaerobic
0	43.0	25	7.0	5.0×10^4	0
12	48.0	30	7.2	2.4×10^6	0
24	54.0	42	7.5	1.2×10^6	1.8×10^4
36	56.0	45	8.1	2.5×10^6	3.0×10^5

food fermentations in Africa which are usually in liquid menstrum.[2] The physicochemical and microbiological changes that occur during fermentation are presented in Table 1.

A. Microorganisms

Only bacteria have been reported to be associated with the dawadawa fermentation. Fungi have not been found in the fermenting mash. Ikenebomeh[33] reported a few fungi in the dawadawa fermentation as contaminants. Campbell-Platt[24] reported that *Penicillium* fermenting oval budding yeasts, and the film yeast *Candida* were incidental contaminants. They constituted 3% of the total microbial isolates. The majority of the bacteria found in dawadawa is aerobic while approximately 10% are anerobic after 36 hr of fermentation (Table 1).

Some of the first nutritional studies on dawadawa were reported in 1907,[34] however, microbiological analysis have only recently been published. Odunfa[21] first reported that the predominant fermentation microorganism was a bacillis, possibly *Bacillus subtilis* and other species. Most recently Odunfa and co-workers confirmed the presence of *Bacillus pumilis, B. licheniformis,* and *B. subtilis* in the fermentation. They found that *B. subtilis* was the predominant species in the fermentation. Similar findings have been reported elsewhere.[27,33] In some Nigerian samples, pediococcus and two varieties of *Staphylococcus saprophyticus* were detected.[21] The two varieties of staphylococcus differ in pigmentation (one yellow and one white) and in their ability to hydrolyze starch.

Campbell-Platt[27] collected numerous dawadawa samples from different countries and identified the associated microorganisms. About 31% of the isolates were *B. subtilis* group, while 13% were other species. In some samples *B. subtilis* constituted 61 to 69% of the total isolates, while all bacillis species were 83 to 93% of all isolates. Lactic acid bacteria, especially lactobacillus and pediococcus constituted 22% of the isolates. Bacteria belonging to the *Bacillus subtilis* group was consistently present in every sample of dawadawa. Adewuyi[32] confirmed the presence of *Bacillus subtilis* strains in dawadawa fermentation. He reported that *B. subtilis* is responsible for production of acceptable dawadawa. The ability of the bacillus strains to bring about dawadawa fermentation is probably due to the proteolytic activity of the organism. All isolates of the *Bacillus subtilis* group from dawadawa are proteolytic, and only a few strains are amylolytic. Campbell-Platt[27] found that Bacillus species composed 95% of both proteolytic and amylolytic isolates, and 76% of lipolytic isolates.

In addition to their proteolytic activity, *Bacillus subtilis* strains are thermotolerant. Many grow at 50°C with an optimum at 35 to 40°C. They are facultative anaerobes and grow over a wide range of pH.[29] They grow very rapidly and reach mid-log phase in 3 to 4 hr. Their rapid growth rate, proteolytic activity, and thermotolerance would enhance their predominance in the fermenting locust beans. *B. subtilis* and *B. licheniformis* use ammonia as a nitrogen source and do not require growth factors.[35] The dominance of *B. subtilis* has also been ascribed to its ability to produce antibiotics which are active against *Streptococcus lactis, S. faecium, Saccharomyces uvarum,* and *B. megaterium*.[27]

Packing clean, dehulled locust beans into fermentation trays while they are hot provides a high initial incubation temperature. This can selectively favor the growth of *Bacillus* spp. and *Lactobacillus plantarum*.[27] The significance of *Lactobacillus plantarum* in dawadawa fermentation is doubtful. It was not isolated from half of the samples analyzed, but was abundant in a few samples of dawadawa. *L. plantarum* lacks the enzyme complement necessary for the major biochemical changes observed in fermenting locust beans, i.e., proteolysis and amylolysis. When a starter culture containing a high proportion of *Bacillus* spp. and *Lactobacillus* spp. is used, the proportion of *Lactobacillus* spp. decreases significantly during fermentation.

Starter cultures are not generally used in dawadawa fermentation. The microorganisms are naturally present in the fermenting beans. *Bacillus subtilis* is a common contaminating microorganism of ubiquitous distribution, and it may come from the air or the calabash trays on which cooked beans are spread. *Staphylococcus* spp. may originate from handling as the cooked beans are spread in the calabash trays. The lactobacilli are often naturally present in plant materials.

B. Physico-Chemical Changes

Several physico-chemical changes occur during fermentation of locust beans. The temperature and moisture content of the fermenting beans increase during fermentation. The pH increases to about 8.1 during the first 30 hr of fermentation, due in part to the production of ammonia.

The most significant biochemical change that occurs during dawadawa fermentation is protein hydrolysis. This is due to the high proteinase activity which results in rapid amino acid production.[37] The free amino acid content increases by over 500% during fermentation. Crude proteinase extract of *Bacillus subtilis* has an optimum temperature range of 40 to 45°C; therefore, the proteinase would be most active at the terminal temperature of the fermenting beans.

Odunfa[36] studied the extracellular enzymatic activities of the microorganisms in fermenting locust beans (Figure 5). The enzymes α-galactosidase, β-galactosidase, sucrase, proteinase, amylase, and lipase were detected in fermenting dawadawa. Glycosidases (α- and β-galactosidases, sucrase, and amylase) hydrolyze locust bean oligosaccharides (stachyose, raffinose, and sucrose) and other complex sugars during fermentation. The high temperature which develops in locust beans during fermentation hastens enzymatic activities and bioconversion. The optimum activity of the α- and β-galatosidases occurs at 24 and 12 hr of fermentation, respectively. Invertase activity is highest at 36 hr. The amylase activity in the fermenting dawadawa is fairly low, and evident only in the first 24 hr of fermentation.[37] Presumably, all the starch is hydrolyzed during this period. Starch is not detected in fermented dawadawa.[38] The sugars produced during fermentation through amylolysis provide easily utilizable substrates for the microorganisms.

Although oil constitutes 31 to 40% of the locust bean, lipase activity is low in fermenting locust beans, and there is a fluctuation in activity throughout the fermentation.[37] Ikenebomeh[33] did not detect lipase in a laboratory fermented dawadawa. Of 95 bacterial isolates from 10 dawadawa samples, only 7% were lipolytic, while most of them were proteolytic.[39] Of the six strains of *Bacillus subtilis* characterized in our laboratory, only one strain had lipolytic activity. The lipase in the fermenting locust beans is possibly from the *Staphylococcus* sp. and *B. licheniformis*.[40,41] Campbell-Platt[27] suggested that lipolytic activity occurred in the later stages of locust bean fermentation. This observation was based on the increase in the number of lipolytic microorganisms.

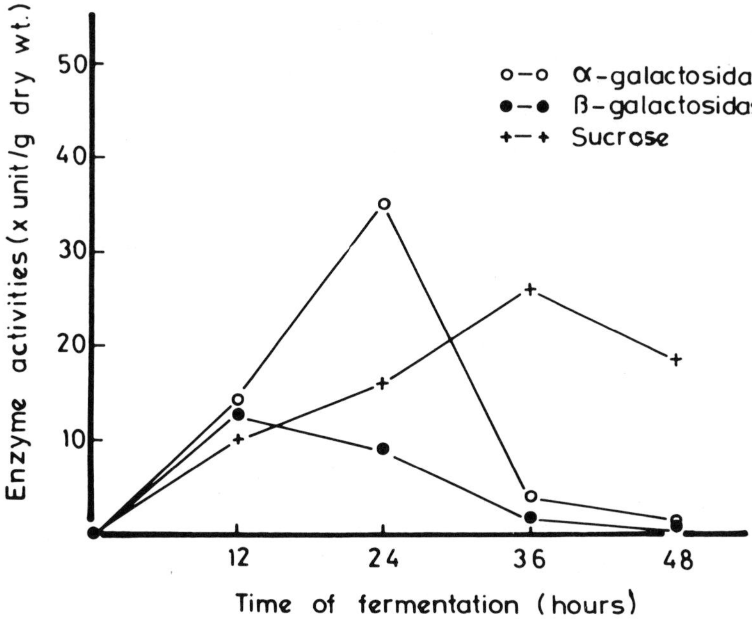

FIGURE 5. Activity of glycosidases in fermenting locust beans.

Another significant change during the fermentation of locust beans is a decrease in percentage of free fatty acids, i.e., from 0.6% in the cooked unfermented locust beans to 0.1% in dawadawa.[39] The decrease in free fatty acids is desirable since large amounts of free fatty acids in foods can result in an objectionable taste and cause rancidity. However, in some fermented foods, free fatty acids produce characteristic flavors.

IV. NUTRITIONAL COMPOSITION AND QUALITY

The nutrient composition of dawadawa is presented in Tables 2 through 5. Campbell-Platt[27] reported that on a moisture-free basis, dawadawa contains 38.5% protein, 31.2% fat, and 23.6% carbohydrate, compared to unfermented locust beans, which have 30% protein, 15% fat, and 49% carbohydrate. The higher protein in dawadawa may be contributed by the microorganisms involved in the fermentation of locust beans. The protein content of *B. subtilis* (the predominant fermentative microorganism) is high, constituting 63% of its biomass.[42] The amino acid content of unfermented and fermented dawadawa is shown in Table 3. The amino acid content of unfermented and fermented dawadawa appears to be very similar. Fetuga et al.[43] observed a small decrease in sulfur-containing amino acids, and a greater decrease in aspartic and glutamic acids as a result of fermentation of locust beans. Like many dry beans, locust beans are low in the sulfur-containing amino acids cysteine and methionine.[43] Eka[28] reported that dawadawa is low in the essential amino acids leucine, isoleucine, phenylalanine, and tryptophan. The deficiency of dawadawa in some of the essential amino acids detracts from the value of dawadawa as a source of high quality protein. However, dawadawa is not consumed alone, but added to stew and other vegetables as a flavoring agent. The essential amino acids in the main meal help to complement the low levels in dawadawa.

Dawadawa contains significant amounts of essential fatty acids.[44] Unsaturated fatty acids account for about 60 to 80% of total lipids in dawadawa.[9,45] Linoleic acid is the major fatty acid in dawadawa lipids. Other fatty acids found in appreciable amounts in dawadawa include palmitic, stearic, oleic, and linolenic acid.[37,45]

Table 2
NUTRITIONAL COMPOSITION OF DAWADAWA

Component	Amount
Crude protein (%)	40.0 — 47.4
Fat (%)	31.4 — 42.9
Carbohydrate (%)	15.0
Crude fiber (%)	3.1 — 7.7
Ash (%)	3.3 — 5.6
Calcium (mg/100 g)	309.0 — 880.0
Iron (mg/100 g)	7.0 — 51.0
Phosphorus (mg/100 g)	517.0 — 584.0
Metabolizable energy (kcal/100 g)	480.0 — 546.0
Gross energy (kcal/100 g)	517.0 — 618.0

Note: On a moisture-free basis.

Table 3
AMINO ACID PROFILE OF UNFERMENTED AND FERMENTED DAWADAWA[28,43]

Amino acid	Unfermented	Fermented
Lysine	6.79	6.17
Histidine	3.14	2.54
Arginine	6.58	6.68
Tryptophan	0.83	0.82
Aspartic acid	8.42	3.49
Threonine	2.45	2.48
Serine	3.66	3.89
Glutamic acid	14.90	10.82
Proline	7.90	7.92
Glycine	3.44	4.66
Alanine	3.79	3.86
Cystine	1.80	1.68
Methionine	0.83	0.58
Isoleucine	3.41	3.62
Leucine	6.22	6.44
Tyrosine	1.98	1.94
Phenylalanine	3.84	3.97

Note: Expressed as g per 16 g nitrogen.

Unfermented locust beans contain high (up to 49%) amounts of carbohydrates. Much of the available reducing sugars and other carbohydrates are utilized by microorganisms during fermentation. Unfermented locust beans contain the raffinose family of oligosaccharides (raffinose and stachyose) and sucrose. The raffinose family of oligosaccharides are reported to be responsible for flatus production in humans and animals. These sugars and sucrose are decreased significantly during fermentation[36] (Table 4). Fully fermented dawadawa contains very little or low amounts of reducing sugars. Watson[38] found less than 0.1% glucose, 0.3% fructose, and no starch or sucrose in fermented dawadawa. This was later confirmed by chromatographic studies.[36]

Changes in the vitamin content occur during fermentation of dawadawa (Table 5). Watson[38] found that dawadawa contained a higher amount of riboflavin than the raw locust bean cotyledons (0.80 vs. 0.51 mg riboflavin per 100 g dry matter). Several other

Table 4
CHANGES IN OLIGOSACCHARIDES DURING FERMENTATION OF LOCUST BEANS[36]

	Oligosaccharides (mg/g dry weight)		
Ferment. time	Sucrose	Raffinose	Stachyose
Unfermented (0 hr)	31.0 ± 4.0	12.0 ± 2.0	28.0 ± 4.0
Fermented (24 hr)	5.5 ± 0.0	—	5.0 ± 0.8
Fermented (36 hr)	—	—	—

Note: Mean value with plus or minus standard deviation.

Table 5
VITAMIN AND TOXICANT CONTENT OF UNFERMENTED AND FERMENTED DAWADAWA[28,47,48]

Component	Unfermented dawadawa	Fermented dawadawa
Thiamin (mg/100g)	0.65	1.35
Riboflavin (mg/100g)	0.45	1.30
Niacin (mg/100g)	7.50	5.30
Oxalate (g/100g)	0.21	0.12
Phytic acid P (g/100g)	15.00	7.50

researchers also reported similar or greater increases in riboflavin content. Up to 1.30 mg riboflavin per 100 g dry matter was reported in the dawadawa samples collected from Zaria province of northern Nigeria.[17] The results reported by Eka, [28] Whitby,[46] Platt,[47] and Leung et al.[48] showed an increase of up to three-fold in riboflavin content during dawadawa fermentation. The increase in riboflavin content may be due to *Bacillus subtilis* and many of its strains of which are unique in producing high levels of riboflavin in commercial fermentations.[49,50]

There are conflicting reports on the thiamin content of dawadawa. Whitby[26] and Leung et al.[48] reported a decrease of up to six-fold in thiamin of dawadawa during fermentation. In contrast, Eka[28] reported a two-fold increase in the thiamin content of dawadawa during fermentation. Dawadawa contains about 5.20 mg vitamin C (ascorbic acid) per 100 g dry matter, which is lower than the raw beans. However, other researchers have reported much higher levels of vitamin C in dawadawa, varying from 57.0 to 117.2 mg per 100 g dry matter.[51,52] These figures appear to be high. It is generally observed that vitamin C is reduced by heat treatment. Dawadawa also contains appreciable amounts of folate.[53,54] Of 24 foodstuffs analyzed, dawadawa is the second highest in folate content after sweet potatoes.[54] Folate content in dawadawa ranges from 0.89 to 0.95 μg/g on a dry weight basis. Significant amounts of folate can be derived from dawadawa to satisfy the adult RDA requirement for folate.

Dawadawa contains most of the important minerals. With the exception of calcium, which is deficient in the diet of many West Africans, the amount of minerals present in dawadawa and other foods is adequate to meet the RDA requirements.[55]

Umoh and Oke[56] reported the fermented locust beans (fermented for 3 days) were more digestible than raw locust beans. The digestibility of fermented beans averages 97.6% with a range of 93.2 to 99.4.[56] The digestibility, protein efficiency ration (PER), net protein utilization (NPU), and biological value (BV) of fermented locust beans is higher than that of raw seeds[43,56,57] when rats are used as test animals. Fetuga et al.[57,58] showed that both raw and cooked beans have a negative PER, NPU, and BV with rats and pigs. Their results indicated that cooking and fermentation improved the nutritive values. Supplementation of the diets with methionine alone results in positive growth and a very substantial increase in protein values.

Nicol[16] compared the nutrition of grain-eating peasants and yam-eating peasants. In the grain-eating area, tryptophan was the first limiting amino acid with a score of 77, while sulfur-containing amino acids were second limiting with a score of 80. In the yam-eating areas, sulfur-containing amino acids were the first limiting with a score of 50. In both populations, dawadawa form a part of the dietary protein intake. The significance of this study is that cereals are more complementary than yams which have a very low amino acid content.

V. TOXICOLOGICAL ASPECTS

There are few reports on the toxicology of dawadawa. The levels of toxic substances such as oxalic acid, phytic acid, and hydrocyanic acid are high in unfermented locust beans (Table 5). However, some of these toxic substances are reduced during cooking and fermentation of dawadawa (Table 5). Hydrocyanic acid is present, but below a dangerous level, in locust beans.[59,60] Soluble oxalate is toxic to animals. It is present in low levels in dawadawa. Phytic acid and oxalic acid form insoluble salts with mineral elements and prevent their utilization. The decrease observed in the phytic acid and oxalic acid content during fermentation would, however, make more mineral elements available.[28]

Various species of *Parkia* contain nonspecific phytohemagglutinins.[61] These are not found in fermented dawadawa. Phytohemagglutinins are normally destroyed by cooking.

Alozie et al.[62] studied the production of aflatoxin in many Nigerian foodstuffs. No aflatoxin was detected in dawadawa. *Aspergillus flavus* failed to grow when inoculated onto dawadawa. Further, dawadawa extract was found to inhibit several fungi including *A. flavus.* The inhibitory factor in the dawadawa extract was nullified when the pH of the extract was reduced from 8.1 to 6.0. The inhibitory factor is attributed to the alkaline pH, which is unfavorable to mold growth.

VI. CONCLUSIONS AND FUTURE RESEARCH NEEDS

The production of dawadawa offers a method to utilize locust beans as a food, otherwise the locust beans are inedible. Dawadawa is generally used as a flavoring agent rather than as a source of dietary protein or calories. It is commonly substituted for food flavors in urban populations. In Nigeria, about $200 million was spent on importation of food flavors in 1981, with a projection of a 15% annual future increase.[63] This underscores the importance of soup flavors in Nigeria. In order to increase the acceptability of dawadawa as a flavoring agent and production, it is essential to modernize the production process and present the product in the market in better forms. With this improvement, dawadawa has a great potential as a key protein source and as a flavoring condiment.

More research is needed to develop the production of dawadawa to the same standard as that of soy sauce and miso in Japan or cheese in Europe. Past research has concentrated on nutritional aspects. Future research should be aimed at translating the research knowledge into developing modern small-scale technology appropriate for rural areas. Small-scale technologies are preferable as they require less capital to initiate, conserve foreign exchange, and generate more employment. Some of the aspects of production that need to be upgraded are the boiling and dehulling processes, and the postfermentation treatment. In traditional production, boiling takes several hours and is not fuel efficient. Future availability of firewood could be a serious constraint. It is projected that firewood will be in short supply by the turn of the century. There is a need for more quantitative information on the length of pressure cooking to produce beans with a texture comparable to the firewood-cooked beans. The dehulling stage in the traditional method is time-consuming, labor intensive, and requires large quantities of water. Dehulling could be mechanized by use of a roller mill.

There is need for more information on the flavor attributes of dawadawa. Also, the conditions of fermentation that will eliminate or reduce objectionable odors due to ammonia, volatile amines, etc. should be investigated. Effective starter cultures need to be developed to eliminate undesirable microorganisms. The use of starter cultures should yield a product with consistent taste and quality. The appearance of dawadawa could be more uniform from batch to batch. This would improve its marketability. Sun-drying of dawadawa causes the product color to change from dark brown with a greyish covering to black. Research is needed to find suitable drying methods that will retain color and minimize loss of flavor components and vitamins.

With the present rapid expansion in educational opportunities, there is an increasing preference for hygienically-produced, well-packaged food products. Hence, in order to compete effectively with the imported flavor cubes, dawadawa should be dried and well-packaged. The nutritional value of dawadawa, especially its contribution to dietary protein, should be more widely publicized.

The production of locust beans is enough to meet present dawadawa demands, but will be a serious constraint if improved technology is widely adopted. A long-term research need is to produce locust bean varieties that are early-maturing, high yielding, and of shorter height. These attributes are more attractive for plantation agriculture. An alternative is to further develop dawadawa made from the local soybean varieties. This research needs to be done as well on other dry beans as substitutes for locust beans in preparation of dawadawa.

REFERENCES

1. Sizaret, A., The shortage of fruit in Niger, *Fruits,* 27, 485, 1972.
2. Odunfa, S. A., African fermented foods, in *Microbiology of Food Fermentations,* Wood, B. J. B., Ed., Applied Science Publishers, London, in press.
3. Park, M., *Travels in the Interior Districts of Africa,* Nicol and Co., London, 1799.
4. Hagos, T. H., A revision of the genus *Parkia* R. Br. (Mim.) in Africa, *Acta Bot. Neerl.,* 11, 231, 1962.
5. Lowe, J., personal communication, 1983.
6. Anon., *Parkia biglobosa, Kew Bulletin of Miscellaneous Information,* 314, 1908.
7. Oliver, B., Nigeria's useful plants. V. Plants yielding dyes, *Nig. Field,* 26, 170, 1961.
8. Cobley, L. S. and Steel, W. N., *An Introduction to the Botany of Tropical Crops,* 2nd ed., Longman, London, 1976, 73.
9. Busson, F., *Plantes Alimentaires de l'Quest African,* Direction de la Copperation Culturelle et Technique, Paris, 1965, 272.
10. Anon., The composition of dawadawa pods from the Gold Coast, *Bull. Imp. Inst.,* 20, 461, 1922.
11. Anon., Technical memorandum on consumer survey of flavouring condiments in Nigeria, Federal Institute of Industrial Research, Oshodi, Lagos, 1984.
12. Perisse, J., *La Consommation des Legumineuses au Togo,* Cote de classement no 4019, Office de la Recherche Scientifique et Technique Outre-Mer, Paris, 1958.
13. Lawson, R. M., The consumption approach to measuring agricultural production in foodstuffs, *Food Res. Inst. Stud. Agric. Econ. Trade Dev. (Stanford),* 5, 205, 1965.

14. Dema, I. S., *Nutrition in Relation to Agricultural Production,* Food and Agriculture Organization, Rome, 1965, 129.
15. Nicol, B. M., Nutrition of Nigerian peasant farmers with special reference to the effect of vitamin A and riboflavin deficiency, *Br. J. Nutr.,* 3, 25, 1949.
16. Nicol, B. M., The protein requirements of Nigerian peasant farmers, *Br. J. Nutr.,* 13, 307, 1959.
17. Simmons, E. B., Calorie and Protein Intakes in Three Villages of Zaria Province, May 1970 — July 1971, Sumaru Miscellaneous Paper 55, Institute of Agric. Res., Ahmadu Bello University, Zaria, Nigeria, 1976.
18. Ladipo, O. O. and Adegboye, R. O., The economic cost of malnutrition and the price of protein and calories in some selected towns of Oyo, Ogun and Ondo States of Nigeria, *Niger. Nutr. Newsl.,* 1, 100, 1976.
19. Ogunbunmi, E. M. and Bassir, O., Protein and amino acid contents of some Nigerian food condiments, *Nutr. Rep. Int.,* 22, 497, 1980.
20. Anon., *List of Foods Used in Africa,* 2nd ed., Nutr. Inf. Doc. Ser. No. 2, Food and Agriculture Organization, Rome, 1970.
21. Odunfa, S. A., Microorganisms associated with fermentation of African locust bean during *iru* preparation, *J. Plant Foods,* 25, 11, 1981.
22. Nakao, S., *Tanaka's Encyclopedia of Edible Plants of the World,* Keigaku Publishing, Tokyo, 1976.
23. Oke, O. L. and Union, I. B., Lesser know oilseeds, a preliminary note on their nutritive value as determined by *in vitro* digestion, *Ecol. Food Nutr.,* 4, 87, 1975.
24. Campbell-Platt, G., African locust bean (Parkia species) and its West African fermented food product, *dawadawa, Ecol. Food Nutr.,* 9, 123, 1980.
25. Oyenuga, V. A., *Nigeria's Food and Feedstuffs: Their Chemistry and Nutritive Value,* 3rd ed., Ibadan University Press, Ibadan, Nigeria, 1968, 50.
26. Dodo, F. E., Traditional Methods of Processing and Preserving Food in Ghana, Food Res. Inst., Accra, Ghana, 1967.
27. Campbell-Platt, G., Microbiology of Dawadawa — A West African Fermented Food Made From African Locust Beans *(Parkia* species), Ph.D. thesis, University of Strathclyde, Glasgow, U.K. 1980.
28. Eka, O. U., Effect of fermentation on the nutrient status of locust beans, *Food Chem.,* 5, 303, 1980.
29. Preston-Holder, V. O., Factors that Affect the Growth and Protease Activity of the Bacillus Species Isolated from Fermenting Locust Bean (iru), Project Rep., Department of Botany, University of Ibadan, Nigeria, 1981.
30. Mossel, D. A. A., *Microbiology of Foods, the Ecological Essentials of Assurance and Assessment of Safety and Quality,* 3rd ed., University of Utrecht, the Netherlands, 1982, chap. 2.
31. Osinowo, F. A. O., Studies on Nigerian Flavouring Materials; Processing of Fermented African Locust Beans *(Parkia clappertoniana),* Staff seminar paper, Federal Institute of Industrial Research, Oshodi, Nigeria, 1981.
32. Adewuyi, E. Y., Studies on the Optimization of Process Conditions for Locust Bean *(Parkia biglobosa)* Fermentation, Masters dissertation, Department of Botany, University of Ibadan, Nigeria, 1983.
33. Ikenebomeh, M. J., The Solid Substrate Fermentaton of African Locust Beans *(Parkia filicoidea),* Ph.D. thesis, McGill University (McDonald Campus), Montreal, 1982.
34. Fincke, H., Uber die Samen von Parkia Africana R. Br. and den Daraus Hergestellt Daug-Daug-Kase, *Z. Unters, Nahr. u. Genusm. Gebrauchsgegenstaende,* 14, 515, 1907.
35. Knight, B. C. J. G. and Proom, H. I., A comparative survey of the nutrition and physiology of mesophilic species in the genus *Bacillus, J. Gen. Microbiol.,* 4, 508, 1950.
36. Odunfa, S. A., Carbohydrate changes in fermenting locust bean *(Parkia filicoidea)* during *iru* preparation, *Qual. Plant. Plant Foods Hum. Nutr.,* 32, 3, 1983.
37. Odunfa, S. A., Biochemical changes in fermenting African locust bean (*Parkia biglobosa*) during iru preparation, *J. Food Technol.,* in press.

38. Watson, J. D., Investigation on the nutritive value of some Gahanian foodstuffs, *Ghana J. Agric. Sci.*, 4, 95, 1971.
39. Odunfa, S. A. and Adesomoju, A. A., Effects of fermentation on the free fatty acids of African locust bean during *iru* preparation, *J. Plant Foods*, in press.
40. Johnson, V. and Snygg, B. G., Lipase production and activity as a function of incubation time, pH and temperature of four lipolytic microorganisms, *J. Appl. Bacteriol.*, 37, 571, 1974.
41. Vadehra, D. V. and Harmon, L. G., Factors affecting staphylococcal lipase, *J. Appl. Bacteriol.*, 32, 147, 1969.
42. Washen, C. I. and Steinkraus, K. H., The potential of microbial cells as protein for men, *BioScience*, 30, 397, 1980.
43. Fetuga, B. L., Babatunde, G. M., and Oyenuga, V. A., Protein quality of some Nigerian foodstuffs. I. Chemical assay of nutrients and amino acid composition, *J. Sci. Food Agric.*, 24, 1505, 1973.
44. Chevassus-Agnes, S. and Pascaud, M., Fatty acid composition of some cooked dishes of the northern Cameroons (Adamova), *Ann. Nutr. Aliment.*, 16, 7, 1932.
45. Girgit, P. and Turner, T. D., Lesser known Nigerian edible oils and fats. III. Fatty acid composition as determined by gas-liquid chromatography, *J. Sci. Food Agric.*, 23, 259, 1972.
46. Whitby, P., Foods and Ghana, Food Res. Inst. Bull., Accra, 1, 1, 1968.
47. Platt, B. S., Biological ennoblement: improvement of the nutritive value of foods and dietary regimes by biological agencies, *Food Technol.*, 18, 68, 1962.
48. Leung, W. T. W., Busson, F., and Jardin, C., Food Composition Tables for use in Africa, U.S. Department of Health/Food and Agriculture Organization, Bethesda, Maryland.
49. Bacher, A., Baur, R., Eggers, V., Hardens, H. D., Otto, M. K., and Schnepple, H., Riboflavin syntheses of *Bacillus subtilis:* purification and properties, *J. Biol. Chem.*, 255, 632, 1980.
50. Ajinomoto Co., Inc., Fermentative Production of Riboflavin, Great Britain Patent 1, 434, 299, 1976.
51. Nicol, B. M., The nutrition of Nigerian children with particular reference to their ascorbic acid requirements, *Br. J. Nutr.*, 10, 255, 1956.
52. Addo, A. A., Ascorbic acid contents of foods commonly consumed in the northern states of Nigeria, *Niger. Food J.*, 1, 128, 1983.
53. Keshinro, O. O., The free and total folate activity in some commonly available foodstuffs, *Food Chem.*, 11, 87, 1983.
54. Hug, R. S., Abalaka, J. A., and Stafford, W. L., Folate content of various Nigerian foods, *J. Sci. Food Agric.*, 34, 404, 1983.
55. Oke, O. L., A nutrition policy of Nigeria, *World Rev. Nutr. Diet.*, 14, 1, 1972.
56. Umoh, I. B. and Oke, O. L., Nutritive value of some lesser known oilseed in rats, *Nutr. Rep. Int.*, 9, 453, 1974.
57. Fetuga, B. L., Babatunde, G. M., and Oyenuga, V. A., Protein quality of some Nigerian foodstuffs. II. Biological evaluation of protein quality, *J. Sci. Food. Agric.*, 24, 1515, 1973.
58. Fetuga, B. L., Babatunde, G. M., and Oyenuga, V. A., Comparative evaluation with the pig and the rat of the protein quality of some Nigerian protein seeds, *Nig. J. Sci.*, 8, 45, 1974.
59. Oke, O. L., Ascorbic acid content of some Nigerian vegetables, *J. Food Sci.*, in press.
60. Oke, O. L., Oxalic acids in plants used in nutrition, *World Rev. Nutr. Diet.*, 11, 170, 1969.
61. Toms, G. C. and Western, A., Phytohaemagglutinins, in *Chemotaxonomy of the Leguminosae*, Hargone, J. B., Boulter, D., and Turner, B. L., Eds., Academic Press, New York, 1971, 367.
62. Alozie, T. C., Rotimi, C. N., and Oyibo, B. B., Production of aflatoxin by *Aspergillus flavus* (UBMI) in some Nigerian indigenous beverages and foodstuffs, *Mycopathol. Mycol. Appl.*, 70, 125, 1980.
63. Essien, N., Business concord survey of the food seasoning market in Nigeria, *Bus. Concord*, 1, 22, 1983.

Chapter 12

PAPADS

S. R. Shurpalekar

TABLE OF CONTENTS

I. INTRODUCTION

Papads have been popular for centuries as a tasty accessory in the Indian menu. This food is a thin, wafer-like product, usually circular in shape, rolled from a legume-based dough often containing other farinaceous materials, along with added salt, spices, and condiments. It is either toasted or deep fat fried to a crispness which produces a crackle when the papad is eaten. Papads are also known as "appalam" in South India. "Papadam" (popular in Kerala State of India) differs from papad or appalam. Papadam has a higher moisture content and it puffs on frying. Though traditionally confined to households, papad-making, in recent years, has developed into a cottage or small-scale industry. Papad exports have reached a value of more than 10 million rupees per year (Table 1).

Until recently, very little was known regarding the technology of the preparation and preservation of papads. Shelf-life of this perishable product is very important, especially since mold growth poses problems when the product is at sea for several weeks. Extensive studies have been done at the Central Food Technological Research Institute in order to collect the requisite scientific information relating to papads, a very important traditional food item. This chapter reviews the currently available information on papads.

II. THE PAPAD INDUSTRY

A survey of papad samples collected from different areas throughout India has shown a wide variation in shape, size, and thickness, distribution of spices, spice content, taste profile (alkaline taste and saltiness), moisture content, chemical composition, and pH (Tables 2 and 3). Papads differ widely in their physico-chemical characteristics, which in turn influence their acceptability. Packaging of papads has received almost no attention, even though the product is highly susceptible to mold attack. Papads break easily if their moisture content is low. Drying methods are generally empirical and they differ from region to region because of the variation in the climate. This often leads to either an excessively dry product or an inadequately dried product.

Since the methodology and ingredients for preparation differ from region to region, the quality and shelf life of the papads produced vary. Papads from the same production unit may even differ in quality.

Papad khars or *sajji khars* (Figure 1) are traditionally used as additives in the papad formulations. Papad khars are presumably used to improve the rolling properties of the dough and the frying quality of papads. They are ashy in nature and are obtained from saltworts by burning a variety of plant species, or from very alkaline deposits in the soil, e.g., *dhobi's* earth. A few hundred tons of papad khars are produced annually in the Punjab and Sind districts and the west coast of India. They contain mainly carbonates, chlorides, and sulfates. The chemical composition of papad khar is presented in Table 4.

With all the variations in methods and ingredients used in preparing papads, producers still encounter difficulties in obtaining a dough with consistent rolling properties and papads of uniform quality attributes. It is desirable to produce papads with acceptable, uniform quality attributes and satisfactory shelf-life.

III. PREPARATION

A. Ingredients

It is necessary to understand the functional attributes of the main ingredients in papad preparation in order to produce the best product. Black gram flour is the most

Table 1
EXPORT PATTERN OF PAPADS FROM INDIA

Country	1967—1968 Quantity exported (metric tons)	1967—1968 Value (million rupees)	1975—1976 Quantity exported (metric tons)	1975—1976 Value (million rupees)	1979—1980 Quantity exported (metric tons)	1979—1980 Value (million rupees)
U.K.	129	0.52	550	3.20	1284	13.23
Malaysia	149	0.48	85	0.57	69	0.56
Singapore	47	0.15	84	0.57	171	1.49
U.S.	10	0.06	64	0.61	122	1.34
United Arab Emirates	—	—	28	0.20	116	1.03
Total	570	1.89	983	6.52	2206	22.19

Table 2
PHYSICO-CHEMICAL CHARACTERISTICS OF LABORATORY PREPARED PAPADS AND 31 COMMERCIAL PAPAD SAMPLES[1]

Characteristics	Laboratory sample	Commercial samples (range)
Weight (g)	4.3	3.1—24.6
Diameter (cm)	11.5	10.2—23.6
Thickness (mm)[a]		
Minimum	0.27	0.21—0.82
Maximum	0.47	0.48—1.25
Moisture (%)	12.4	9.3—17.1
Ash (%)		
Total	9.8	5.3—13.2
Acid-insoluble	0.6	0.2—1.7
Chloride (as NaCl)	6.7	3.0—9.1
Alkalinity (as Na_2CO_3)	1.6	0.9—4.4
Alkalinity (as $NaHCO_3$)	0.6	0.4—2.1
Protein (N × 6.25) (%)[b]	20.8	18.0—22.0
pH	6.9	5.2—9.1
Fat (dry basis) (%)	0.8	0.6—4.5

[a] Thickness represents the average of 10 papads.
[b] Protein content for three papad samples (sago, jackfruit, and sweet potato) ranged between 0 to 4%.

indispensible constituent in papad dough, because of the mucilaginous substance it contains. The role of each of the main ingredients was studied by Shurpalekar et al.[3] Their results are presented in Table 5 and discussed below.

1. Water

About a 40 to 45% water content and a kneading time of 3 to 4 min are satisfactory conditions for obtaining papad doughs of desired consistency and rolling characteristics, and papads of desired quality attributes.

2. Salt

About 8% salt is optimum for obtaining a dough with the desired characteristics and

Table 3
GRADING OF 31 COMMERCIAL PAPAD SAMPLES BASED ON QUALITY PARAMETERS[1]

Quality parameter	Grading (number of samples)				
	Excellent	Good	Satisfactory	Fair	Poor
Color					
Before frying	6	8	7	7	3
After frying	6	10	9	4	2
Size, shape, and thickness	7	10	7	7	—
Pliability	8	16	4	3	—
Spicing[a]	3	3	8	3	2
Blister formation	2	20	4	2	3
Appearance	6	9	10	6	—
Texture	2	11	14	4	—
Taste	3	4	10	14	—
Expansion on frying	4	3	3	11	8
Overall score	4	4	12	11	—

Note: Grading based on scoring system: excellent, 76—100%; good, 66—75%; satisfactory, 56—65%; fair, 41—55%, and poor, 0—40%. The quality parameter appearance has a maximum score of 40 consisting of scores for color, size, pliability, spicing, and blister formation. The quality parameter texture has a maximum score of 20; taste 20; expansion on frying, 20, with an overall score of 100.

[a] Spices were used in 19 samples; the spices used include chili, cumin seeds, pepper, garlic, and asafetida.

FIGURE 1. Papad (sajji) khar samples of varying colors and hardness.

Table 4
CHEMICAL COMPOSITION OF 13 COMMERCIAL SAMPLES OF PAPAD KHAR[2]

Constituent	Range
Specific gravity/bulk density (g/mℓ)	0.68—1.74
Moisture (%)	71.7—96.5
Ash[a]	
Total ash (%)	0.6—26.3
Acid-insoluble ash (%)	0.0—28.4
Ash as carbonate (Na_2CO_3) (%)	1.6—58.3
Ash as bicarbonate ($NaHCO_3$) (%)	0.0—24.7
Ash as chloride (NaCl) (%)	1.2—72.0
Phosphorus (mg/100 g)[a]	0.0—60.0
Iron as ferric (mg/100g)[a]	0.0—5.3
Total sulfates as sulfur (g/100g)[a]	0.0—2.5
Volatile sulfides as sulfur (g/100g)[a]	0.0—1.6

[a] Dry weight basis.

papads having a balanced and acceptable taste when fried. Excessive addition of salt causes salt bloom on raw papads during storage, while inadequate salt yields an insipid and unacceptable product.

3. Sodium Carbonate

Analytical data on different samples of papad khar or sajji khar led to the choice of sodium carbonate for standardizing the dough and producing papads of desired taste and crispness. Sodium carbonate in the range of 1.0 to 1.5% in the dough gives optimum results. Sodium carbonate in excess of 1.5% yields a product with alkaline taste and a foamy or oily surface when fried. Using an inadequate level of sodium carbonate adversely affects the uniformity of product color and yields a papad that tends to turn leathery.

4. Flour Quality

The particle size of flour is important in order to obtain papad dough of homogeneous consistency that will, in turn, yield papads of uniform taste and texture. Black gram flour with a particle size of 85 mesh is most desirable for uniform absorption of water and development of dough during kneading.

A mixture of 100 parts flour of 85 mesh particle size, 45 parts water, 7 to 8 parts of common salt, and 1.0 to 1.5 parts sodium carbonate, yields the best dough and papads.

B. Method of Preparation

Based on the mixture given above, the recommended method of preparation of papads is as follows.

Black gram flour or a black gram flour blend, is placed in a mixer (Hobart Model C-100) with water, salt, and carbonate. The mixture is kneaded at minimum speed into a homogeneous lump of dough. An optimally kneaded dough does not stick to the sides of the mixing bowl.

The dough is left to rest for about 30 min, then divided into small balls of about 2 cm diameter, each weighing about 5 to 6 g. Fermentation occurs during this resting period. There have been no published reports on this fermentation other than with papadams. The balls are rolled into thin circular discs of about 1 mm thickness, using

Table 5
EFFECT OF INGREDIENTS ON QUALITY CHARACTERISTICS OF DOUGH AND PAPADS BASED ON BLACK GRAM[3]

Ingredient	Hand feel	Rolling	Appearance	Texture	Taste	Diametrical expansion on frying (%)
Water						
35%	Somewhat tough	Difficult	Dull	Somewhat hard, brittle	Unbalanced	28
40%	Soft	Easy	Bright and attractive	Somewhat hard, brittle	Unbalanced	40
45%	Soft	Easy	Bright and attractive	Crisp and brittle	Balanced and acceptable	45
50%	Soft and somewhat sticky	Difficult	Bright and attractive	Crisp and brittle	Balanced and acceptable	45
Common salt						
0%	Tough	Very difficult	Bright and attractive	Crisp and brittle	Unacceptable, bland and floury	22
4%	Somewhat tough	Difficult	Bright and attractive	Crisp and brittle	Unacceptable and somewhat floury	28
8%	Soft	Easy	Bright and attractive	Crisp and brittle	Balanced and acceptable	44
10%	Soft	Easy	Bright and attractive, but salt bloom	Crisp and brittle	Balanced and acceptable	45
12%	Soft	Easy	Bright and attractive, but salt bloom	Crisp and brittle	Salty and unacceptable	44
Sodium carbonate						
0.0%	Soft	Easy	Dull, brown patches and nonuniform	Crisp and brittle, but turns soggy	Unbalanced pulsy	42
0.5%	Soft	Easy	Normal, somewhat patchy, nonuniform	Fairly crisp and brittle	Somewhat pulsy	42
1.0%	Soft	Easy	Normal, somewhat patchy nonuniform	Fairly crisp and brittle	Balanced and acceptable	45
1.5%	Soft	Easy	Uniform, bright and attractive	Crisp and brittle	Slightly alkaline and acceptable	45

2.0%	Soft	Easy	Uniform, bright foamy and oily	Crisp and brittle	Alkaline and unacceptable	43
Flour (various mesh sizes)						
44—60	Fairly soft, but granular	Very difficult	Normal, but coarse	Somewhat hard, brittle and gritty	Floury and unacceptable	25
61—85	Fairly soft	Easy	Bright and attractive	Crisp and brittle	Balanced and acceptable	35
86—100	Soft	Easy	Bright and attractive	Crisp and brittle	Balanced and acceptable	42
More than 100	Soft	Easy	Bright and attractive	Crisp and brittle	Balanced and acceptable	40

Note: Kneading time was 3—4 min for all samples, except for the 35% water, and the more than 100 mesh flour samples. Kneading time for these samples was 5 min. Papads prepared with high salt (10 and 12%) had salt bloom after 6 weeks storage.

a wooden rolling pin. Cornstarch is used as dusting material to prevent stickiness during rolling. Papads are then cut into uniform circular discs and any adhering starch is brushed off. Papads are dried to 14 to 15% moisture in an air conditioned room maintained at 25 to 27°C and 65% RH. (The drying procedure is discussed in detail in the next section.) Refined groundnut oil, heated to 195 ± 5°C is used for deep-fat frying of papads.

Common salt gives taste to the product; it softens the papad dough and helps in rolling it out into thin sheets. Salt also facilitates uniform distribution of flour components such as protein, mucilages, and starch, and contributes to the blooming or expansion of papads on frying. Food-grade carbonate contributes to the uniform and bright color as well as the desirable crisp and brittle texture of papads upon frying. Carbonates also mellow the undesirable beany flavor, typical of legumes. The traditionally used papad khars or sajji khars have widely varying composition and yield a product of varying quality attributes — taste in particular. In contrast, use of sodium carbonate yields papads of uniform taste and alkalinity.

In the traditional method of preparation, black gram flour or blend is hand mixed with a small quantity of oil, common salt, and *papad khar* or *sajji khar* and then beaten or pounded with a cylindrical stone device (20 cm long, 6.5 cm in diameter). The dough is then left to rest 1 to 2 hr during which time fermentation occurs. However, the fermentation process has not been studied. The small dough balls are rolled into papads of circular shape and generally dried under shade. In households, the papad requirement for the whole year is met by preparing the product during the summer months.

When papads are manufactured on either a small scale or a commercial scale, a mechanical mixer-cum-kneader is essential for uniform distribution of ingredients throughout the mass of dough. Papad dough can then be rolled into thin sheets of desired thickness by passing it through a series of teflon coated metallic rollers. The sheets can be cut into circles of desired diameter by using a sharp punching mechanism. This is helpful in packaging operations on a commercial scale, when uniformity in size, shape, and thickness is very important. Papads of different diameters are presented in Figure 2.

Papadams are prepared from dehulled black gram flour. A paste made with black gram flour is mixed with spices (asafetida, caraway, cardamom, cloves, red pepper, and salt) and the mixture is inoculated with inoculum from a previous batch and left to ferment for 4 to 6 hr. The fermented dough is then rolled into flat tortilla-like wafers (10 to 20 cm wide and 0.5 mm thick), sun-dried, and stored. *Candida krusei* and *Saccharomyces cerevisiae* are involved in the papadam fermentation.[4] These microorganisms are naturally present in black gram.

C. Drying

Traditionally, papads are dried in the shade, under widely fluctuating conditions of atmospheric humidity and temperature. The nonuniform thickness of papads also influences their drying characteristics. Papads have two levels of critical moisture: (1) a maximum level of 17.5%, above which they become susceptible to fungal spoilage, and (2) a minimum level of 12%, below which their pliability is affected and they tend to warp.

The commercial papads exhibit a variation in their moisture content ranging from 9 to 17%. This may affect their shelf-life. Therefore, it is desirable to dry papads under controlled conditions. Venkatesh and Shurpalekar[5] reported on the drying of papads under atmospheric conditions and under controlled humidity and temperature. They set the optimum level of moisture at 14.5% to make adequate allowance for either ingress or loss of moisture due to vagaries of atmospheric conditions during storage.

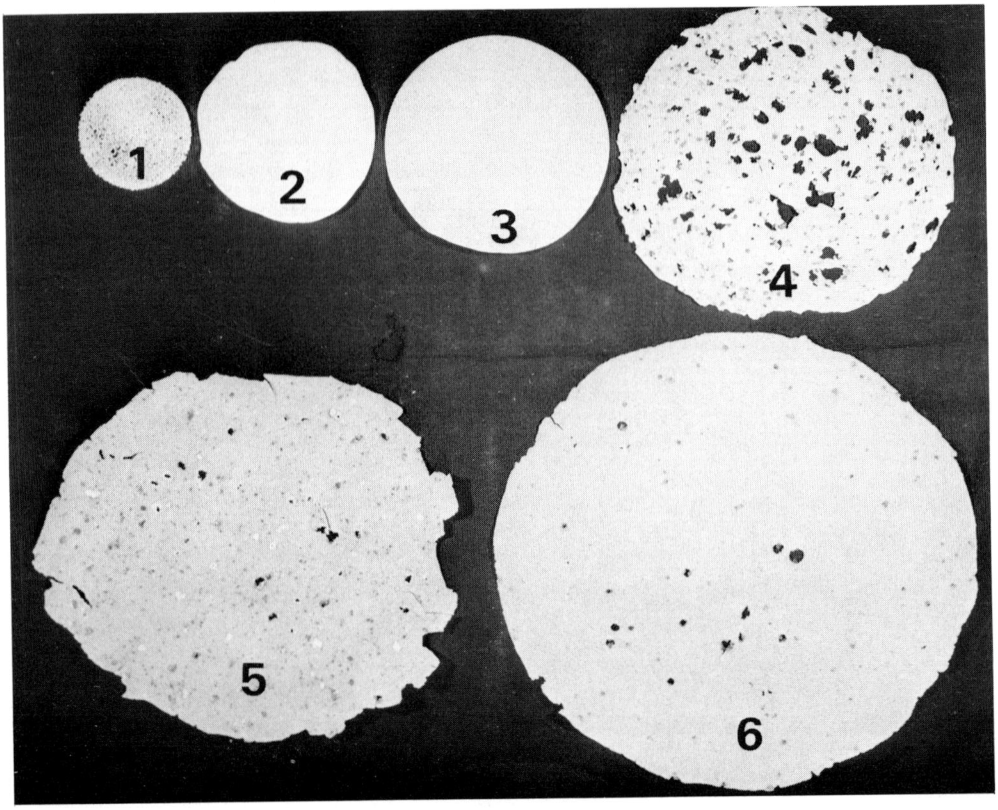

FIGURE 2. Papads of different diameters (6 to 24 cm). Note frayed edges (5); irregular shape (2 and 5); degree of brightness (2, 3, and 5); too many holes seen as black spots (4); and uniform thickness and diameter (1, 3, and 6).

The results of Venkatesh and Shurpalekar's study are presented in Figure 3. On days of low relative humidity (38 to 43% RH), papads could be dried from 30% moisture to 14.5% in just 2 1/2 hr at 24.5°C. Papads dried for 4 hr warped, and became brittle and unsuitable for packing. Under desirable atmospheric conditions of 66 to 70% RH and 25°C (somewhat comparable to controlled conditions in a humidity cabinet, i.e., 64 to 66% RH at 31°C) warping was prevented because of the steady decrease in the loss of moisture over a 6 hr period. At a high RH of 78 to 82% and at 24.5°C, the drying rate was slower; even after 6 hr papads did not dry adequately, and contained 18% moisture. This renders them susceptible to fungal attack, especially when they are packed and sealed in polyethylene bags where a very high humidity can build up.

Under controlled conditions in a humidity cabinet, papads can be dried in about 5 hr to the desired moisture content of 14.5% without warping. To avoid overdrying or underdrying during storage at 64 to 70% RH and 25 to 31°C, papads should have a moisture content of 14.5%. This can be obtained with a drying period of 5 to 6 hr. Simple facilities can be constructed for artificial drying of papads under controlled conditions by using an electric heater and water spray within an enclosed room or space.

D. Packaging and Storage

In order to develop a proper packaging for papads, shelf-life for both export transhipment and the inland marketing must be considered. Balasubramanyam et al.[6] stud-

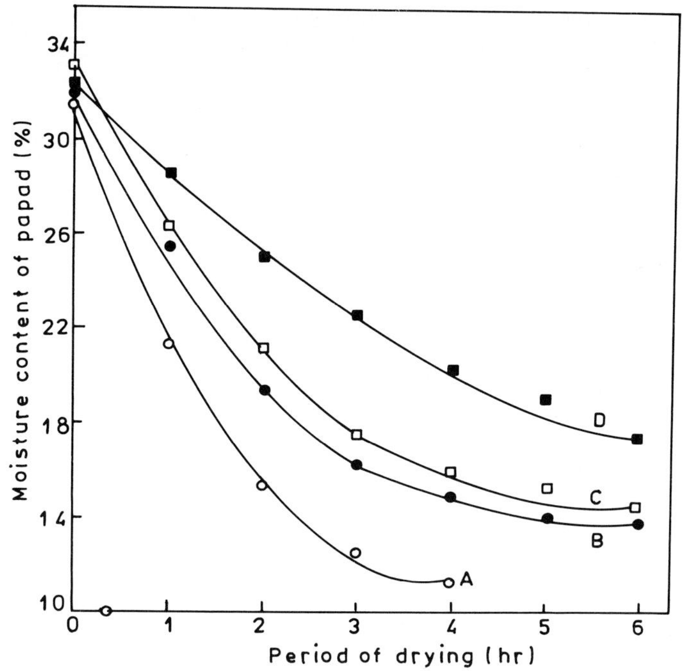

FIGURE 3. Effect of relative humidity on drying of papads: (A) atmospheric drying at 38 to 43% relative humidity (RH) and 24.5°C, (B) cabinet drying at 64 to 66% RH and 31.0°C, (C) atmospheric drying 66 to 70% RH and 25.5°C, and (D) atmospheric drying at 78 to 82% RH and 24.5°C.

ied the moisture-absorption characteristics of papads under varying climatic conditions, as well as the effect of additives (common salt and sodium carbonate) and the use of polyethylene bags on shelf-life.

The following formulae for papads were studied:

1. Black gram flour (BGF) only
2. BGF + 1% Na_2CO_3
3. BGF + 1% Na_2CO_3 + 7% salt
4. BGF + 8% salt + 0.66% Na_2CO_3 or 3.6% $NaHCO_3$

Moisture absorption behavior of papads equilibrated to 66% RH at 27°C was studied using desiccators maintained at a relative humidity ranging from 32 to 86%. The period required for the appearance of fungal attack was the main criterion used.

In these packaging studies, 200 and 400 gauge polyethylene, and 350 gauge high-density polyethylene, were used for packing papads based on the following formula: BGF — 100 parts; salt — 7 parts; sodium carbonate — 1 part. The samples equilibrated to 66% RH at 27°C had a moisture content of 14.8%. Packages containing 170 g (35 papads) and 340 g (70 papads) were used. 250 g unit packages of commercial samples, purchased locally, were also included for comparison. Shelf-life studies were conducted under accelerated storage conditions at 92% RH and 38°C. When the packages indicated the moisture pickup of about 3%, they were withdrawn from the humidity chamber.

Inclusion of 1% sodium carbonate in the formula has little effect on the equilibrium moisture content (EMC) of black gram papads. EMC was 14.6% at 70% RH in case

Table 6
EQUILIBRIUM MOISTURE CONTENT (EMC) OF PAPADS AT 27°C

Ingredients	EMC at different relative humidities (% RH) 32	64	75	86	Initial moisture (%)
BF + common salt only	8.9	13.1	15.8	21.8	10.8
BF + Na_2CO_3 (1%) + common salt (0%)	8.8	13.5	18.0	22.3	18.0
BF + Na_2CO_3 (1%) + common salt (4%)	9.2	18.0	22.9	31.4	23.1
BF + Na_2CO_3 (1%) + common salt (7%)	8.9	17.1	22.7	32.5	16.6
BF + Na_2CO_3 (1%) + common salt (8%)	9.2	15.8	25.7	38.7	17.4
BF + Na_2CO_3 (1%) + common salt (12%)	8.9	14.4	35.1	50.8	17.6
BF + Na_2CO_3 (1%) + common salt (16%)	8.8	14.4	42.4	61.8	16.2
BF + Na_2CO_3 (0.66%) + common salt (8%)	7.3	17.5	27.3	39.9	25.4
BF + common salt (8%) + $NaHCO_3$(3.6%)	8.7	19.5	28.8	44.1	22.7
Commercial papads (5 number)					
Range	8.0 to 8.6	14.3 to 17.5	23.7 to 25.8	—	12.3 to 17.2
Average	8.3	16.3	24.8	—	15.8

Note: EMC is expressed as percent on dry basis. The recipe for papads consisted of black gram flour (100 parts), common salt (7 to 8 parts), sodium carbonate (1 part). When required, sodium carbonate or any other alkaline salt was altered keeping the other ingredients the same. BF = black gram flour.

of papads based on black gram flour alone. Inclusion of 7% salt increases the EMC to 20.3%, indicating that salt is the main factor contributing to the increase in EMC. Increasing the level of salt in papads beyond 4% (up to 16%) does not significantly increase the EMC of 20.5%. The moisture level in papads should not exceed a critical limit of 21% on a dry weight basis, or 17.5% based on finished product, for preventing onset of fungal spoilage.

Eight percent salt is the critical limit in papads, since there is a large increase in their EMC at RH values higher than 75%. At lower humidities, salt bloom poses a problem in papads containing more than 8% salt. Considering these parameters and the taste factor, a level of 7 to 8% salt is most desirable. As long as the salt level in papads is maintained at 8%, use of different alkaline salts at comparable levels does not significantly alter the EMC of papads.

Extrapolation of the data in Table 6 also indicates that the EMC of commercial papads ranges between 21 and 22.6% at 72.5% RH. Commercial samples containing traditionally used *papad khars* or *sajji khars* closely resemble papads containing 7% salt and 1% sodium carbonate.

Thus, for obtaining papads of uniform quality and taste, 1% food-grade sodium carbonate can easily replace *papad khar* which has a widely varying composition.

Sorption studies have shown that papads equilibrate to 17.5% moisture (finished product) at 70% RH. Thus, in papads having initial moisture content of 14.8%, an uptake of about 3% moisture will be permissible before the onset of fungal growth during storage.

Extrapolation of the data presented in Table 7 for papads stored at 38°C and 92%

Table 7
MOISTURE PICKUP BY PAPADS STORED IN POLYETHYLENE BAGS AT 92% RELATIVE HUMIDITY AND 38°C TEMPERATURE

Polyethylene bags			Package	% Moisture pickup during storage (days)			
Film type	Gauge	WVTR ($g/m^2/24$ hr)	weight (g)	29	64	99	128
LDPE	200	6.4	170	1.31	2.84	—	—
			340	0.61	1.36	2.02	2.57
LDPE	400	4.5	170	0.86	1.81	2.63	3.23
			340	0.49	1.04	1.52	1.93
HDPE	350	1.9	170	0.35	0.76	1.13	1.35
			340	0.24	0.52	0.78	1.00
LDPE	250		250[a]	1.61	3.69	5.66	—

Note: WVTR = Water vapor transmission rate under 90% relative humidity gradient (RH outside polybags, 92%; RH inside, 2%; gradient, 90%) at 38°C temperature; LDPE = Low density polyethylene; HDPE = High density polyethylene.

[a] Commercial samples with initial moisture content of 14.4 and 16.9% on fresh and dry weight basis, respectively.

RH shows that after 78 days storage, the moisture pickup in packs of 170 g in 200 gauge low-density polyethylene (LDPE) probably exceeds 3%. Though no visible fungal attack was observed, the samples developed an unpleasant musty odor, rendering them organoleptically unacceptable. Packs of 340 g had a moisture pickup of just 2.6%, even after 128 days. A similar trend was observed in case of 400 gauge LDPE. For 350 gauge high-density polyethylene (HDPE) low moisture pickups of 1.4% and 1.0% were recorded at the end of 128 days storage in 170 and 340 g unit packs, respectively. On the same gauge basis, HDPE with nearly one third the water vapor transmission rate of LDPE provides excellent protection against ingress of moisture.

The shelf life was assessed under accelerated storage conditions of 38°C and 92%. As such, the actual shelf-life according to Painc's correlation may be nearly 3 times longer, i.e., about 1 year, when 400 gauge polyethylene is used.

E. Color and Acidity

During the storage of papads under different conditions of packaging, temperature, and humidity, the main quality attributes that are affected are color of dried product, appearance, and sour taste in the fried product due to the development of acidity[7] (expressed as pH). Studies have been reported for black gram papads packaged in low-density polyethylene, high-density polyethylene, and an aluminum foil laminate consisting of paper/aluminum foil/polyethylene, and stored under various conditions: (1) refrigerator (4°C and 40 to 50% RH); (2) 65% RH at 27°C, and (3) 92% RH at 38°C. The data on the changes in moisture, pH, and color in papads stored for 140 days are summarized in Table 8.

Under accelerated storage conditions (92% RH and 38°C), the color changes significantly from a reflectance value of 26.5 to 16%, irrespective of the moisture content in the final product. Color changes in papads are primarily due to temperature. Neither

Table 8
CHANGES IN MOISTURE, pH, AND COLOR OF PAPADS PACKED DIFFERENTLY AND STORED AT VARIOUS CONDITIONS FOR 140 DAYS

	Initial values	At 92% RH and 38°C			At 65% RH and 27°C			In refrigerator
		LDPE	HDPE	AFL	LDPE	HDPE	AFL	LDPE
		Papads containing 14.0% moisture						
Moisture[a] (%)	13.5	19.1	18.5	13.5	13.7	13.8	13.5	13.9
pH	7.6	5.7	5.9	5.9	6.9	6.9	6.9	7.6
Color (% reflectance) average[b]	26.5	16.0	16.0	17.0	26.0	27.0	26.0	27.0
		Papads containing 16.0% moisture						
Moisture[a] (%)	15.9	20.1	20.0	16.4	16.2	15.9	16.2	15.9
pH	7.2	5.6	5.7	5.7	6.5	6.5	6.4	7.2
Color (% reflectance) average[b]	29.5	16.5	16.5	16.5	28.0	28.0	27.0	29.0

Note: Values for moisture, pH, and color for the papads packed in HDPE and AFL and stored in refrigerator were almost the same as that of LDPE values. LDPE = low density polyethylene; HDPE = high density polyethylene; AFL = aluminum foil laminate.

[a] Moisture content on fresh weight basis.
[b] Average of 50 papads.

packaging material nor humidity significantly affects color. Practically no discoloration is observed for samples packed in different films and stored at 65% RH and 27°C, or under refrigerated conditions.

At 92% RH and 38°C, a significant decrease in pH from 7.6 to 5.6 was observed; consequently, these papads had a slightly sour taste after frying. Under other conditions of storage, however, there was less change in the pH values.

Acceptability studies on papads with a 14% moisture level stored at 65% RH and 27°C showed that the product could be kept well for 300 days. Papads containing 16% moisture, however, indicated a slightly sour taste and unpleasant odor.

F. Evaluation of Dough

Many quality parameters relating to papad dough are based on subjective assessment such as soft or tough when felt by hand, easy or difficult to roll, etc. The Brabender Farinograph was used by Shurpalekar and Venkatesh[9,10] to measure the strength of papad dough.

For the evaluation of the papad dough, black gram, green gram, rice, horse gram, corn, gelatinized cornstarch, and wheat flour (maida) were ground to pass through 80 mesh sieves. For every 100 parts of flour, 7 parts of salt and one part of food-grade sodium carbonate were used. Papads rolled from doughs mixed in a Hobart mixer with an arm in planetary motion were compared with papads rolled from doughs mixed in a Farinograph mixer (Sigma type blades).

Since black gram flour is considered to be an indispensable component of papad dough, the influence of additives such as sodium carbonate and/or common salt on black gram dough was studied (Figure 4). The Farinograph curve was obtained by running the mixer in fast position for a total period of 10 min. The maximum consistency, maintained within 600 Brabender units (BU) by adjusting the level of water addition (45 to 55%) to the dough, fell at the end of 10 min. The width and the pattern of the curve were taken as criteria for evaluating the dough characteristics.

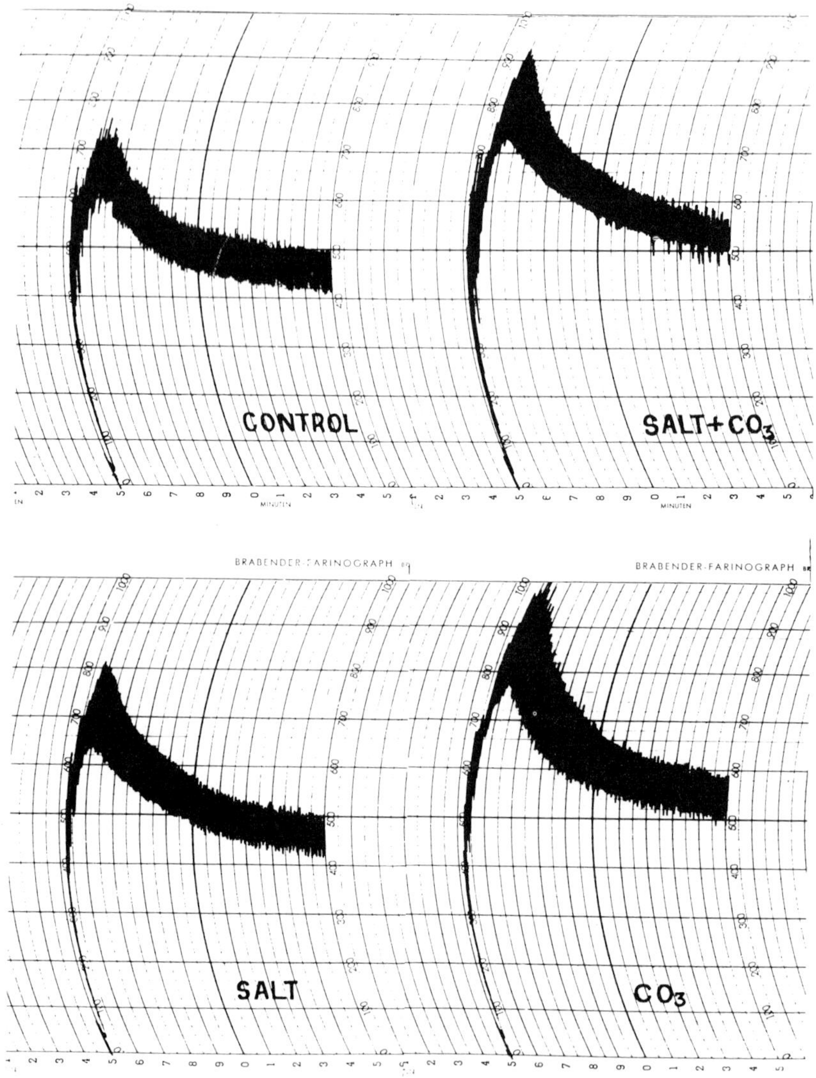

FIGURE 4. Farinograms depicting effect of salt, with or without carbonate, on the consistency of papad dough made from black gram.

The effect of additives on black gram dough were (Figure 5) (1) instead of 45% water normally used for hand-rolled papad dough, 50% water was essential for obtaining the curves within 1000 BU; (2) 3% extra water was necesary for papad dough containing sodium carbonate and no salt. These observations corroborated the subjective observations such as tough and noncohesive nature of papad dough containing carbonate. The toughness of the dough and the consistency were improved significantly by inclusion of 7% common salt which improved the rolling property of the dough. Use of both carbonate and common salt yielded a dough having the desired strength, softness, and rolling property.

The results presented in Table 9 (Series I) indicate no correlation between the maximum dough consistency and diametrical expansion of papads on frying. The diametrical expansion (45 to 50%) of papads based on doughs containing salt, with or without carbonate, was significantly higher (10 to 20%) than those of papads based on black gram, with or without carbonate. A dough consistency of about 900 BU is desirable for obtaining maximum diametrical expansion in black gram papads.

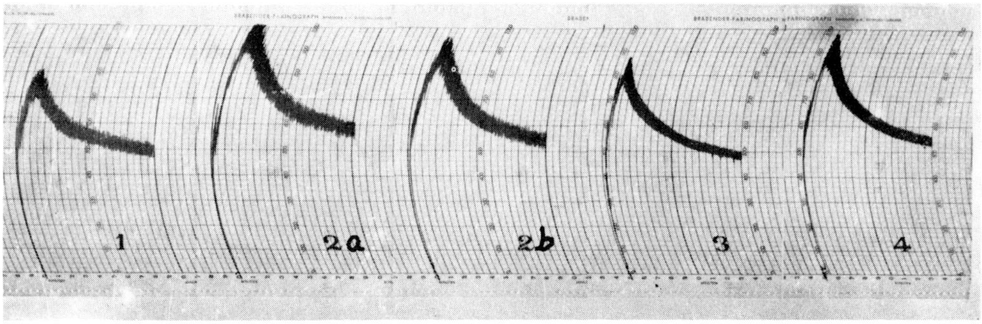

FIGURE 5. Farinograms of black gram dough containing: (1) 50% water; (2a) 52% water and 1% sodium carbonate; (2b) 53% water and 1% sodium carbonate; (3) 50% water and 7% salt, and (4) 50% water, 1% sodium carbonate, and 7% salt.

For doughs prepared from other flour blends, water addition should be between 40 and 55%. This will give the desired consistency in a Farinograph mixer.

The papads based on a wheat blend had an excellent diametrical expansion of 55%, while the papads made from a corn blend had an expansion of only 33% (Table 9). Though the maximum dough consistency was comparable in the case of rice and horse gram blends, the diametrical expansion of the papads made from a rice blend was almost three times that of the horse gram blend (Table 9).

The fall in the consistency measurements of black gram doughs with different additives or with starch was in the range of 290 to 370 BU, while in case of doughs based on blends of black gram with green gram, rice, or horse gram, the low values ranged from 80 to 150 BU (Table 9). The weakening effect on the dough was maximum in the black gram-wheat blend (445 BU).

The black gram dough containing only carbonate had a large curve width corroborating, thereby, the toughness of the dough indicated in Table 10. Taking into consideration different attributes, after subjective as well as objective evaluation, it has been inferred that salt renders black gram dough soft, while carbonate imparts the desired toughness and strength to the doughs. As such, both these additives are necessary to obtain the dough of desired strength, consistency, and rolling property.

Uniform consistency of doughs for black gram papads is very important from the standpoint of rolling characteristics. A spreader-roller device has been successfully used for objective assessment of rolling characteristics of different papad doughs containing black gram.[11]

Papad doughs based on blends of black gram flour with flours of green gram, horse gram, corn, wheat, rice, and different starches were prepared in a Hobart mixer, using the general recipe which included 1.0% sodium carbonate and 7.0% salt. An inclined plane of wood of 900 mm length and 100 mm width was adjusted at an angle of 1.4°, so that a solid steel cylinder, weighing 3.8 kg and having a diameter of 90 mm and a length of 80 mm, could just start rolling down the inclination on its own (Figure 6). The top surface of the incline was covered with a glass sheet, beneath which centimeter graph paper was placed. Strips of different doughs 10 mm long, 40 mm wide and 1 mm thick were placed centrally along the incline, 400 mm away from the top end. The roller was then allowed to run down over the strip along the incline resulting in an increase in the area of the strip. The time taken for the roller to travel along the incline was also recorded. Table 11 shows the increase in area for doughs based on different flour blends.

Black gram dough, with or without 1% sodium carbonate had the same rolling prop-

Table 9
FARINOGRAPH MEASUREMENTS, DOUGH CHARACTERISTICS, AND EXPANSION OF PAPADS BASED ON BLENDS OF LEGUMES, CEREALS, AND STARCHES

Ingredients	Water added (mℓ)	Dough consistency: Max. (BU)	Dough consistency: Time to reach max. (sec)	Dough characteristics: Hand feel	Dough characteristics: Rolling	% Diametrical expansion
Series I						
BF (50 g) only	25	750	78	Fairly soft, not cohesive	Somewhat difficult, edges crack	10[a]
BF (50 g) + Na_2CO_3 (0.5 g)	26	950	112	Tough but cohesive	Somewhat easy	20[a]
BF (50 g) + Na_2CO_3 (0.5 g)	26.5	900	92	Tough but cohesive	—	—
BF (50 g) + salt (3.5 g)	25	820	56	Soft and cohesive	Easy	50
BF (50 g) + salt (3.5 g) + Na_2CO_3 (0.5 g)	25	910	68	Soft, cohesive, and elastic	Easy	45
Series II[b]						
BF (50 g) + green gram flour (40 g)	22.5	480	37	Soft but slightly sticky	Easy	45
BF (25 g) + rice flour (25 g)	25	340	48	Fairly soft	Easy	50
BF (40 g) + corn flour (10 g)	25	640	44	Soft, somewhat sticky, less cohesive	Easy	33
BF (40 g) + cornstarch (10 g)	25	660	35	Soft, somewhat elastic	Easy	64
BF (35 g) + cornstarch (12.5 g) + gelatinized starch (2.5 g)	25	760	49	Fairly soft only after resting	Somewhat difficult	55

BF (40 g) + gelatinized starch (10 g)	27.5	940	74	Somewhat tough	Somewhat difficult	50
BF (25 g) + wheat flour (25 g)	25	770	39	Very soft, elastic	Very easy	55
BF (10 g) + horse gram (40 g)	25	400	—	Soft but slightly sticky	Easy	17[a]

Note: Moisture content of papads was between 13.0 and 15.0%. The stickiness of some doughs could easily be overcome by using dusting material. BU = Brabender units. BF = black gram flour.

[a] Refers to papads made from doughs kneaded in Hobart mixer.
[b] In addition, each dough contained 3.5 g of salt and 0.5 g of Na_2CO_3 except the dough made with black gram and horse gram flours which had only salt and no sodium carbonate.

Table 10
CHANGES IN THE FARINOGRAPH CONSISTENCY OF PAPAD DOUGHS BASED ON DIFFERENT LEGUMES, CEREALS, AND STARCHES

Ingredients	Dough consistency (BU)		Width of farinogram (mm)		Time taken to reach max. curve width (sec)
	Final	Fall	Maximum	Minimum	
Series I					
BF (50 g) only	495	255	15	9	120
BF (50 g) + Na_2CO_3 (0.5 g)	585	365	18	10	75
BF (50 g) + Na_2CO_3 (0.5 g)	545	355	16	10	90
BF (50 g) + salt (3.5 g)	485	335	11	6	60
BF (50 g) + salt (3.5 g) + Na_2CO_3 (0.5 g)	540	370	15	7	60
Series II[a]					
BF (10 g) + green gram flour (40 g)	330	150	9	3	60
BF (25 g) + rice flour (25 g)	260	80	11	3	30
BF (40 g) + corn flour (10 g)	370	270	14	5	30
BF (40 g) + cornstarch (10 g)	335	325	9	4	45
BF (35 g) + cornstarch (12.5 g) + gelatinized starch (2.5 g)	395	365	10	5	150
BF (40 g) + gelatinized starch (10 g)	650	290	14	11	120
BF (25 g) + wheat flour (25 g)	325	445	13	5	60
BF (10 g) + horse gram (40 g)	260	140	9	2	45

Note: Final dough and fall in dough consistencies were measured at the end of 10 min. BU = Brabender units. BF = black gram flour.

[a] In addition, each dough contained 3.5 g of salt and 0.5 g of Na_2CO_3 except the dough made with black gram and horse gram flours which had only 3.5 g of salt.

erty. However, dough containing sodium carbonate was tough, while the dough without carbonate was fairly soft. Both the doughs were neither cohesive nor easy to roll. In contrast, the black gram dough containing 7% salt, with or without 1% carbonate, was cohesive and showed an increase of 100 to 110% in area.

When black gram flour in the dough was replaced to the extent of 20 to 80% by other flours, the increase in area varied over a wide range of 70 to 150%. Most of the doughs were soft and easy to roll. When starches were used to replace black gram flour, the water requirement of the dough of desired consistency was more, while the increase in area was comparatively smaller (70 to 90%). Further, the doughs tended to be slightly tough and difficult to roll. The maximum increase in area of 150% was observed for dough based on 50:50 blend of black gram and wheat.

FIGURE 6. The steel roller and inclined plane used for recording increases in area of the strip of papad dough, seen in the center of the inclined plane.

Table 11
ROLLING CHARACTERISTICS OF PAPAD DOUGH BASED ON BLENDS OF DIFFERENT LEGUMES, CEREALS, AND STARCHES

Ingredients	Water added (mℓ)	Final area (mm^2) Range	Average
BF	45	673—738	706
BF + salt (7%)	45	781—835	804
BF + Na_2CO_3 (1%)	45	691—725	705
BF + salt (7%) + Na_2CO_3 (1%)	45	809—854	833
BF (20) + green gram flour (80)	40	830—919	873
BF (50) + wheat flour (50)	45	971—1066	1015
BF (80) + corn flour (20)	45	864—896	883
BF (70) + cornstarch (25) + gelatinized starch (5)	50	752—792	766
BF (80) + cornstarch (20)	45	703—750	724
BF (50) + rice flour (50)	45	934—964	949
BF (80) + gelatinized starch (20)	60	674—708	689
BF (20) + horse gram flour (80)	45	913—948	927

Note: Rolling characteristics were measured as the roller traversed a distance of 900 mm along the inclined plane in 4.4 to 4.6 sec. Final area is the average of 10 trials. Initial area of the dough strip was 400 sq. mm. BF = black gram flour. The figures in the parentheses indicate the proportions of the main ingredients used in the papad dough. In addition these doughs contained 7 parts of salt and 1 part of sodium carbonate. The dough made with black gram and horse gram flours contained only 7 parts of salt.

Table 12
EFFICACY OF FUMIGANTS IN CONTROLLING MOLD GROWTH IN PAPADS OF VARYING MOISTURE CONTENTS

Treatment	Fumigant dosage (mg/ℓ)	Mold growth at moisture content of 18.0%	20.0%
Control	0	Heavy growth	Heavy growth
Acrylo-nitrile	32	No growth	Moderate growth
	64	No growth	No growth
Methyl-formate	32	Slight growth	Heavy growth
	64	Slight growth	Heavy growth
Methyl-iodide	32	No growth	Moderate growth
	64	No growth	No growth
Ethylene dibromide	32	Moderate growth	Heavy growth
	64	Moderate growth	Heavy growth
Ethylene oxide	32	No growth	No growth
	64	No growth	No growth

Note: No mold growth was observed on the papads containing 14.0 and 16.0% moisture content.

IV. PREVENTION OF INSECT AND MOLD GROWTH

Spoilage in improperly dried papads due to mold growth and insect infestation is the common problem faced by the producers. The ingredients used in preparation can carry an inoculum of microorganisms and insect pests like the spice beetle. The microclimate within a package containing inadequately dried papads favors growth of the microorganisms and beetles.

Fumigants are successful in eliminating or preventing contamination or spoilage in different foods. The advantages of these fumigants are their diffusibility through containers of plastic films or paper; the ability to sterilize, *in situ,* in packets; high penetrability into bulk commodity; and their nonpersistent nature. The efficacy of fumigants like methyl formate, methyl bromide, methyl iodide, ethylene oxide, ethylene dibromide, and acrylo-nitrile in controlling infestation and preventing mold attack in black gram papads has been investigated.[8]

In experimental studies, papads of 12.5 cm diameter were dried at 66% relative humidity for different periods, to obtain a moisture content of 20, 18, 16, and 14%. They were packed in 250 gauge polyethylene bags in lots of 10 papads. Adult flour beetles (*Tribolium castaneum*) and spice beetles (*Stegobium panicium*) were introduced into one set of packages before heat sealing. Packages were placed in desiccators. Fumigants were then introduced into the desiccators.

The results presented in Table 12 suggest that polyethylene is not a barrier to organic fumigants. At a dosage of 32 mg/ℓ, 100% mortality of test insects was observed. From the photographs (Figure 7), it is evident that visibly heavy mold growth occurs in papads containing 18 and 20% moisture. Fumigated as well as unfumigated control papads, with 14 to 16% moisture, were free from mold growth. This also confirmed the findings reported by Balasubramanyam et al.[6] that a moisture content less than the

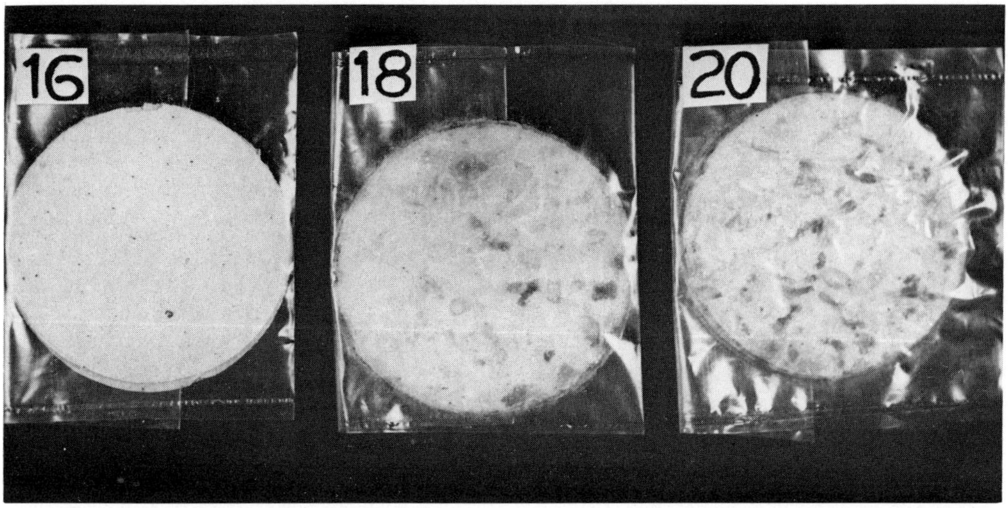

FIGURE 7. Control group of papads containing 16, 18, and 20% moisture stored at 25 to 28°C.

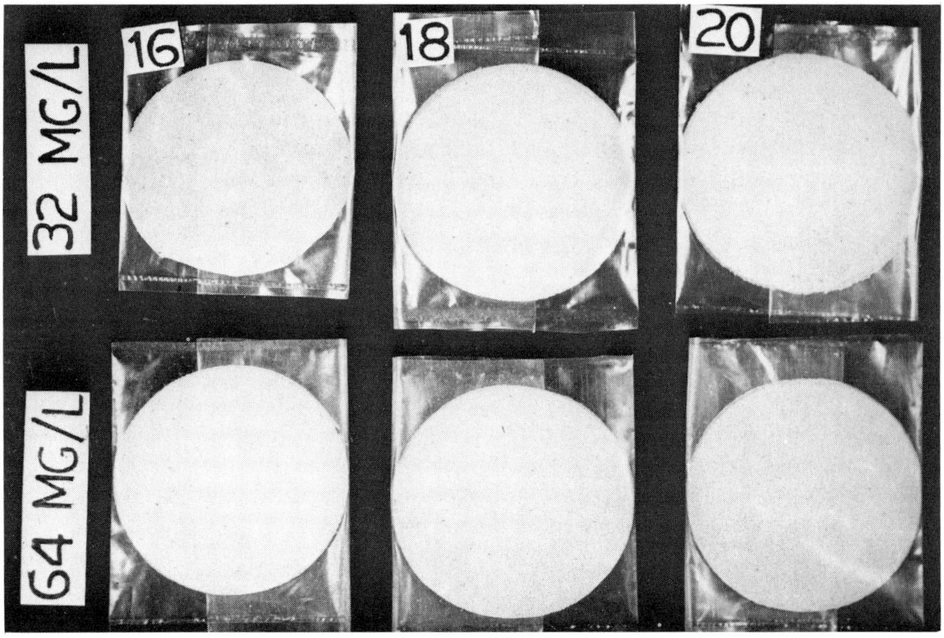

FIGURE 8. Papads containing 16, 18, and 20% moisture treated with ethylene oxide (32 and 64 mg/ℓ). Exposure was for 48 hr at 25—28°C. Note the absence of mold attack.

critical value of 18% is necessary for preventing mold growth. All fumigants, other than ethylene dibromide and methyl formate, prevented mold growth in papads with 18% moisture, even at the lower dose of 32 mg/ℓ. In papads with 20% moisture, only acrylonitrile, methyl iodide, and methyl bromide could prevent mold growth. Ethylene oxide was found to be the most effective fumigant, even at the lower dose of 32 mg/ℓ (Figure 8).

The final choice of fumigants depends, however, upon the cost, availability, permis-

Table 13
DOUGH CHARACTERISTICS AND QUALITY OF PAPADS BASED ON BLENDS OF BLACK GRAM AND GREEN GRAM FLOURS

Flour blend		Dough characteristics		Fried papads			
					% Diametrical expansion after storage (days)		
Black gram	Green gram	Hand feel	Rolling	Taste	1	14	60
100	0	Soft elastic	Easy	Balanced and acceptable	49	56	60
80	20	Soft elastic	Easy	Balanced and acceptable	43	54	58
60	40	Soft elastic	Easy	Balanced and acceptable	40	53	56
40	60	Soft	Easy	Balanced and acceptable	36	52	57
20	80	Soft	Easy	Balanced and acceptable	35	50	56
0	100	Very soft	Difficult	Beany taste	27	31	32

Note: Water added to flour blends ranged between 40 and 45%; kneading time was between 1.5 and 2.5 min. All fried papads were bright, attractive, crisp, and had brittle texture.

sibility of possible residues, and ease of handling, as in the case of liquid fumigants like acrylonitrile and methyl iodide. As papads are fried in oil at temperatures of 190 to 195°C, there is little likelihood of finding any fumigant residues in the fried product that is consumed. All the fumigated papads which were free from mold growth were organoleptically acceptable upon frying.

V. USE OF OTHER FLOURS AND FLOUR BLENDS

In many regions of India, papads are prepared from blends of black gram with cereals such as rice, wheat, and corn or legumes, like green gram or horse gram, and corn or tapioca starches. In northern India, green gram (*Phaseolus aureus*) papads are preferred to black gram papads, while the latter are more popular in southern India. Green gram papads are made in combination with black gram, which contains a unique mucilaginous substance which gives the desired consistency and rolling property to the dough.

Shurpalekar and Venkatesh[9] reported on: (1) the minimum level of black gram necessary in the recipe of green gram papads; (2) the levels of spices to be used; and (3) maximum levels of substitution of other cereals, legumes, and starches in place of black gram for obtaining papads of desirable quality attributes.

Different blends of green gram and black gram flours (100 to 0, 80 to 20, 60 to 40, 40 to 60, 20 to 80, and 0 to 100) were used in the normal recipe of papads containing 7% salt and 1% sodium carbonate. Water addition was maintained at 40 to 45%, to obtain the desired dough consistency and rolling property. Different combinations of spices like pepper, chili, asafetida, cumin, and garlic were added at various levels in the 80 to 20 blend of green gram and black gram flours, either in the form of thin flakes, or in the form of liquid oleoresins extracted from different spices.

The results reported in Table 13 indicate that at least 20% of black gram flour is essential in making green gram papads having desired quality attributes, and a diametrical expansion of about 45%. Papads prepared with green gram flour have a poor

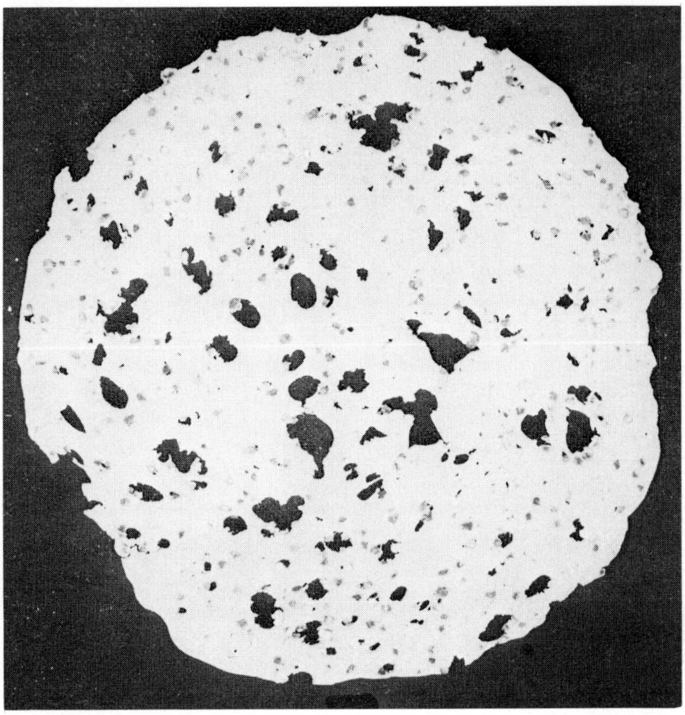

FIGURE 9. Spiced papad with holes of different sizes due to the falling off of spice particles.

diametrical expansion of 27%, and a bean-like taste. The green gram dough was difficult to roll, and had a tendency to crack at the edges.

Acceptable spice levels were 1% chili powder, 2% cumin flakes, 2% black pepper flakes, 0.2 to 0.3% asafetida powder, and 0.2% garlic when used alone or in combination. Inclusion of spices does not affect the normal range of diametrical expansion of papads. The falling out of pepper particles, leaving undesirable holes in papads, is one of the serious defects (Figure 9) that was overcome by increasing the desirable thickness of spiced papads to about 0.8 mm and using flakes of pepper and cumin of 0.8 and 0.3 mm thickness, respectively. This facilitated uniform distribution of spice particles on the surface of the papad (Figure 10) and upon frying, gave the characteristic spicy taste. Papads containing chili or pepper oleoresins (0.15 and 0.2%, respectively), instead of the natural spices, had a spotless, bright, creamy yellow color, and pleasing appearance, which are desirable parameters for the export market.

Recipes for acceptable papads with maximum possible substitution of black gram by relatively cheaper raw materials, and the evaluation of the doughs as well as papads made therefrom, are indicated in Table 14. For arriving at the acceptable recipes, dough consistency, rolling property, texture, taste, and expansion characteristics of papads, were considered as the main criteria.

Black gram flour in papads could be substituted to the extent of 30% by corn flour, 50% by either rice or wheat flour, 20% by cornstarch, and 25% by gelatinized starch, or by a combination of 5% gelatinized starch and 25% cornstarch. The water requirement of the doughs for desired consistency ranged between 45 to 50% in case of blends with cereals, and 57.5% to 60% in case of corn flour and cornstarch blends. Most of the papads based on different blends had highly acceptable color, appearance, taste, and crispness.

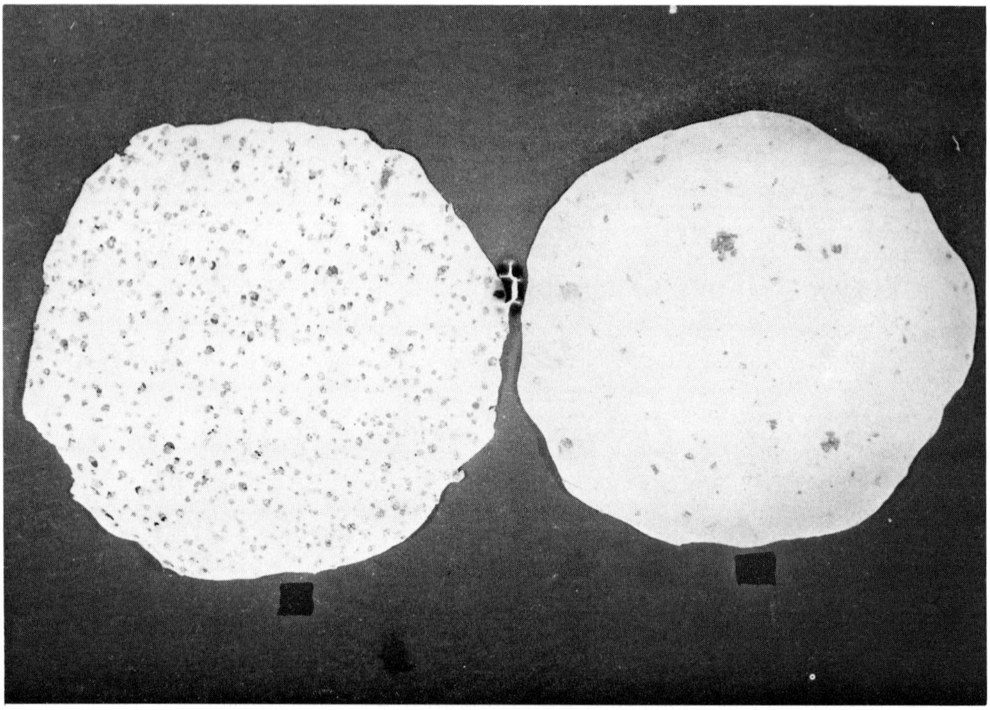

FIGURE 10. Papad (left) with uniform size and distribution of bold spice particles. Papad (right) with spice particles of varying sizes and boldness distributed unevenly. Note the variation in thickness as indicated by brightness.

In the case of horse gram papads, only 20% black gram flour was adequate to get a dough of desired consistency. However, the fried papads had an unacceptable color and gritty taste. Horse gram papads are consumed mostly in toasted or baked form.

VI. NUTRITIONAL QUALITY

Very little information is available on the nutritive value of papads.[12] Papads based on black gram flour or its blend with other legumes contain generally more than 20% protein. The lysine content often exceeds 6%, hence in India, papads form a good food adjunct to the cereal-predominant diet which is deficient in lysine. Papads fried in oil naturally offer more calories than papads toasted or baked. However, in the light of high cost and shortage of vegetable oils, more and more consumers are preferring toasting or baking of papads over frying. Systematic research studies on the nutritive value of papads are needed.

VII. SPECIFICATIONS

The results of the studies reported in this chapter played an important role in the development of an Indian Standard for papads (IS: 2639-1972).[13] Specifications included in this standard are presented in Table 15.

VIII. SUMMARY

Papads are a thin wafer-like, crisp product usually made from black gram flour.

Table 14
DOUGH CHARACTERISTICS AND QUALITY OF PAPADS BASED ON DIFFERENT BLENDS OF BLACK GRAM WITH OTHER LEGUMES, CEREALS, AND STARCHES

	Blends used								
	I	II	III	IV	V	VI	VII	VIII	IX
Recipe (%)									
Corn flour	30	—	—	—	—	—	—	—	—
Rice flour	—	—	—	50	—	—	—	—	—
Wheat flour	—	50	60	—	—	—	—	—	—
Horse gram flour	—	—	—	—	—	—	—	—	80
Gelatinized starch	—	—	—	—	25	—	—	5	—
Cornstarch	—	—	—	—	—	20	30	25	—
Black gram flour	70	50	40	50	75	80	70	70	20
Water added (mℓ)	57.5	45.0	45.0	47.5	60.0	50.0	50.0	57.5	45.0
Dough characteristics	Soft	Soft, elastic	Soft, elastic	Soft	Soft	Soft	Soft	Soft after resting	Soft
Rolling of papads	Easy	Easy	Easy	Easy but cracked edges	Easy	Easy	Easy	Easy	Easy
Expansion (%)	37	55	24	50	51	64	47	55	17

Note: All the papads were bright and attractive with the exception of dull appearance of papads containing rice or horse gram. Except horse gram, the remaining papads were crisp to bite and had a brittle texture and acceptable taste. Texture and taste attributes of fried horse gram papads were unacceptable.

Table 15
INDIAN STANDARD
SPECIFICATIONS FOR PAPADS
(IS: 2639-1972)

Characteristics	Requirements
Moisture content (%)	12.5—15.0
Total ash[a] (maximum) (%)	12.0
Acid-insoluble ash[a] (maximum) (%)	0.2
Alkalinity ash[a] (as Na_2CO_3) (%)	2.2
pH of water extract	8.0
Fat[a] (maximum) (%)	3.0
Crude fiber[a] (maximum) (%)	1.0

[a] On dry weight basis.

Until recently, there was little reported research on the technology of papad preparation.

A standardized recipe for papad dough based on black gram is: 100 parts black gram flour (85 mesh sieve size); 45 parts water; 7 to 8 parts salt, and 1 part sodium carbonate. Sodium carbonate is recommended as a replacement for the traditional papad khar or sajji khar.

Part of the black gram flour can be replaced by up to 30% corn flour, 50% by rice or wheat flour, or 30% by cornstarch. This substitution can reduce cost without adversely affecting the dough properties or the desired quality attributes of papads. Papads based on green gram should be made from a flour blend containing at least 20% black gram. The desirable levels at which spices may be incorporated are: chili powder 1%, cumin flakes 2%, black pepper flakes 2%, and asafetida and garlic powders, 0.2% each. The fermentation during the resting stage for the dough has not been studied with papads; however, *C. krusei* and *S. cerevisiae* are known to be involved in the fermentation of papadams.

Papads have two critical limits for moisture: a lower limit of 12.0%, below which the product tends to warp and break, and an upper limit of 17.5% (finished product base), above which it becomes susceptible to attack by molds and insects. The following optimal drying conditions have been established: (1) 6 hr at 66 to 70% RH and 25°C in air, or (2) 5 hr at 64 to 66% RH and 31°C in a drying cabinet. Undesirable changes in color and taste of papads during storage can be prevented by ensuring a pH value of about 7.5 and a moisture content of about 14.0% in the packaged product.

Fumigants effective against molds and insects include: ethylene oxide at 32 mg/ℓ and acrylonitrile, methyl bromide, or methyl iodide at doses of 64 mg/ℓ.

The Brabender Farinograph can be used to make objective measurements. The Farinograph can also be used to numerically express dough consistency (in Brabender Units).

The rolling, characteristics of papad doughs can be evaluated with a metallic roller device. The increase in the area of the dough strip after rolling provides a quantitative estimate of the dough softness and ease in rolling.

The scientific data obtained have been used in the formulation of an Indian Standard for papads (IS: 2639-1972).

ACKNOWLEDGMENTS

Grateful thanks are due to Mr. K. M. Dastur, Scientist, Central Food Technological Research Institute, for critical reading of the manuscript.

REFERENCES

1. Shurpalekar, S. R., Venkatesh, K. V. L., Prabhakar, J. V., and Amla, B. L., Physico-chemical characteristics and quality assessment of commercial papads, *J. Food Sci. Technol.*, 7, 100, 1970.
2. Venkatesh, K. V. L., Shurpalekar, S. R., Prabhakar, J. V., and Amla, B. L., Physico-chemical characteristics of papad (sajji) khar, *J. Food Sci. Technol.*, 7, 106, 1970.
3. Shurpalekar, S. R., Prabhakar, J. V., Venkatesh, K. V. L., Vibhakar, S., and Amla, B. L., Some factors affecting the quality of black gram (*Phaseolus mungo*) papads, *J. Food Sci. Technol.*, 9, 26, 1972.
4. Batra, L. R. and Millner, P. D., Asian fermented foods and beverages, in *Developments in Industrial Microbiology,* Underkoefler, L. A., Ed., American Institute of Biological Sciences, Washington, D.C., 1976, 117.
5. Venkatesh, K. V. L. and Shurpalekar, S. R., Effect of atmospheric humidity on drying of papads, *J. Food Sci. Technol.*, 18, 66, 1981.
6. Balasubramanyam, N., Shurpalekar, S. R., and Venkatesh, K. V. L., Moisture sorption behaviour and packaging of papads, *J. Food. Sci. Technol.*, 10, 20, 1973.
7. Balasubramanyam, N., Venkatesh, K. V. L., and Shurpalekar, S. R., Changes in color and acidity of black gram (*Phaseolus mungo*) papads during storage, *J. Food. Sci. Technol.*, 11, 79, 1974.
8. Shurpalekar, S. R. and Venkatesh, K. V. L., Studies on papads based on blends of blackgram with cereals, pulses and starches, *J. Food. Sci. Technol.*, 12, 32, 1975.
9. Shurpalekar, S. R. and Venkatesh, K. V. L., Brabender Farinograph as a tool in the objective evaluation of papad dough, *J. Food. Sci. Technol.*, 12, 36, 1975.
10. Shurpalekar, S. R. and Venkatesh, K. V. L., A spreader-roller device for objective evaluation of rolling characteristics of papad dough, *J. Food. Sci. Technol.*, 12, 84, 1975.
11. Narasimhan, K. S., Shurpalekar, S. R., and Venkatesh, K. V. L., Use of fumigants in prevention of insect infestation and mold growth in papads, *J. Food. Sci. Technol.*, 9, 134, 1972.
12. Shurpalekar, S. R. and Amla, B. L., A production and marketing proposal for papads, Proceedings of Workshop organized by Protein Foods Assoc. India on "Better Foods for Better Nutrition", Hyderabad, India, 1974.
13. Shurpalekar, S. R., Prabhakar, J. V., and Amla, B. L., Quality control in papad industry, *Indian Stand. Inst. Bull.*, 23, 81, 1971.

Chapter 13

OTHER LEGUME-BASED FERMENTED FOODS

M. D. Pierson, N. R. Reddy, and S. A. Odunfa

TABLE OF CONTENTS

I. INTRODUCTION

This chapter provides a brief description of some of the less common fermented legumes. Rather than attempting to provide an exhaustive listing of all the variations and cultural types of these foods, several selected examples are presented. These fermented legumes include: ugba, African fermented oil beans; inyu and kecap, types of soy sauce produced in Taiwan and Indonesia, respectively; waries, a condiment of India that is produced by fermentation of black gram paste; kenima, a fermented soybean food of Nepal, Sikkim and India; meitauza, fermented food produced from the solid waste of tofu and Chinese cheese manufacture; Philippine tao-si, mold fermented soybeans; and fermented chickpeas and cowpeas. Most of these foods have received little or no scientific study or significant technological development. Thus, the potential of these products as basic foods, ingredients, or condiments has not been fully explored.

II. UGBA

Nigerian fermented oil beans are called *ugba* by the Ibos and *ukana* by the Efiks. These fermented foods are consumed throughout western and central Africa,[1] but predominantly by the Ibos and other smaller ethnic groups of eastern Nigeria. Ugba is one of the lesser known foods of Africa. It is a traditional food generally prepared in homes as a small family business. Ugba is consumed as a basic food or used as a flavoring agent. The quantity of ugba produced is not known, since the seeds are collected by individuals and sold in the market to ugba producers.

Oil bean seeds for ugba are obtained from a perennial legume tree, *Pentaclethra macrophylla* Benth, commonly called the oil bean tree, Congo Acacia, or Atta bean tree. These trees are found growing throughout the forest areas of west and central Africa, from Senegal to the Cameroons.[2,3] In eastern Nigeria, oil bean plants are cultivated around communities as a cash crop and along roadsides as shade trees. The seeds of the oil bean tree are large, brown, glossy, and flat. They measure 5 to 7 cm in diameter and weigh between 15 and 20 g. The seeds are commonly called owala seeds, opachelo nuts, paracaxy kernel,[4] or okpagha in different countries. Possible commercial uses of oil bean seeds were suggested as early as 1902.[5]

A. Preparation

The flow diagram for the preparation of ugba is shown in Figure 1. Oil bean seeds (Figure 2A) are boiled in water in metallic pots over an open fire for several hours, varying from 4 to 12 hours.[3,6] The cotyledons (kernels) are separated from the cooked seeds by removing the seed coats and washing. The separated cotyledons are boiled again in water overnight over a low flame, allowed to cool, drained, and washed many times with water to remove bitter components in the cotyledons. The bitter components of the oil bean seeds are water soluble. The washed cotyledons are cut into long thin slices (Figure 2B). These slices are mixed with salt, put in a clean pot without water, covered, and fermented for up to 5 days at room temperature. Often after one or two days of fermentation, the sliced cotyledons are wrapped in banana leaves (which have been "browned" over firewood flame) and tied tightly. Wrapped ugba at different stages of fermentation is sold directly to customers (Figure 2C). The customer is often told of the length of fermentation at the time of purchase.

Ugba has a short shelf-life (less than one week). It is consumed in one of two ways: (1) beans fermented for less than three days are used as a delicacy or hors d'oeuvre by mixing them with oil, pepper, salt, and fish to make a so-called native salad, or (2) well-fermented ugba is added to soups as a flavoring condiment.[3]

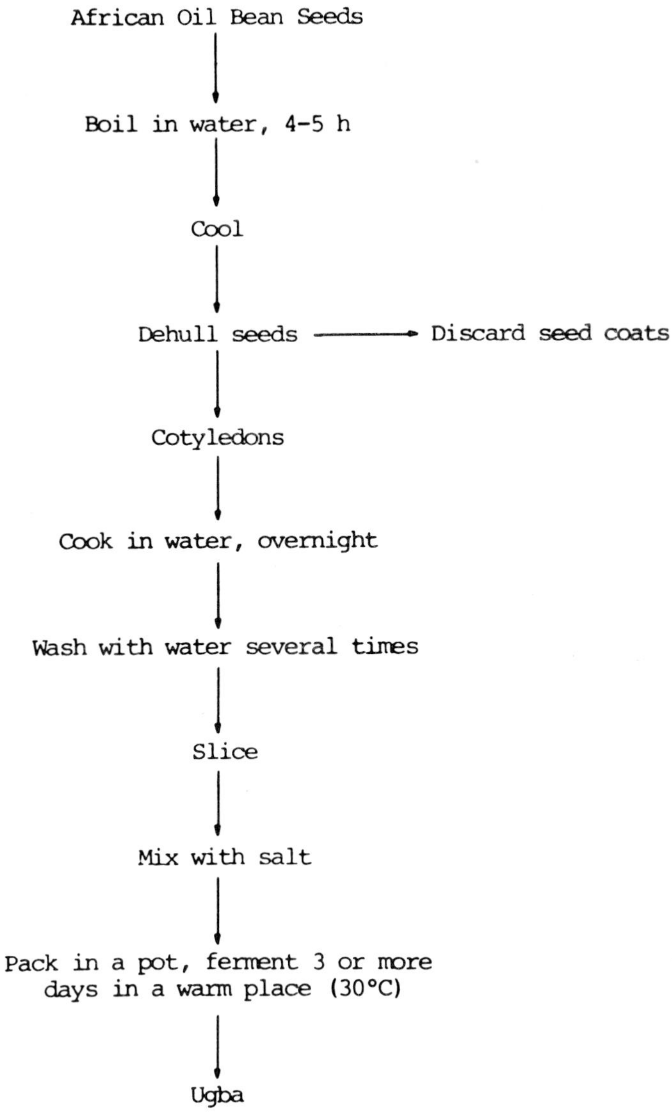

FIGURE 1. Flow diagram for the preparation of ugba.

B. Microorganisms Involved

The microorganisms isolated at different stages of the fermenting ugba are mainly proteolytic *Bacillus subtilis* and coagulase-negative *Staphylococcus* spp.[6,7] *B. subtilis* is reported to be more predominant and its population increases throughout the fermentation while that of *Staphylococcus* decreases. Obeta[8] confirmed the presence of *B. subtilis* as well as the presence of various other bacteria including *Micrococcus luteus, Micrococcus roseus, Bacillus licheniformis, Leuconostoc mesenteroides, Bacillus cereus* var. *mycoides, Bacillus circulans,* and *Bacillus macerans* in the ugba fermentation. *Bacillus* spp. was reported to be responsible for softening of cotyledon slices and production of the characteristic aroma of ugba.[8] The microorganisms involved in ugba fermentation are natural contaminants from the air, from the banana leaves used for wrapping, from handling, and from the utensils used in slicing the cooked cotyledons. *Staphlococcus* spp. is associated with handling since it commonly occurs on the skin. The salt added before the fermentation selects for *Staphylococcus* and *Micrococcus* spp.

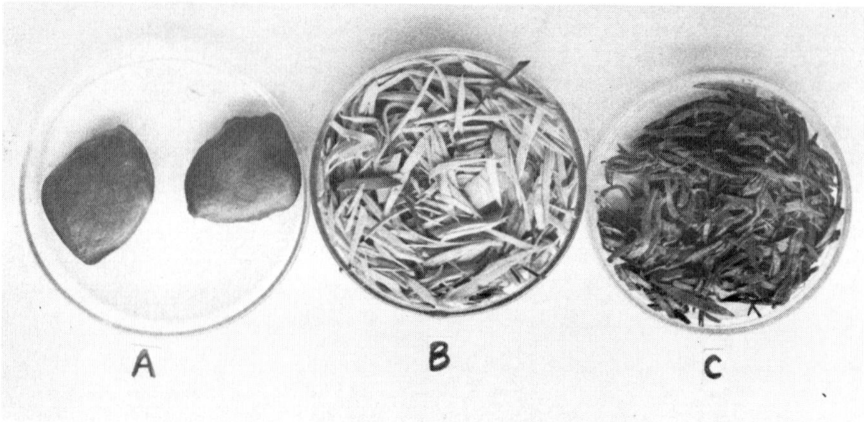

FIGURE 2. (A) Oil bean seeds, (B) cooked, sliced and unfermented beans, and (C) fermented ugba.

Table 1
PROXIMATE COMPOSITION OF UNFERMENTED AND FERMENTED UGBA[12,20]

Component	Unfermented ugba	Fermented ugba	Fermented ugba[a]
Moisture (%)	6.2	46.8	46.8
Crude protein (%)[b]	22.6	11.7	43.9
Fat (%)	46.3	36.1	18.4
Ash (%)	2.3	0.6	3.2
Carbohydrate (%)[c]	22.6	4.8	20.8
Crude fiber (%)	2.5	2.0	6.5
Energy (cal/100g)	558.0	362.0	—

[a] Fermented ugba prepared by using defatted beans.
[b] Protein content was determined after defatting the unfermented and fermented ugba.
[c] Obtained by difference.

C. Chemical Changes

Changes that occur during the fermentation of ugba were studied by Oyeyiola.[6] The pH of the sliced seeds increases from 5.0 to 8.7 during fermentation.[6] A rise in pH is attributed to the hydrolysis of proteins and the abundant production of ammonia by various biochemical pathways.[9] The increase in pH would encourage the growth of *Bacillus* spp. and *Staphylococcus* spp., both of which grow well at pH 7.0 to 8.0. The moisture content of the beans increases from 52% to 74% during the 5 day fermentation.

D. Proximate Composition

The unfermented seeds have an oil content of 38 to 47%.[10,11] The composition of fermented and unfermented seeds is presented in Table 1. The comparison between fermented and unfermented seeds should be viewed with caution since they have different levels of moisture. Unfermented, defatted seeds contain about 46.8% protein.[10]

Table 2
AMINO ACID CONTENT (g/100g protein) OF AFRICAN OIL BEAN SEEDS[12,13,16]

Amino acid	Content
Aspartic acid	7.95—10.30
Threonine	3.27—4.17
Serine	4.80—5.54
Glutamic acid	9.32—11.60
Proline	2.90—5.77
Glycine	3.84—4.62
Alanine	3.81—4.70
Cysteine	1.10—4.80
Valine	4.90—6.60
Methionine	0.90—1.80
Isoleucine	3.30—4.88
Leucine	5.30—6.68
Tyrosine	1.80—5.58
Phenylalanine	5.01—7.00
Lysine	5.46—6.97
Histidine	1.53—2.44
Arginine	4.70—6.53
Tryptophan	1.15—1.78

Table 3
MINERAL AND VITAMIN CONTENT OF UNFERMENTED AND FERMENTED UGBA[20]

Component (mg/100g)	Unfermented ugba	Fermented ugba
Minerals		
Calcium	190.0	110.0
Iron	16.0	3.3
Phosphorus	172.0	—
Vitamins		
Thiamin	0.07	0.07
Riboflavin	0.32	0.30
Niacin	0.90	0.30

The unfermented seeds are low in sulfur-containing amino acids.[12-15] The high content of other essential amino acids makes these seeds a potential source of protein (Table 2).[12,13,16] Some rare amino acids, including dicarboxylic amino acids, have been found in the unfermented seeds.[17-19] The vitamin content of ugba is low.[21] The content of niacin and riboflavin decreases during fermentation (Table 3). Like many other plant materials, the unfermented seed is a poor source of calcium and phosphorus.[12,20] Since ugba is eaten with fish or added as a condiment to stew containing animal proteins, much of the needed calcium and protein may be obtained from other sources.

The unfermented seeds are a potential source of edible oil.[11,22] The oil has a pleasant aroma and burning taste, possibly due to the presence of an alkaloid. Kar and Oke-

Table 4
FATTY ACID COMPOSITION OF AFRICAN OIL BEAN SEEDS

	Ref.				
	23[a]	24[b]	16[b]	11[b]	22[b]
Yield of oil (%)	29.6	46.3	31.0	44.5	45.0
Saturated fatty acids					
Lauric acid			0.48		
Myristic acid			Trace		Trace
Palmitic acid	5.2	3.4	5.2	0.62	6.6
Stearic acid	4.3		1.8	7.23	2.5
Arachidic acid	2.6				3.0
Behenic acid	5.8	5.2			5.0
Lignoceric acid	13.1	12.0			12.0
Unsaturated fatty acids					
Oleic acid	20.3	29.0	15.6	89.83	30.3
Linoleic acid	48.7	42.8	73.0	2.30	38.0
Linolenic acid		3.2	3.9		
Gadoleic acid		0.28			2.4
Free fatty acid (%)	4.8			2.7	

[a] As percent weight.
[b] As percent of total oil.

chukwu[10] reported that the unfermented seed oil contains about 75% saturated fatty acids and 25% unsaturated fatty acids. The free fatty acid content varies from 2.7 to 4.8%.[11,23] The fatty acid composition of oil bean seeds is presented in Table 4. The presence of appreciable amounts of behenic and lignoceric acids is not desirable for edible oils.

Stachyose, galactose, and fructose are the major sugars in the unfermented seed. The saponins make up 2.1% in unfermented seed.[24] These saponins yield oleanolic acid, hederagenin, glucose, arabinose, and rhamnose on hydrolysis.[25]

E. Nutritional Value

Achinewu[26] studied the protein quality of raw and 3-day fermented beans. Both the raw and fermented beans had a negative protein efficiency ratio (PER) and food conversion efficiency (FCE) in rats. Several others found similar results[15,27,28] even after supplementing the diets with deficient amino acids. Digestibility improved after heat treatment and fermentation, indicating that there may be other factors which could have hindered effective utilization of amino acids.[28,30] Mbadiwe[15] attributed the poor nutritional quality to the presence of growth-depressing factors.

F. Toxicology

The unfermented oil seeds contain a number of antinutritional and/or toxic factors including saponins, alkaloids (alkaloid paucine), sterols, glycosides, and growth depressant caffeolyputrescine, but no hemagglutinins.[29,31] The presence or absence of these toxic substances in the fermented beans has not been investigated. It is possible that they are eliminated or reduced during fermentation.[26] Although saponins have been reported to be toxic, they may be beneficial since they have been found to lower plasma cholesterol.[32]

G. Conclusions

Ugba will probably remain as a food of local importance in central and west Africa. In order to improve its nutritional quality, more information is needed on its toxic components. There appears to be no published information on the microbiological safety of ugba. Since ugba is often consumed without prior heating, there need to be more studies of food-borne pathogens such as *Clostridium botulinium* and *Clostridium perfringens*. Also, *Staphylococcus aureus* growth and toxin production during fermentation need to be examined. There is no published information on the flavor components of ugba. Information is needed on processing techniques, especially washing procedures, and means for complete removal of the bitter compounds from oil bean seeds. Finally, appropriate preservation techniques for extending keeping-quality of ugba needs to be investigated.

III. INYU AND KECAP

A. Inyu

Inyu is a fermented black soybean sauce produced in Taiwan, mainland China, Hongkong, and other Chinese cultures.[33] It is used as a condiment in Chinese cuisines and as a flavor enhancer in meat and vegetable products.[34] The flavor of inyu becomes more intense following cooking, thus differing from soy sauce. The production of inyu usually is done using conventional methods, which are low efficiency and have adequate quality control. There have been very few studies on the microbiological and chemical changes during inyu fermentation or on improvements in processing.[33]

The preparation of inyu is outlined in Figure 3. Black soybeans are used as the substrate for inyu production. The beans are washed and then soaked in water for 4 to 7 hr at room temperature, cooked at 117°C for 30 min, and cooled to room temperature. The cooled beans are inculated with koji mold, *Aspergillus oryzae* or *Aspergillus soyae*, mixed well, and incubated aerobically at room temperature for 5 days. The mold mycelia will completely cover the surface of the black soybeans in koji production. Heavy sporulation of the koji mold may result in the development of off-flavors in the final product and may interfere in water absorption. Therefore, the spores on the inyu-koji are partially removed by washing with water. Sodium chloride is then added to the washed beans to give a final salt content of 25 to 30% w/v of the original black soybeans. The mixture is sealed in a earthen container and allowed to undergo anaerobic fermentation for up to 90 days. This stage of fermentation is usually called "aging". The mash undergoes a lactic acid and a yeast fermentation.[34] After the completion of aging, the mash is pressed, cooked, pasteurized, and bottled.

For the industrial manufcture of inyu, the inoculated black beans are distributed to about a 3 cm depth in small flat trays (about 100 cm diameter and 5 cm in depth). The trays are stacked on top of each other but separated by a gap of 5 to 15 cm to allow heat to escape. The koji is then washed. The aging time of anaerobic fermentation can be shortened by preincubation of the washed koji at an elevated temperature (45°C) for 4 hr before mixing it with salt for fermentation. The fermented product is processed as described above.

There are very few published reports on the composition and flavor components of inyu. This product is like soy sauce in many ways (manufacturing process, use, aging, and fermentation).[33] Inyu production requires about 45 days, compared to the 180 days needed for soy sauce production.[34] A major difference between soy sauce and inyu is the increase in the flavor of inyu when the food is cooked.

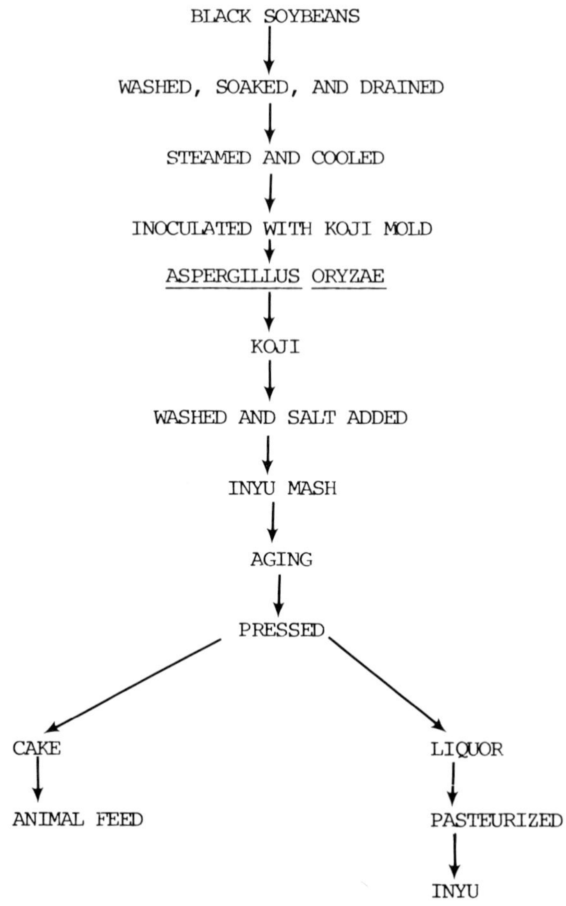

FIGURE 3. Flow diagram for the preparation of inyu.

B. Kecap

Kecap, Indonesian soy sauce, is a dark brown liquid with a meat-like flavor. There are two types of kecap: a sweetened sauce called *kecap manis*, and a salty sauce called *kecap asin*. Both of these sauces are used as condiments to add flavor and color to foods such as soybean curd, rice, meat, and vegetable products.[34] In Indonesia, it is thought that consumption of good quality kecap mixed with *Curcuma xanthorhiza* tuber extract will relieve stomach and/or liver disorders.[34]

To prepare kecap, black soybeans are cleaned and soaked in water for 12 hr, boiled for 4 to 6 hr, drained, and then cooled on bamboo trays. The soybeans become overgrown with mold after the cooled beans are either inoculated with ragi tempe or contaminated naturally. This step requires 3 to 12 days at room temperature. During incubation, the trays containing the beans are covered with gunny sacks or polyethylene sheets. The gunny sacks serve as a source of mold inoculum since they are reused. The mold-overgrown beans are crumbled, sun-dried, and winnowed to remove seed coats or hulls and excess mold. The winnowed beans are placed in earthenware jars, wooden barrels, or other containers, covered with a 20 to 30% brine, and allowed to ferment for 14 to 120 days at room temperature. The fermented material is boiled and filtered.[34]

To prepare kecap manis, palm sugar and spices are added to the filtrate, mixed well and boiled for 4 to 5 hr. To make kecap asin, a small amount of palm sugar is added to the filtrate and boiled for 1 hr. The boiled filtrate is decanted or filtered again, cooled, and bottled. There are three quality grades of kecap. First-quality kecap has a protein content ≥ 6%, second-quality is 4 to 6% protein, and third-quality is 2 to 4% protein.[34]

IV. WARIES

Waries are used as condiments in India and Pakistan.[35,36] They are prepared from dehulled black gram paste. Waries are spicy hollow balls, 3 to 8 cm in diameter, brittle, and friable. Each ball weighs about 15 to 40 g and contains many spices and salt.

To prepare waries, black gram cotyledons are thoroughly washed, soaked in water for 6 to 12 hr at room temperature, and wet ground on a stone mortar to a smooth paste. The ground smooth paste is fermented with or without spices and salt, and used for preparation of two types of foods: (1) waries and (2) wada and pakorian.[37]

The ground smooth paste of black gram is often mixed with a small portion of paste from a previously fermented batch, salt, and spices and then fermented at room temperature for 2 to 8 days. Some typical spices used are asafetida, caraway, cardamom, cloves, fenugreek, ginger, red pepper, and salt.[36] After fermentation, the paste is hand molded into small balls and air dried in the sun on bamboo or palm mats. During drying, a mucilaginous exterior seal forms and encloses gases and moisture within the ball. After 6 to 8 days of drying, the waries look hallow and spongy-like with yeast spherules inside. Dried waries are stored in a container for subsequent use as condiments in cooking vegetables, pulses, or rice. *Candida krusei* (cast.) Berkhout and *Saccharomyces cerevisiae* Hansen are reported to be responsible for warie fermentation. The role of the bacteria present is not yet known.[36] Recently Batra[37] reported that the unfermented paste (pH 6.20) with amendments for waries collected from Amritsar, Punjab, India, had a bacterial count of 3.6×10^6 per g (Gram-positive cocci and rods and acid forming bacteria) and 3.2×10^3 per g yeasts. The fermented paste (pH 4.20) had a bacterial count of 6×10^5 per g (Gram-positive cocci and rods and acid forming bacteria) and 2.9×10^9 per g yeasts.

Waries are also prepared by substituting bengal gram and mung beans for black gram. Village households in the Pakistan districts of Mianwali, Dehara-Ghazi-kan, Multan, and Jhang prepare waries by using bengal gram. Such waries are called *channa ke waries*, the bengal gram waries.[37] Bengal gram seeds are inexpensive and easily available in these regions. The use of bengal gram in the preparation of waries offers a simple means for effective utilization of a local commodity at the peak of grain production. No report is available regarding the composition of waries.

The unsalted and unspiced black gram paste can be fermented for 6 to 24 hr and used in the preparation of wada and pakorian. The fermented paste is pan- or deep-fried as small flat round donut shaped balls and then soaked briefly in water. The soak water is drained away and the wadas are marinated in *dahi* (yogurt) for a few hours before consumption. During marination, fermentation by lactic acid bacteria occurs. The role of specific bacteria and yeasts in wada and pakorian preparation, particularly during the second phase of fermentation in yogurt, is not known.[37]

V. KENIMA

Kenima is a soybean based fermented food that is produced and consumed mainly in Nepal, Sikkim, and Darjeeling districts of India.[36] This product looks much like tempe. It is salted, deep-fat fried, and used as an adjunct to foods such as rice and vegetables.

Kenima is prepared by soaking soybeans overnight. The soaked beans are dehulled and cooked in water for 2 to 3 hr. After cooling, cooked dehulled beans are inoculated

FIGURE 4. Kenima product. (Courtesy of Dr. Lekh R. Batra, Plant Protection Institute, U.S. Department of Agriculture, Beltsville, Maryland.)

with inoculum from a previous batch of kenima and wrapped in banana leaves or other suitable leaves. The wrapped beans are incubated at 22 to 23°C for 24 to 48 hr. During fermentation the beans become mucilaginous due to bacterial growth. The microorganisms involved in the fermentation have not been identified. However, Batra and Millner[36] recovered two rod-shaped, acid-producing bacteria from three samples of kenima from Darjeeling. These bacteria were present at levels of 10^6 to 10^7 per g of kenima. Kenima has a pleasant nut-like taste when it is deep-fried and salted. Uncooked kenima has an unacceptable taste (Figure 4). There appear to be no published reports on the composition, nutritional quality, or toxicological aspects of kenima. This product is produced in individual homes and its commercial production has not been explored.

VI. MEITAUZA

Meitauza is a traditional fermented food of China.[38,39] It is produced by the fermentation of the solid waste material (okara) from the manufacture of tofu or Chinese cheese. To prepare meitauza, solid waste residue from tofu preparation is pressed into small cakes and inoculated with a mold (*Mucor meitauza* or *Actinomucor elegans*). The inoculated cake is allowed to ferment for 10 to 15 days. The cakes become covered with white mycelia of the mold as the fermentation progresses. The fermented cakes are sundried for a few hours. Dried cakes are fried in oil or cooked with vegetables before eating. This product is somewhat similar to tempe.[39]

Recently Kronenberg and Hang[40] modified the fermentation time by using a standardized inoculum, acidifying the cake, and controlling incubation at 15°C. They produced well-molded firm cakes of meitauza in 70 hr, a marked decrease from 10 to 15 days required using traditional methods. Organoleptic evaluation of deep-fried meitauza indicated desirable changes in appearance, odor, and texture of the fermented residue.

VII. PHILIPPINE TAO-SI

Philippine tao-si is soybean fermented food, wherein soybeans are overgrown with a mold, *Aspergillus oryzae* and heated to boiling in a brine solution (18% w/w).[34] The partially digested, mold-covered, salty soybeans are eaten without further preparation.[34]

To prepare tao-si soybeans are soaked overnight in running water at room temperature. The soaked beans are boiled for 1 hr, drained, and cooled. The cooled beans are coated with either raw or roasted wheat flour and inoculated with *Aspergillus oryzae*. Inoculated beans are spread on bamboo trays, covered with banana leaves, and incubated for 2 to 3 days in a warm place. The soybeans become overgrown with white mold mycelia. The fermented soybeans are placed in a brine solution (18% w/v) and heated to boiling to prevent further mold growth and to "inactivate" enzymes.[34]

VIII. FERMENTED COWPEAS AND CHICKPEAS

Zamora and Fields[41-43] conducted studies on the nutritive value of naturally fermented cowpeas and chickpeas. They fermented cowpeas and chickpeas for 4 days at 25°C and used the fermented flours to prepare soup and chip-like products. The products were evaluated for their nutritional, microbiological, and toxicological properties. The organisms isolated from fermented cowpeas and chickpeas were *Lactobacillus casei*, *Lactobacillus leichmanii*, *Lactobacillus plantarum*, *Pediococcus pentosaceus*, and *Pedicoccus acidilactici*. No pathogenic organisms were found in the fermented batters. Fermentation increased the relative nutritive value, limiting amino acids, and significantly reduced niacin content, trypsin-inhibiting activity, and flatus-producing factors.[42] No toxic substances were detected in the fermented chickpea and cowpea flours.[43] They further observed that chips and soups made from the fermented bean flours were in a "neither like nor dislike category" even though the products had a relatively high nutritive value and content of limiting amino acids. The color and acid taste appeared to be the main reasons for low sensory panel scores for the product.

REFERENCES

1. Anon., List of Foods Used in Africa, 2nd ed., Nutr. Inf. Doc. Ser. No. 2, Food and Agriculture Organization, Rome, 1970.
2. Keay, R. W. J., Onochie, C. F. A., and Stanfield, D. P., *Nigerian Tree*, Vol. 2, Nigerian National Press, Lagos, 1964, 119.
3. Oliver, R., Nigeria's useful plants part 1, *Niger. Field*, 23, 147, 1958.
4. Nakao, S., *Tanakas Encyclopaedia of Edible Plants of the World*, Keigaku Publishing, Tokyo, 1976.
5. Heckel, E., *Les graines grasses nouvelles*, Paris, Challamef, 1902.
6. Oyeyiola, G. P., Studies on the Microorganisms Isolated from *Ugba*, a Nigerian Indigenous Food, Master's dissertation, University Ibadan, Ibadan, Nigeria, 1981.
7. Oyeyiola, G. P. and Odunfa, S. A., Microbiological study of the fermentation of *ugba*, a Nigerian indigenous fermented flavouring condiment, *J. Plant Foods*, in press.
8. Obeta, J. A. N., A note on the microorganisms associated with the fermentation of seeds of the African oil bean tree (*Pentaclethra macrophylla*), *J. Appl. Bacteriol.*, 54, 433, 1983.
9. Whitaker, J. R., Biochemical changes in fermented protein foods, *J. Food Technol.*, 12, 163, 1978.
10. Kar, A. and Okechukwu, A. D., Chemical investigations on the edible seeds of *Pentaclethra macrophylla* Benth., *Qual. Plant. Plant Foods Hum. Nutr.*, 28, 29, 1978.
11. Nwokedi, G. I. C., *Pentaclethra macrophylla*, Benth, — a potential oil seed, Tech. Rep. No. 14, Nigerian Stored Products Research Institute, 1975.
12. Mba, A. V., Njjke, M. C., and Oyenuga, V. A., Proximate chemical composition and amino acid content of Nigerian oil seeds, *J. Sci. Food Agric.*, 25, 1547, 1974.

13. Achinewhu, S. C., Composition and food potential of African oil bean (*Pentaclethra macrophylla)* and velvet bean *(Mucuna uriens), J. Food Sci.,* 47, 1736, 1982.
14. Njike, M. C., Mba, A. V., and Oyenuga, V. A., Chick bioassay of available methionine and sulphur amino acids (SAA), assay of African oil bean *(Pentaclethra macraphylla),* conophor seed meal (*Tetracarpidium conophorum*) and groundnut meals (*Arachis hypogea*), *J. Sci. Food Agric.,* 26, 807, 1975.
15. Mbadiwe, E. I., Nutritional evaluation of seeds of *Pentaclethra macrophylla, Qual. Plant. Plant Foods Hum. Nutr.,* 28, 261, 1978.
16. Ikediobi, C. O., Amino acid and fatty acid composition of *Pentaclethra macrophylla* and *Treculia Africana* seeds, *J. Am. Oil Chem. Soc.,* 58, 30, 1981.
17. Krauss, G. Y. J. and Reinbothe, H., Free amino acids in Mimosaceae seeds, *Phytochemistry,* 12, 125, 1973.
18. Mbadiwe, E. I., A new dicarboxylic amino acid from seeds of *Pentaclethra macrophylla, Phytochemistry,* 14, 1351, 1975.
19. Welter, A. M., Jadot, J., Dardeene, G., Marlier, M., and Casimr, J., 3(R) <1(S)-aminocarboxy methyl-2-pyrollidone-5-5-5-carboxylic acid from the seeds of *Pentaclethra macrophylla, Phytochemistry,* 14, 1347, 1975.
20. Duke, J. A., *Handbook of Legumes of World Economic Importance,* Plenum Press, New York, 1981, 185.
21. Okwoli, N., Adesoye, K., and Oluyomi, B. A., Microbiological Assay of Vitamins During the Fermentation of Some Traditional Vegetable Proteins, Bachelor of Science Rep., Department of Botany, University of Ibadan, Nigeria, 1984.
22. Ngiefu, C. K., Paguot, C., and Vieux, A., Oil Plants of Zaire II Botanical families producing oils of relatively high nonsaturation, *Oleagineux,* 32, 535, 1977.
23. Hilditch, T. P., Meara, M. I., and Patel, C. B., The component acids and glycerides of *Pentaclethra* (*Leguminosae*) and *Lophira* (*Ochnaceae*) seed fats, *J. Sci. Food Agric.,* 2, 142, 1951.
24. Achinewhu, S. C., The saponis content of some Nigerian oil seeds, *Qual. Plant. Plant Foods Hum. Nutr.,* 33, 3, 1983.
25. Welter, A. and Delande, C., Contribution of the chemical study of saponins of an African mimosaceae *Pentaclethra ectvelaleana, Bull. Soc. R. Sci.* (Liege), 44, 687, 1975.
26. Achinewu, S. C., Protein quality of African oil bean seed (*Pentaclethra macrophylla*), *J. Food Sci.,* 48, 1374, 1983.
27. Njike, M. C., Mba, A. U., and Oyenuga, V. A., The nutritive value of some Nigerian feeds and foods, *East Afr. Agric. For. J.,* 40, 138, 1974.
28. Njike, M. C. and Oyenuga, V. A., The nutritive value of some Nigerian feeds and foods. II. Evaluation with growing chicken, *Ghana J. Sci.,* 15, 187, 1975.
29. Mbadiwe, E. I., Caffeoylputrescine from *Pentaclethra macrophylla, Phytochemistry,* 12, 1246, 1979.
30. Mears, J. A. and Mabry, T. J., Alkaloids in the Leguminosae, in *Chemotaxonomy of the Leguminoseae,* Harborne, J. B., Boulter, D., and Turner, B. L., Eds., Academic Press, New York, 1971, 73.
31. Toms, G. C. and Western, A., Phytohaemaglutinins, in *Chemotaxonomy of the Leguminoseae,* Harborne, J. B., Boulter, D., and Turner, B. L., Eds., Academic Press, New York, 1971, 367.
32. Oakenfull, D. G., Fenwick, D. E., Hood, R. L., Topping, D. L., Illman, R. L., and Storev, G. B., Effects of saponins on bile acids and plasma lipids in the rat, *Br. J. Nutr.,* 42, 209, 1979.
33. Hua, J., Wang, I. H., Chen, S. K., and Hsu, W. H., Fermented sauce made from black soybean (Inyu), in *Advances in Biotechnology,* Vol. 2, Young, M. M. and Robinson, C. W., Eds., Pergamon Press, New York, 1981, 517.
34. Steinkraus, K. H., Ed., *Handbook of Indigenous Fermented Foods,* Marcel Dekker, New York, 1983.
35. Batra, L. R. and Millner, P. D., Some Asian fermented foods and beverages and associated fungi, *Mycologia,* 66, 942, 1974.
36. Batra, L. R. and Millner, P. D., Asian fermented foods and beverages, in *Developments in Industrial Microbiology,* Underkofler, L. A., Ed., American Institute of Biological Sciences, Washington, D.C., 1976, 117.
37. Batra, L. R., Fermented cereals and grain legumes of India and vicinity, in *Advances in Biotechnology,* Vol. 2, Young, M. M. and Robinson, C. W., Eds., Pergamon Press, New York, 1981, 547.
38. Beuchat, L. R., Traditional fermented food products, in *Food and Beverage Mycology,* Beuchat, L. R., Ed., AVI Publishing, Westport, Conn., 1978, 224.
39. Hesseltine, C. W., A millennium of fungi, food, and fermentation, *Mycologia,* 57, 149, 1965.
40. Kornenberg, H. J. and Hang, Y. D., Studies on meitauza — a fermented food of China, presented at 43rd Annu. Meet. of Inst. of Food Technol., New Orleans, 1983, 168.
41. Zamora, A. F. and Fields, M. L., Nutritive quality of fermented cowpeas, (*Vigna sinensis*) and chickpeas (*Cicer arietinum*), *J. Food Sci.,* 44, 234, 1979.

42. Zamora, A. F. and Fields, M. L., Microbiological and toxicological evaluation of fermented cowpeas (*Vigna sinensis*) and chickpeas (*Cicer arietinum*), *J. Food Sci.*, 44, 928, 1979.
43. Zamora, A. F. and Fields, M. L., Sensory evaluation and nutritive value of soups and chips made from fermented and non-fermented beans, *J. Food Sci.*, 44, 930, 1979.

Chapter 14

FUTURE OF LEGUME-BASED FERMENTED FOODS

S. K. Sathe, N. R. Reddy, and M. D. Pierson

Fermentation is one of the oldest food preservation processes known to mankind. It was originally used to extend the shelf-life of highly perishable foods. Fermentation is a dynamic process during which several catabolic and anabolic reactions proceed simultaneously, depending on several conditions (including substrate, microflora, and environmental factors). From the viewpoint of human diet, fermentation serves several important functions including enhancement or creation of unique flavors, change in textural properties of the food, and improvement in nutritional quality and digestibility of foods. The contribution of fermented foods to the human diet is crucial in underdeveloped countries where numerous fermented foods (especially cereal-based and legume-based) are an integral part of daily food intake. Many populations depend on these foods to satisfy their nutritional needs.[1-5]

Fermented foods can be classified into several categories:

1. Cereal-based fermented foods such as bread, fermented crackers, distilled spirits, beer, etc.
2. Legume-based fermented foods such as soy sauce, miso, tempe, natto, sufu, idli, etc.
3. Vegetable-based fermented foods such as sauerkraut and pickled cucumbers
4. Fruit-based fermented drinks such as wines
5. Milk-based fermented foods such as yogurt, fermented milks, etc.

Legume-based fermented foods serve as an important source of protein, calories, and certain vitamins to the people of Southeast Asia, the Neareast, and parts of Africa. The prospects for increased consumption of legume-based fermented foods is good for both developed and underdeveloped countries.

Most legume-based fermented foods are introduced to the western world by immigrants from the Orient. Some of fermented foods such as miso, soy sauce, and tempe are becoming popular in the Western Hemisphere as a result of: (1) their nutritional and organoleptic characteristics, (2) awareness of relationship between food and health, which favors the use of plant products and low-processed foods, and (3) beneficial aspects of legume-based fermented foods. For instance, fermented soy sauce is consumed widely among the U.S. population due to its excellent organoleptic characteristics. With the increasing awareness of the nutritional value of legume-based fermented foods, popularity and consumption of these foods is on the rise in the western world.[4,6] This is helped to some extent by an increase in research efforts in western countries on the traditional legume-based fermented foods of the Orient. Further, the research efforts of western countries have helped in understanding the fermentation process, nutritional quality, and microorganisms involved of traditional fermented foods. Due to the diversity of indigenous fermented foods, much remains to be learned. Undoubtedly, the consumption of legume-based fermented foods will continue to increase in underdeveloped and developing countries in the future. Improvements in the organoleptic characteristics of legume-based fermented foods may increase their production and consumption in the western world. However, a number of social and scientific factors, such as:

1. Scientific interest in fermented foods
2. Prevention of food poisoning
3. Fermentation and increased shelf-life
4. Improvement of the physical properties of the product
5. Interest in a variety of natural products of plant origin
6. Modification of the substrate
7. Interest in a more healthy food
8. Necessity for increased consumption of plant materials as the population increases
9. Cultural and religious grounds
10. Movement of people

will affect wider use of fermented foods including legume-based fermented foods in the West, and in the developing countries.[1] Several areas related to legume-based fermented foods need further research in order to increase the consumption and production of these foods in the west as well as in the countries where they are popular. The following are areas for future research.

Microorganisms — Identification and characterization of the microorganism(s) responsible for fermentation. In order to understand the nature of fermentation, it is important as a first step to determine which microorganism(s) is (are) responsible for a particular fermentation system. There are several legume-based fermented foods such as kenima, dawadawa, papad, waries, etc. where the microorganism(s) responsible for fermentation are not clearly identified and need to be characterized and identified. Such efforts would help the understanding of the contribution of microorganism(s) towards the final product.

Biochemical changes — The biochemical changes such as protein and carbohydrate hydrolysis need to be critically evaluated, and the relationship between the extent of these changes and product acceptability also need to be evaluated. These efforts would help to establish the desirable extent of certain changes during fermentation, i.e., establish the end point of fermentation. Further, this may help in developing appropriate cultures to achieve the desired organoleptic and nutritional changes as well as the optimum fermentation time which subsequently will result in reduction of fermentation time. Isolation, identification, and characterization of enzymes should provide important insights into the specific fermentative changes accompanying a particular fermentation. Understanding these processes may help in development of processes without the use of microorganisms, i.e., bring about specific changes needed to achieve final product characteristics. Some of the legume-based fermented foods such as papad, dhokla, dawadawa, khaman, etc. need detailed studies on their nutritional and organoleptic characteristics.

Final products — Identification and characterization of end products, especially flavor components, of legume-based fermented foods are important. Flavor components are the primary contributors to product acceptability. Changes in flavors can be classified as (a) desirable, such as destruction of beany flavors and production of new flavors, and (b) undesirable, like strong ammonia-like flavors if the fermentation goes uncontrolled. Knowledge of both types of changes is essential. Except for a few soy products, not much is known about the chemical nature of other legume-based fermented food flavors. Identification and characterization of flavor components responsible for the final product flavor may help in development of synthetic flavors which may reduce product costs and improve product reproducibility.

Pure cultures — Isolation and characterization of pure cultures aid in the development of uniform quality starter cultures. This has been accomplished for only a few legume-based fermented foods such as miso, soy sauce, and tempe. Improvement in cultures for fermentation, particularly with respect to selection of strain(s), would permit reduction in fermentation time. This is important for products that require longer fermentation times. Genetic engineering undoubtedly has a potential to contribute in this regard.

Keeping quality — Except for a few products such as soy sauce, the keeping quality of fermented foods is lacking and merits scientific attention.

Byproduct disposal — Development of acceptable methods to dispose of byproducts of fermentation is required.

Safety — Safety of legume-based fermented foods needs thorough scientific documentation. There is a need to carefully evaluate potential chemical and microbiological hazards of these fermented foods. The potential hazard of mycotoxins in fermented foods exists; therefore, control measures should be aimed at preventing the development of toxins in raw materials and during fermentation process.

Benefits — Nutritional and medical benefits of legume-based fermented foods need to be more thoroughly established.

New foods — Development of new foods such as barbecue sauce and salad dressings from miso, soy sauce, and other fermented foods should be investigated. Better production methods and modification of the foods for western taste should be studied.

Consumer education — Consumers need to be educated on the health benefits of legume-based fermented foods as well as the limitation(s) so that they can make an intelligent choice based on scientific facts.

The above list of research needs is not intended to be complete, but serves to point out many broad areas where research is needed. The use of legume-based fermented foods undoubtedly will continue to increase in the Orient, other parts of Asia, and Africa, since these are traditional foods in these parts of the world. If popularity of soy sauce and tempe in the western world is any indication, then clearly there is a considerable potential for development of new markets for traditional legume-based fermented foods in the western world.

REFERENCES

1. Hesseltine, C. W., The future of fermented foods, *Nutr. Rev.*, 41, 293, 1983.
2. Reddy, N. R., Pierson, M. D., Sathe, S. K., and Salunkhe, D. K., Legume-based fermented foods: their preparation and nutritional quality, *CRC Crit. Rev. Food Sci. Nutr.*, 17, 335, 1982.
3. Hesseltine, C. W. and Wang, H. L., Fermented foods, *Chem. Ind.*, 12, 393, 1979.
4. Steinkraus, K. H., *Handbook of Indigenous Fermented Foods*, Marcel Dekker, New York, 1983.
5. Wang, H. L. and Hesseltine, C. W., Use of microbial cultures: legume and cereal products, *Food Technol.*, 35, 79, 1981.
6. Beuchat, L. R., Fermented soybean foods, *Food Technol.*, 38, 64, 1984.

INDEX

A

C

D

E

F

G

H

I

J

K

L

M

N

R

U

V

W

X

Y

Z